SPACE SCIENCE, EXPLORATION AND POLICIES

MARS

EVOLUTION, GEOLOGY AND EXPLORATION

SPACE SCIENCE, EXPLORATION AND POLICIES

Additional books in this series can be found on Nova's website
under the Series tab.

Additional e-books in this series can be found on Nova's website
under the e-book tab.

SPACE SCIENCE, EXPLORATION AND POLICIES

MARS

EVOLUTION, GEOLOGY AND EXPLORATION

ALBERTO G. FAIRÉN
EDITOR

nova publishers
New York

For permission to use material from this book please contact us:
Telephone 631-231-7269; Fax 631-231-8175
Web Site: http://www.novapublishers.com

Library of Congress Cataloging-in-Publication Data

Mars : evolution, geology and exploration / editor, Alberto G. Fairén.
p. cm.
Includes index.
ISBN 978-1-62618-102-1 (hardcover)
1. Mars (Planet)--Geology. 2. Mars (Planet)--Exploration. 3. Exobiology. I. Fairén, Alberto G.
QB643.G46M36 2013
523.43--dc23

2013004724

Published by Nova Science Publishers, Inc. †New York

CONTENTS

PREFACE: EARLY MARS, A COLD AND WET PLANET

This book is about Mars, its history and traits, and is also about us, scientists trying to understand the planet from home and sending bold surveyors to the dusty Martian surface, so far robotic explorers and hopefully soon humans. Let us start the book defining graphically the place where this volume stands in time and in space. In this book, Mars is described through the formal stratigraphic systems (Noachian, Hesperian, and Amazonian) devised by Scott and Carr (1978), the series (upper, middle, and lower divisions of systems) defined by Tanaka (1986), and the absolute age estimates for the surface of Mars based on crater statistics refined by Hartmann and Neukum (2001) (Table 1). The major geologic provinces described are (1) the ancient southern highlands, including Arabia, Cimmeria, Sirenum, Noachis, Aonia, Tempe and Xanthe terrae, the mountains ranges of Thaumasia highlands and Coprates rise, and the largest canyon in the Solar System, Valles Marineris; (2) the northern lowlands, including the large flats of Utopia, Isidis, Amazonis, Chryse, Elysium, Arcadia and Acidalia planitiae; (3) the major volcanic provinces such as Tharsis and Elysium rises, Syrtis Major, and Hadriaca, Tyrrhena and Apollinaris montes; and (4) the gigantic impact basins Hellas, Argyre and Isidis. The northern lowlands and the southern highlands of Mars are separated by a large topographic escarpment up to 5 km in height, known as the dichotomy boundary (Figure 1). Once we have settled where and when we are, we can have a glimpse to the contents of the book.

This volume starts with a detailed summary of the geological evolution of Mars as can be envisioned today. Mars, like other terrestrial planets, should have evolved through accretion, differentiation, and final lithospheric tectonism controlled by mantle convection and expressed as either plate tectonics or stagnant-lid tectonics. A main characteristic of its evolution is that Mars has been a water-rich terrestrial planet up to recent times, but especially during the Noachian. Geologic evidence indicative of flowing and ponding liquid water on the surface of ancient Mars appears abundantly across most of the Martian landscape, indicating that liquid water has been present in variable amounts and for long periods of time on and/or near the surface at different moments of Mars' early history, the Noachian period. Early Mars appears to have been "wet". However, the presence of liquid water on the surface of early Mars is difficult to reconcile with the reduced solar luminosity at 3.8 Ga. and before (with a relative solar constant $S/S0 \cong 0.75$ so that the Sun was about 25% fainter than today), which would have imposed mean temperatures below freezing all over the planet. Atmospheric greenhouse gases and carbon dioxide ice clouds in the upper troposphere are suspected to provide the means for over freezing temperatures in early Mars, explaining some of this discrepancy, but these solutions have been probed to face numerous problems. So, early Mars was "cold", and therefore it is difficult to explain the early Martian hydrology invoking global "warm" conditions. An alternative and exciting possibility is that the Martian environment was characterized by global mean freezing conditions, as predicted by climate models, and at the same time a vigorous hydrogeological cycle was active during hundreds of millions of years, as confirmed by geomorphological and mineralogical analyses. In this scenario of a "cold and wet" early Mars, a combination of geochemical and climatic traits would help explain the presence of liquid water on the cold surface of early Mars. Some advancements on our understanding of the environmental conditions of a "cold and wet" early Mars are offered in this volume.

Table 1. Absolute age estimates for the surface of Mars

Epoch	Absolute age range (Ga)
Late Amazonian	0.6-0.3 to present
Middle Amazonian	2.1-1.4 to 0.6-0.3
Early Amazonian	3.1-2.9 to 2.1-1.4
Late Hesperian	3.6 to 3.1-2.9
Early Hesperian	3.7 to 3.6
Late Noachian	3.82 to 3.7
Middle Noachian	3.95 to 3.82
Early Noachian	>3.95

The book subsequently deepens in the depiction of particular geomorphologies and mineralogies characterizing specific regions and/or times, for an enhanced understanding of the general picture defining the geology and evolution of Mars under cold aqueous environmental conditions. First, we will find discussions about the presence of icebergs in very cold oceans occupying the northern plains of Mars in the past. The northern plains of Mars appeared in early orbital images as persistently flat and showing few evident topographic features, closely resembling that of the ocean basins on Earth, and as such the lowlands have been previously interpreted as the floor of an ocean that once covered the Martian northern hemisphere. But some differentiated geomorphologies are appreciable in modern high resolution datasets throughout the northern lowlands of Mars, including boulders distributed forming clusters, flat-topped and conical circular fractured mounds, and scattered plough marks and chains of crater-like structures. These features are proposed here to be informative of processes of iceberg rafting, scouring and grounding on ancient cold glacial oceans on Mars.

The book then describes the existence of massive glaciated landscapes rimming the planet's dichotomy boundary, between the northern lowlands and the southern highlands, at the time when the cold ocean was ponded on the plains. Imprints of the ancient glaciers can be found on the Martian fretted terrains. Fretted terrain landforms occur along the topographic dichotomy boundary, within latitudes ranging from 20ºS to 45ºN. They range from small, isolated buttes and knobby terrain, to large, flat-topped mesas and plateaus, with major axes several hundred kilometres in length, and flanked by abrupt escarpments up to 1–2 km high. Here is proposed that the dichotomy boundary would have been glaciated in a restricted sense, i.e. with a cold ocean to the north supplying moisture, but the glaciers unable to extend too far south on the highlands because they would have been further away from the moisture source.

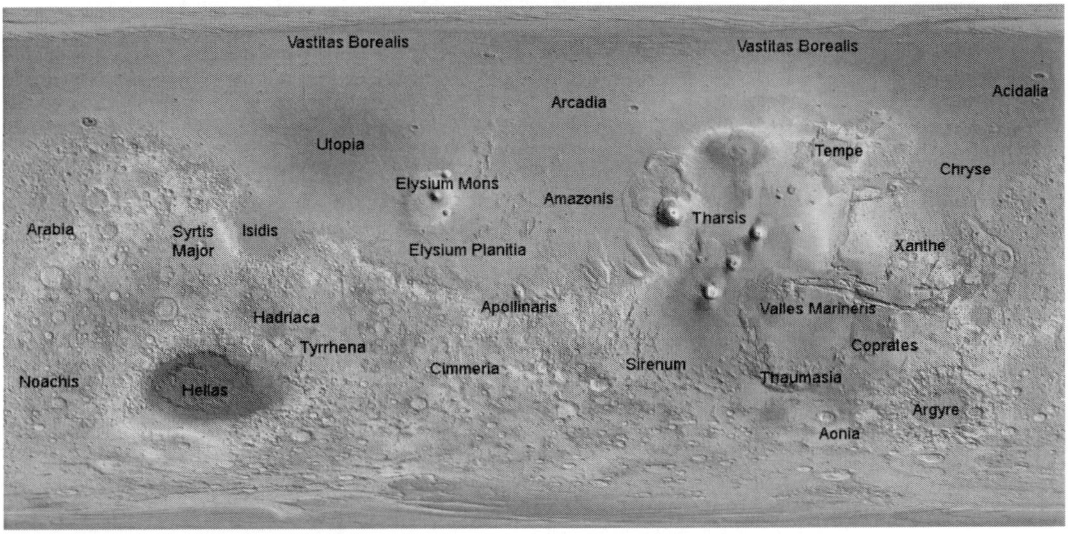

Figure 1. The major geologic provinces on Mars. Base map is a laser altimeter topographic map (NASA).

The following chapter deals with the evolution of trough morphologies on the South Pole. The north and south polar plateaus of Mars consist of CO_2 and H_2O ice intermixed with dust, forming layered strata developed in response to Mars' cyclical obliquity, orbital eccentricity, and precession. The surface age of the South Polar layered strata has been estimated to be younger than 100 Ma, but some of the layered materials may date from times as old as from the beginning of the Amazonian Period, up to 3.5 Ga. These materials can offer clues about the environmental evolution of Mars at the time when the surface of the planet transitioned from "cold and wet" to "very cold and hyperarid" prevalent conditions, and therefore about the fate of the water that characterized the surface of Mars during the Noachian and the Hesperian.

Next, the focus is on the aqueous mineral deposits formed in cold watery environments and eventually identified on the surface and shallow subsurface. Iron oxides and oxyhydroxides, hydrated sulfate salts, hydrated silica, phyllosilicates, carbonates and chlorides have been detected on Mars. Here is proposed that all these major water-derived mineralogies were deposited simultaneously in space and in time, in different aqueous environments, during the "cold and wet" Noachian/Hesperian times. This would be indicative of the presence of coetaneous diverse geochemical conditions over the entire early Martian surface, revealing an active and environmentally heterogeneous early Mars, which would have witnessed changing weathering regimes.

The book then focuses on the possibilities for life origin and development on Mars, particularly analyzing the biological potential of the young Mars. On early Mars, environmental conditions were apparently similar to those of the primitive Earth when life arose, and Mars seems to have had many of the essential conditions that are expected to have intervened in the emergence of life on Earth. In a "cold and wet" early Mars scenario, favorable conditions for habitability would have included a denser and warmer atmosphere than today, volcanism to recirculate gases and water and to produce new heterogeneous environments, permanent bodies of liquid water in which chemical evolution can occur, a global magnetic field to protect the planet from damaging radiation, and availability of chemical elements. In this scenario, Martian hydrothermal systems would have been the best places where physico-chemical processes could have evolved towards higher levels of chemical organization, as occurred on Earth.

An additional discussion follows about the biological communities thriving in the Antarctic Dry Valleys, as the microbiota typical of the polar environments on Earth is the most relevant analog for a possible primeval biosphere on an early "cold and wet" Mars. Aqueous environments with low temperatures and high salt concentrations, such as the Antarctic Dry Valleys, are considered good analogues to study the surface conditions on early Mars. In such environments, as liquid water levels decrease by freezing and/or evaporation, the concentration of dissolved salts increase, creating brine pockets where the freezing point of water is greatly depressed, to temperatures as low as -20°C. The growth of psychrophiles has been measured in the Dry Valleys (and many psychrophiles are also halophiles), demonstrating that metabolism is not constrained to external temperature. The Dry Valleys are therefore an optimal location for testing hypotheses about the limits to which life can survive in very cold conditions, such as those defining the surface aqueous environments of early Mars.

The last part of the book is devoted to the analysis of the exploration of Mars and the search for life on the planet, delving into three complementary and sequential strategies: (1) in situ robotic analyses, (2) bringing Martian samples to Earth for examination, and (3) sending humans to Mars for in situ exploration. Arguments are provided here to support the extended notion that the more promising target in the search for life on Mars is the Martian subsurface. This is so because shallowly buried materials would have been sheltered from the harsh conditions defining the surface of Mars during the last 3 billion years, and therefore these covered materials are the most promising settings to have preserved organic signatures from ancient living forms, or even to maintain subsurface niches for current life. Here is argued that, in order to identify ideal locations for subsurface analyses which would be accessible by in situ robotic exploration, three characteristics of the sampling site need to be determined: the dust and dune coverage, the cap layer distribution, and the location of outcrops. For the sample return-oriented strategies, here is provided an overview on methods which can be used to address the designs of both a generic sample return capsule and an entry sample return vehicle. The sample return capsule would be expected to be able to select, collect and finally return samples to Earth. The entry sample return vehicle would need to withstand high entry velocities, heat fluxes and heat loads. Finally, in the effort to send humans to Mars for exploration, here are investigated different options that could bring important simplifications to a potential manned trip to Mars, and significantly reduce the risks.

All the discussions summarized above, and detailed hereafter in the chapters of this book, configure a general description of our understanding of the geological evolution and the prospects for exploration of Mars at the time of this writing. Indeed, it is challenging to publish a book about Mars when an armada of audacious missions is still operating on the planet (2001 Mars Odyssey, 2003 Mars Express, 2004 Mars Exploration Rover Opportunity, 2006 Mars Reconnaissance Orbiter), and especially after the successful landing (on August 6, 2012) of the more sophisticated and complete spacecraft ever sent to another world: the Mars Science Laboratory (or Curiosity rover). Additional missions are already approved to fly on 2014 (Maven) and 2016 (InSight), as this journey of discoveries continues. Since this book is delivered for printing just at the beginning of Fall 2012, it is actually very possible that some of its contents would demand a certain degree of updating when the volume become available in libraries. But that would be excellent news, and in fact we might wish that the discussions presented henceforth in this book would become obsolete at the earliest possible time: that would mean that we keep on making astounding findings about the geology and the evolution of Mars in this golden age of planetary exploration.

Alberto G. Fairén
Space Science and Astrobiology Division,
NASA Ames Research Center, Moffett Field, CA, US

In: Mars
Editor: Alberto G. Fairén

ISBN: 978-1-62618-102-1

Chapter 1

MARS EVOLUTION

J. M. Dohm*[*1,2], H. Miyamoto[3], S. Maruyama[1], V. R. Baker[2,4], R. C. Anderson[5], B. M. Hynek[6,7], S. J. Robbins[7], G. Ori[8], G. Komatsu[8], M. R. El Maarry[9], R. J. Soare[10], W. C. Mahaney[11], K. J. Kim[12] and T. M. Hare[13]

[1]Earth-Life Science Institute, Tokyo Institute of Technology, Meguro, Tokyo, Japan
[2]Department of Hydrology and Water Resources, University of Arizona, Tucson, AZ, US
[3]The Museum, University of Tokyo, Tokyo, Japan
[4]Lunar and Planetary Laboratory, University of Arizona, Tucson, AZ, US
[5]Jet Propulsion Laboratory, California Institute of Technology, Pasedena, CA, US
[6]LASP, University of Colorado, Boulder, CO, US
[7]Geological Sciences Department, University of Colorado, Boulder, CO, US
[8]International Research School of Planetary Sciences, Università d'Annunzio, Pescara, Italy
[9]Physikalisches Institut, Universität Bern, Bern, Switzerland
[10]Department of Geography, Dawson College, Montreal, Canada
[11]Quaternary Surveys, Toronto, Ontario, Canada
[12]Geological Research Division, Korea Institute of Geosciences and Mineral Resources, Daejeon, South Korea
[13]United States Geological Survey, Flagstaff, AZ, US

ABSTRACT

An overarching geologic hypothesis, GEOMARS, coherently explains many aspects of the geological history of Mars. This includes ancient geologic terrains that comprise mountain ranges, structurally-controlled basins, highly degraded promontories, magnetic anomalies, and basin-containing stacked sequences of sedimentary deposits. In addition,

[*]Corresponding author: Earth-Life Science Institute, Tokyo Institute of Technology, Meguro, Tokyo, Japan, jmd@hwr.arizona.edu.

the theory also clarifies an ever increasingly realized mineralogically diverse planet, marine-, lacustrine-, fluvial-, eolian-, and glacial-sculpted landscapes, and superplumes that have dominated the geologic, hydrologic, and climatic histories for more than 3.5 Ga until the present. The hypothesis comprises eight major stages of martian geological evolution (from oldest to youngest): Stage 1 - shortly after accretion, Mars differentiates to a liquid metallic core, a mantle boundary (MBL) of high-pressure silicate mineral phases, upper mantle, magma ocean, thin komatiitic crust, and convecting steam atmosphere; Stage 2- Mars cools to condense its steam atmosphere and transforms its mode of mantle convection to plate tectonism; subduction of water-rich oceanic crust initiates arc volcanism and transfers water, carbonates, and sulfates to the mantle; Stage 3 - the core dynamo initiates, with the associated magnetosphere and possible photosynthetic production of oxygen; Stage 4 –accretion of thickened, continental crust and subduction of hydrated oceanic crust to the mantle boundary layer and lower mantle of Mars continues; Stage 5 - the core dynamo stops during Noachian heavy bombardment but plate tectonism continues; Stage 6 – large basin-forming impacts result in Hellas, Argyre, Isidis, and Chryse; Stage 7–plate tectonism terminates and Tharsis superplume (~between 4.0 and 3.8Ga) initiates, and Stage 8 – the superplume phase (stagnant-lid regime) of martian planetary evolution with episodic phases of volcanism, water outflows, and related transient hydrological cycling as well as climate and environmental change which interrupts prevailing ice-house conditions.

This portrayal of the geological evolution of Mars, as a testable hypothesis, is highlighted through the geologic provinces of Mars and their special attributes. The provinces are (mostly from oldest to youngest, as there is an overlap in relative age among several of the provinces): the ancient southern highland province, including Terra Cimmeria, Terra Sirenum, Arabia Terra, Zanthe Terra, and the mountains ranges, Thaumasia highlands and Coprates rise, as well as the younger Hellas-Argyre province, Tharsis and other volcanic provinces such as Elysium rise, Malea Planum, Hadriaca and Tyrrhena Mons, Apollinaris Mons, and Syrtis Major, the northern plains, and the Tharsis/Elysium corridor.

Keywords: Mars, Evolution, Geology, Hydrology, Climate, Geologic provinces, superplume, volcanic provinces, Tharsis, Elysium, Terra Sirenum, Terra Cimmeria, Apollinaris, Valles Marineris, Argyre, Hellas, Life

INTRODUCTION

A coherent geologic hypothesis, the geological evolution of Mars and related synthesis (GEOMARS), was proposed to explain many first-order geological observations of Mars (Baker et al., 2007). The main ingredients of this hypothesis are that: (1) Mars has been a water-rich terrestrial planet up to present (<10's Ma), (2) Mars like other terrestrial planets should have evolved through accretion, differentiation, and final lithospheric tectonism controlled by mantle convection and expressed as either plate tectonics or stagnant-lid tectonics, and (3) the early history of Earth affords an analog to the evolution of Mars (Kaula, 1975; Sleep, 2000). The hypothesis leads to the following inferred major stages of Mars evolution (from oldest to youngest): Stage 1 - shortly after accretion, Mars differentiates to a liquid metallic core, a mantle boundary (MBL) of high-pressure silicate mineral phases, upper mantle, magma ocean, thin komatiitic crust, and convecting steam atmosphere (Fig. 1); Stage 2- Mars cools to condense its steam atmosphere and transform its mode of mantle convection to plate tectonism; subduction of water-rich crust initiates arc volcanism and transfers water, carbonates, and sulfates to the mantle (Fig. 2); Stage 3 - the core dynamo initiates, with the associated magnetosphere leads to conditions (Fig. 2); Stage 4 –accretion of thickened, continental crust and subduction of hydrated oceanic crust to the mantle boundary layer and lower mantle of Mars continues(Fig. 2); Stage 5 - the core dynamo stops during Noachian heavy bombardment while plate tectonism continues (Fig. 2); Stage 6 – large impact events, Hellas, Argyre, Isidis, and Chryse occur following the shutdown of the dynamo, but determining the exact age relationship between the shutdown of plate tectonism and the incipient development of Tharsis proves difficult; Stage 7 – plate tectonism terminates and Tharsis superplume initiates (~between 4.0 and 3.8Ga) (Fig. 3); and Stage 8 – the superplume phase (stagnant-lid regime) of martian planetary evolution with episodic phases of volcanism, water outflows, and related transient hydrological cycling as well as climate and environmental change away from prevailing ice-house conditions continues into the present, including the development of Valles Marineris which began to form at least as far back as the Late Noachian (Fig. 3). This portrayal of the evolution of Mars as a hypothesis that may be tested through ongoing and future orbital and in situ investigations is optimally highlighted through the geologic provinces of Mars and their distinct characteristics. The formal stratigraphic systems (Noachian, Hesperian, and Amazonian) devised by Scott and Carr (1978), the series (upper, middle, and lower divisions of systems) defined by Tanaka (1986), and the absolute age estimates for the surface of Mars based on crater statistics (Hartmann and Neukum, 2001) are used in this work, although such information is likely to be continually revised, especially when there is sample return from prime locations (e.g., Dohm et al., 2004; Fairén et al., 2005).

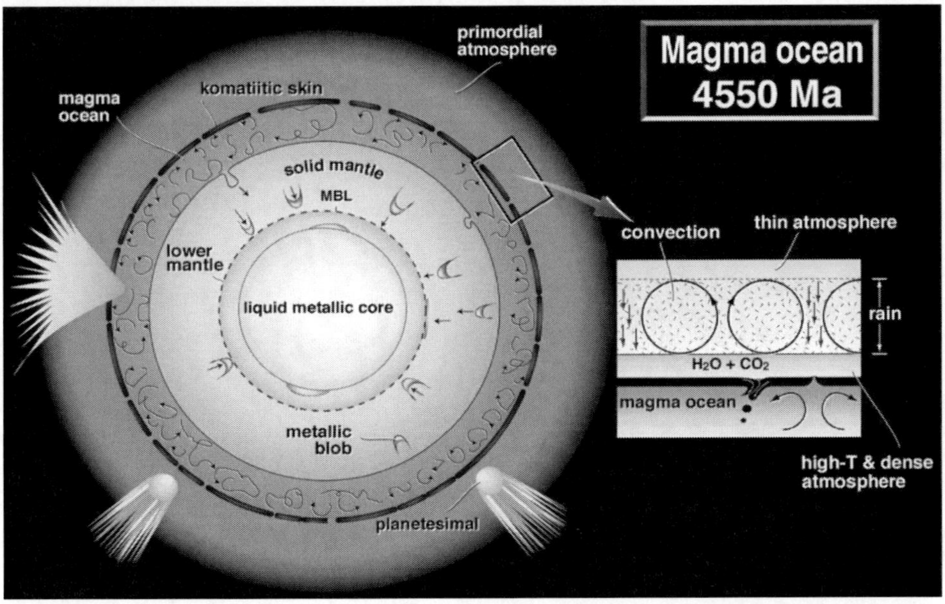

Figure 1. Adapted from Baker et al. (2007), schematic cross section through Mars shortly after accretion, showing differentiation, probably very rapid, to a metallic core, a mantle boundary layer (MBL) of high-pressure silicate mineral phases, upper mantle, magma ocean, thin komatiitic crust, and very hot convecting steam atmosphere (inset). Iron-rich metallic blobs are shown descending from the magma ocean and through the developing mantle in a process that generates the liquid metallic core.

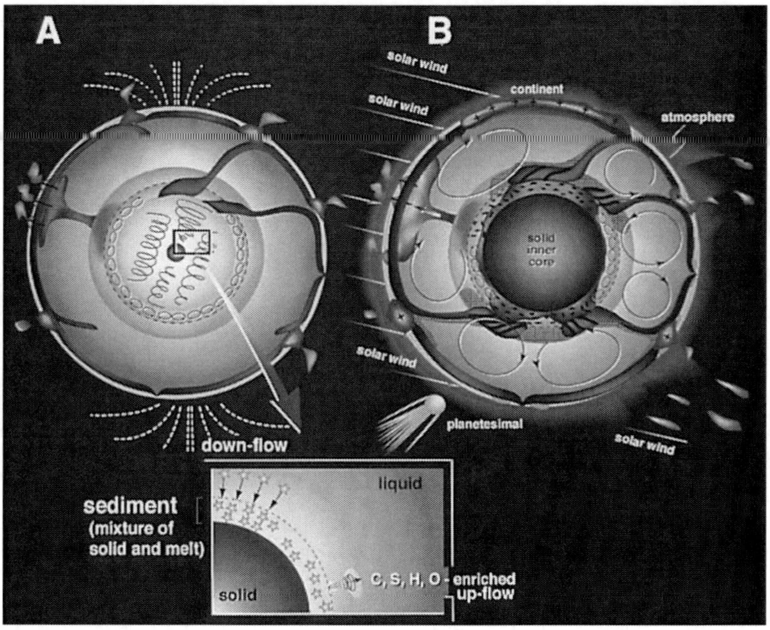

Figure 2.A. Adapted from Baker et al. (2007), possible scenario for the initiation of the martian core dynamo during the period of the early heavy bombardment. The inset detail shows how siderophile elements in the outer liquid core (yellow stars) move to the lowermost portions of the fluid core and chemical convection is promoted as light elements (C, S., H, O) are expelled from the solidifying inner core. The resulting dynamo continued for few hundred million years during a time or rapid accretion of thickened continental crust. The early Noachian ocean that promoted the warm, wet surface conditions conducive to rapid denudation was progressively depleted by the return of seawater through subduction of hydrated oceanic curst (shown as dark layer of surface material intruding into the planetary interior). B. The core dynamo stopped during the heavy bombardment, while plate tectonics continued, when the inner core grew so large that its internal convection could no longer sustain the dynamo.

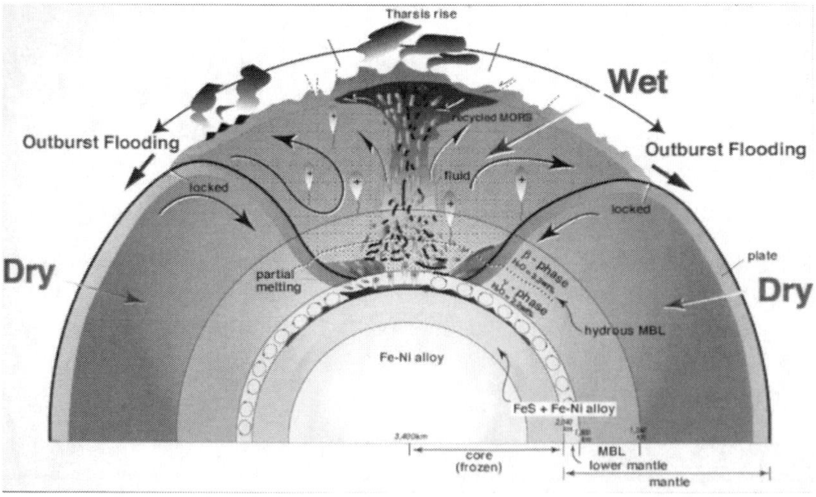

Figure 3. Adapted from Baker et al. (2007), hypothetical initiation of the Tharsis superplume by focused subduction, delivering huge amounts of water-rich oceanic crust to the local MBL. Massive basaltic volcanism at Tharsis is associated with temporary outgassing of water to enhance the atmosphere, and to produce huge outburst floods of water, episodically replenishing the northern plains ocean.

Table 1. Geologic province, GEOMARS stage, estimated age, aspect, and process

Province	Stage	Estimated Absolute age	Aspect	Process
Ancient Southern Cratered Highlands	1-5	> 3.8Ga	Linear crustal magnetization anomalies of surprisingly great intensity occur in portions of Mars' southern highlands; the remnant magnetization of crustal rocks implies that the martian dynamo shut off during the Noachian Epoch, about 4 billion years ago (Acuna et al., 1999), prior to the large impact events, Hellas, Argyre, and Isidis (Louzada et al., 2011)	Some form of plate tectonism (Sleep, 1994; Dohm et al., 2002; Fairen et al., 2002; Fairén and Dohm, 2004; Baker et al., 2007)
Ancient Southern Cratered Highlands	1-5	> 3.8Ga	martian crust showing major variations from thin beneath the northern plains (~30 km) to thick (~60 km) beneath the southern highlands and Tharsis (Zuber et al., 2000), as well as the origin of the great hemispheric dichotomy between the low-lying northern plains and the elevated southern highlands	Some form of plate tectonism, mantle convection associated with core formation (Wise et al., 1979), a colossal impact that formed a basin in the north polar region (Wilhelms and Squyres, 1984), and extensive southward erosional retreat of the heavily cratered highland plateau (Scott, 1978)
Ancient Southern Cratered Highlands	1-5	> 3.8Ga	Ancient mountain ranges, Thaumasia highlands and Coprates rise, which comprise complex structure such as thrust faults, cuestas, hogbacks (Dohm and Tanaka, 1999; Dohm et al., 2001a), as well as presently unknown primary rock compositions; other large structures include Phlegra Montes and Tanuis Fossae, located north of the Elysium rise and within the Tempe Terra plateau, respectively (Dohm et al., 2002),	Some form of plate tectonism, construction of a large volcanic pile that caused flexural bending of the underlying lithosphere (Williams et al., 2008), mega sliding (Montgomery et al., 2009)
Ancient Southern Cratered Highlands	1-5	> 3.8Ga	Structurally-controlled basins among highly eroded promontories and bands of strong magnetic signatures in the extremely ancient provinces of Mars (Dohm et al., 2002, 2005; Davila et al., 2011)	Some form of mobile crust, including possible plate tectonism, as well as growth of Tharsis such as possible formation and reactivation of basement structures in Terra Sirenum (Anderson et al., 2012; Karasozen et al., 2012)
Ancient Southern Cratered Highlands	1-5	> 3.8Ga	Deep crustal rocks possibly granitic in composition (Bandfield et al. 2000, 2004).	Intrusive activity of more felsic rocks (e.g., Dohm et al., 2009a), concentration related to hydrothermal activity (Ruff et al., 2011), heterogeneity in the mantle and crust exposed by processes such as plate tectonism and giant impacts (Baker et al., 2007)
Argyre-Hellas	6	4.2-3.9Ga*	Giant impact events and associated impact-induced basins, rim materials, basin deposits, and basement structures (Scott et al., 1986; Greeley and Guest, 1987; Moore and Wilhelms, 2007; Dohm et al., 2001a, 2011a,b,2012); destruction of remanent magnetization; Harvesting of olivine-enriched lower crustal and upper mantle materials (Louzada et al., 2011; Buczkowski et al., 2010).	Impact, including impact-induced basins and basement structures which would subsequently serve as sinks for sediments and liquid water, water-ice, and sediments and conduits for the migration of magma and other volatiles, respectively (Parker and Gorsline, 1991; Kargel and Strom, 1994; Kargel, 1995; Dohm et al., 2001a, 2011a,b, 2012; Moore and Wilhelms, 2007). Impact-induced basement structures would influence the development of volcanic fields on the margins of Hellas and Isidis(Anderson et al., 2008). Argyre lake sourced Uzboi Valles (Parker and Gorsline, 1991)

Table 1. (Continued)

Province	Stage	Estimated Absolute age	Aspect	Process
Tharsis and other volcanic provinces such as Elysium rise, Malea Planum, Hadriaca Mons and Tyrrhena Mons, Apollinaris Mons, and Syrtis Major	7-8	>3.8-present-day	Tharis and Elysium superplumes, and Apollinaris, Tyrrhena and Hadrica, Malea Planum, and Syrtis Major volcanic fields (Scott and Tanaka,1986; Greeley and Guest, 1987; Mouginis-Mark, 1990; Crown et al., 1992; Dohm et al., 2001b, 2007; Tanaka et al., 2005; Williams et al., 2010a,b). Tharsis is composed of distinct features such as Valles Marineris (Fig. 10), which began to form during the Naochian Period.	Magma-water-basement structure interactions, including volatile release such as related to the development of the circum-Chryse outflow channel system and hydrothermal activity, endogenic-driven perturbations of environmental and global climate change, transient hydrological cycling away from prevailing ice house conditions due to the magmatic-driven outflow channel development and associated ponding to form water bodies ranging from oceans to lakes, and influential on life if existent (Baker et al., 1991, 2007; Dohm et al., 2001b, 2007, 2009c; Fairén et al., 2003)
Northern plains	1-8	>3.8-present-day	Northern plains and its record of diverse materials and far-reaching processes	Far-reaching record of marine and paleolake activity, impact cratering, magmatic- and impact-driven tectonism and related marsquakes, periglacial, glacial, aeolian, fluvial, debris-flow, and volcanic resurfacing, subterranean gas release such as sedimentary volcanism, and extant mantled ice bodies of regional extent from past water inundation phases (Tanaka et al., 2005; Dohm et al., 2009c). It could have originated due to extensive southward erosional retreat of the heavily cratered highland plateau, subcrustal erosion resulting from mantle convection, excavation by one or more large, overturn of unstable cumulates in an initial magma ocean, flexure of the southern highlands in response to vertical loading, or some form of plate tectonism (Scott, 1978; Wilhelms and Squyres, 1984; Sleep, 1994; Hess and Parmentier, 2001; Frey et al., 2002; Watters, 2003; Fairén and Dohm, 2004)
Tharsis/Elysium corridor	7-8	>3.8-present-day	Volcanoes and fissure-fed vent structures of diverse sizes and shapes, lava flows many of which have few if any superposed impact craters, structurally-controlled releases of water and other possible volatiles such as from Cerberus Fossae, channels that dissect relatively young lavas, Medusae Fossae Formation materials possibly ignimbrites, anastomosing dark slope streaks possibly indicative of water in cases, tectonic features such as fractures, faults, graben, and structurally-controlled pit crater chains that cut stratigraphically young rock materials, geologic terrains that display few if any superposed impact craters, and MO Gamma Ray Spectrometer (GRS)-based elevated hydrogen (H) and chlorine (Cl)	Long-lived magmatism, tectonism, and hydrologic activity often interlinked, including activity up to present-day. Amazonis and Elysium basins and connecting Marte Vallis (AME) is a prime target to test whether there is near-surface magma and groundwater, seismic activity, elevated heat flow, and life (Dohm et al., 2008 and references therein, 2012).

*based on Robbins and Hynek (2012).

THE GEOLOGIC PROVINCES OF MARS: WINDOWS INTO THE GEOLOGIC EVOLUTION OF MARS

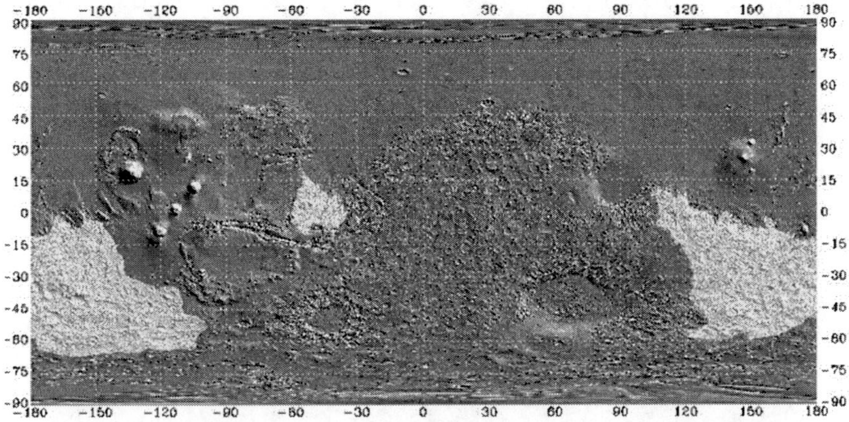

(a)

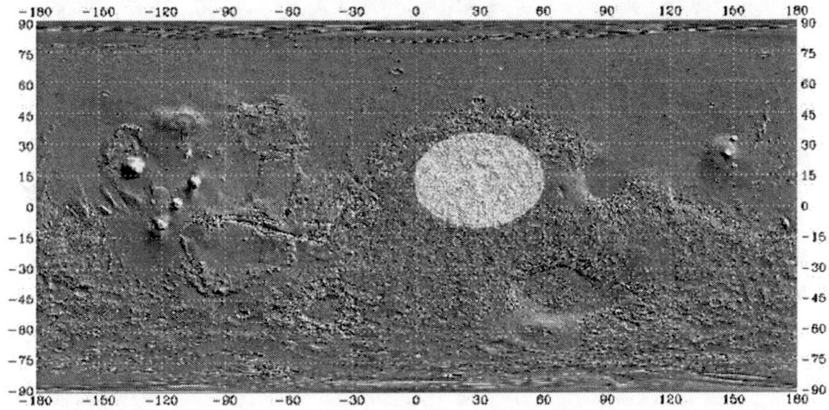

(b)

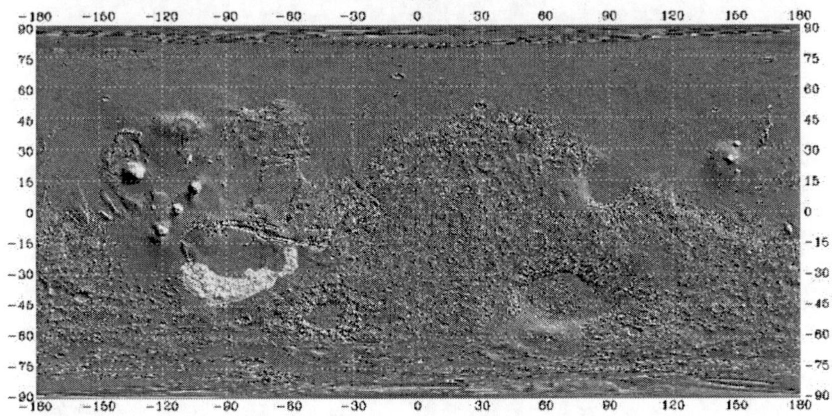

(c)

Figure 4. Ancient southern cratered highlands province, including (a) Terra Sirenum, Terra Cimerria, and Xanthe Terra, (b) Arabia Terra, and (c) the ancient mountain ranges, Thaumasia highlands and Coprates rise.

The geologic provinces, as identified through a synthesis of geologic, paleohydrologic, topographic, geophysical, spectral, and elemental information (Scott and Tanaka, 1986; Greeley and Guest, 1987; Acuña et al., 1999; Connerney et al., 1999; Bandfield et al., 2000; Dohm et al., 2005; Tanaka et al., 2005; Boynton et al., 2007), are windows into the evolution of Mars, including (mostly from oldest to youngest, as there is overlap of relative age among several of the provinces; also see Table 1): (1) the ancient southern highland province, including Terra Cimmeria, Terra Sirenum, Arabia Terra, Zanthe Terra, and the mountains ranges, Thaumasia highlands and Coprates rise (Fig. 4a-c), (2) Hellas-Argyre province (Fig. 5), (3) Tharsis and other volcanic provinces such as Elysium rise, Malea Planum, Hadriaca Mons, Tyrrhena Mons, Apollinaris Mons, and Syrtis Major (Fig. 6), (4) the northern plains province (Fig. 7), and (5) the Tharsis-Elysium corridor province (Fig. 8).

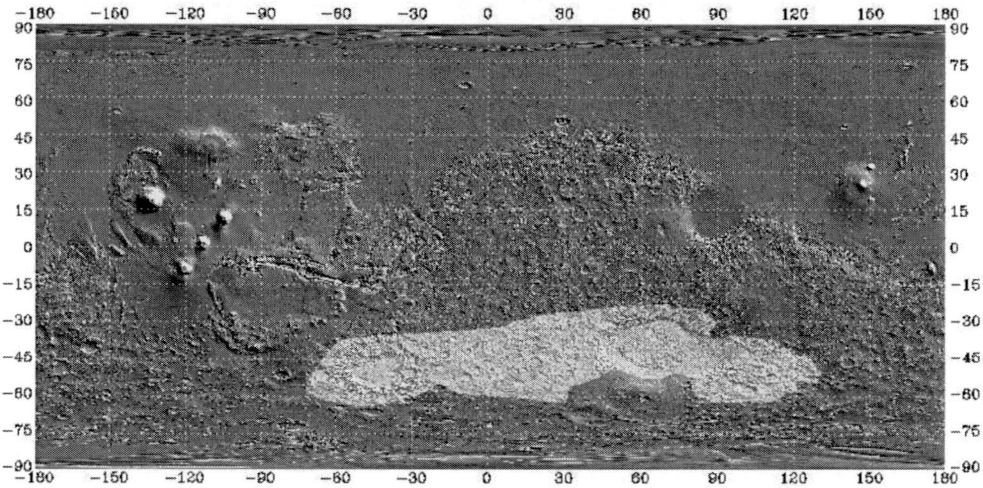

Figure 5. Hellas-Argyre province.

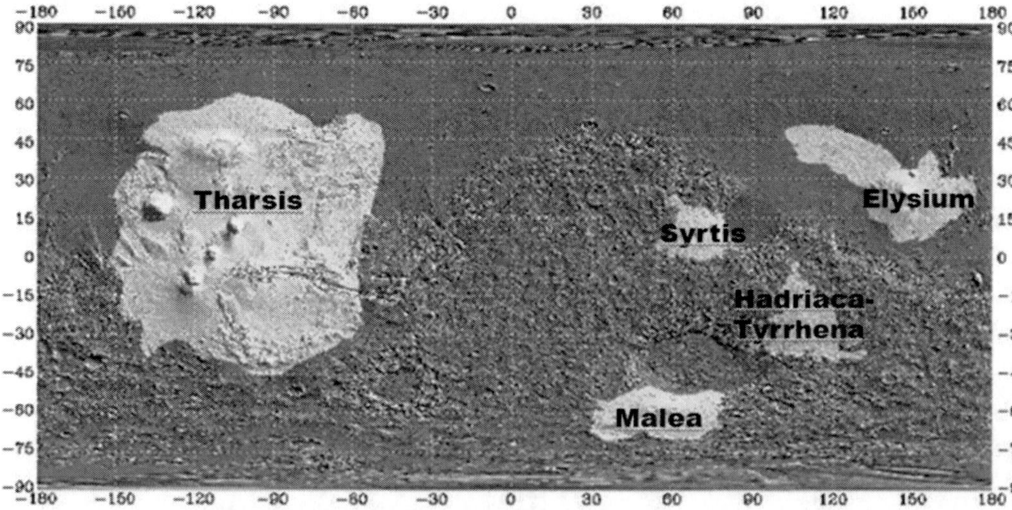

Figure 6. Tharsis and other volcanic provinces such as Elysium rise, Malea Planum, Hadriaca-Tyrrhena Montes, and Syrtis Major.

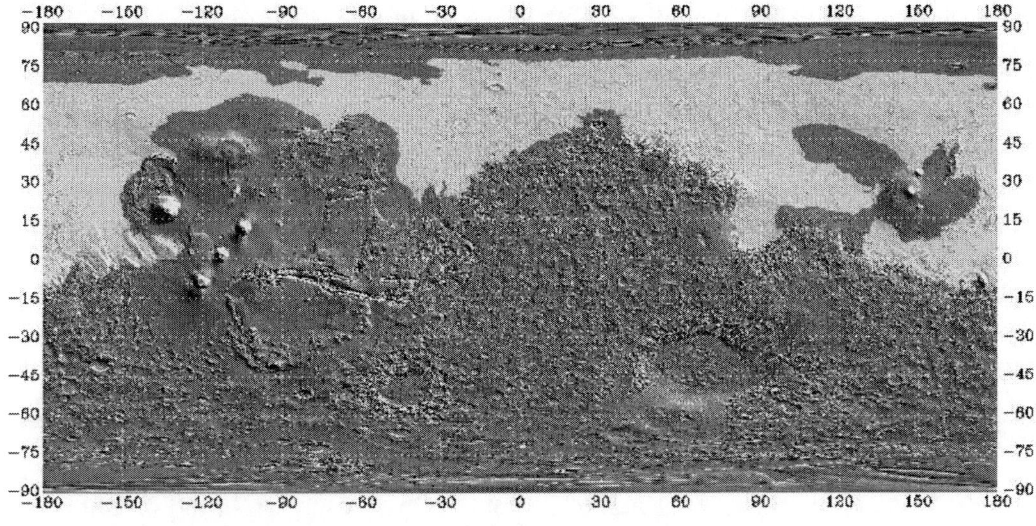

Figure 7. Northern plains province.

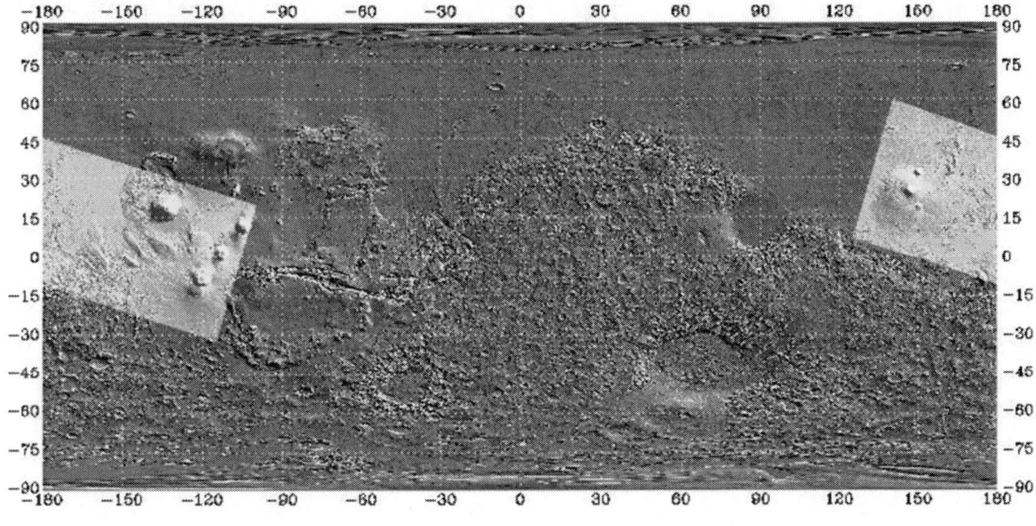

Figure 8. Tharsis/Elysium corridor province.

Ancient Cratered Southern Highland Province (Representative of Stages 2-5; Figs. 1,2,4a-C; Table 1)

The ancient cratered southern highland province (Fig. 1) includes Noachian geologic terrains that are marked by magnetic anomalies (Acuña et al., 1999, 2001; Connerney et al., 1999, 2005; Lillis et al., 2008). The terrains include: (1) Noachian mountain ranges, Thaumasia highlands and Coprates rise, both of which exhibit complex structures such as thrust and normal faults and rift systems, as well as cuestas and hogbacks along their margins (Dohm and Tanaka, 1999; Dohm et al., 2001a), (2) basin and range topography (Anderson et al., 2012; Karasozen et al., 2012), including salt-containing, structurally-controlled basins

(Davila et al., 2011), as exemplified through a detailed geologic investigation of Terra Sirenum (Anderson et al., 2012) and analysis of topographic profiles and crustal thickness models (Karasoen et al., 2012), (3) faults that are tens to thousands of kilometers long (referred to hereafter as macrostructures; Dohm et al., 2002), and (4) degraded promontories, many of which are interpreted as silica-rich volcanoes (e.g., Scott and Tanaka, 1986; Hodges and Moore, 1994) or in some cases, impact crater massifs (Craddock et al., 1990). These terrains can be aptly explained through dynamic endogenic activity, including some form of primitive plate tectonism and/or mobile crust (Dohm et al., 2002; Fairén et al., 2002; Fairén and Dohm, 2004; Yin, 2011; Yin, 2012 in press), as portrayed especially during GEOMARS Stages 2-5 (Fig. 2), and planetary shrinkage due to cooling, rather than impact events. Furthermore, such dynamic activity would result in a complex basement structural fabric that would play a key role in the migration of magma and volatiles such as water through time. Subsequent activity including salt tectonism (Montgomery et al., 2009; Wezel and Baioni, 2010) and Tharsis development (Dohm et al., 2001b, 2007a; Anderson et al., 2012; Karasoen et al., 2012) would contribute to the complex structural history of parts of the ancient cratered southern highland province, as discussed later, including formation and reactivation of pre-existing structures such as within the Thaumasia highlands and Coprates rise mountain ranges and the structurally-controlled basins of Terra Sirenum.

Also during this time, giant impact basins formed, becoming regional aquifers which were subsequently masked by significant erosion and/or burial. This hypothesis is exemplified by the putative Arabia Terra basin (Dohm et al., 2007b). Such regional aquifers interacted with volcanic provinces such as the Arabia Terra basin and the much younger Syrtis Major volcanic field, the marginal zone of which is considered a prime target for future exobiologic investigation (Dohm et al., 2011a). Other ancient terrains in the southern cratered highlands include Xanthe Terra, a highly cratered terrain that occurs along the eastern margin of the north-trending Maja Valles and forms the eastern margin of an ancient Europe-sized basin in the east-central/southeast part of Tharsis (referred to hereafter as the Tharsis basin) (Dohm et al., 2001b, 2007). The central and eastern parts of the circum-Chryse outflow channel system distinctly modify this part of the ancient southern cratered highlands province; thus, the surface expression of a complex basement structural fabric may be subdued when compared to other parts of the highlands.

Hellas-Argyre Province (Representative of Stage 6; Fig. 5; Table 1)

Large impact events, Hellas, Argyre, Isidis, and Chryse, occurred subsequent to the shutdown of the dynamo and even possibly contributing to the shutdown (Roberts et al., 2009; Roberts and Arkani-Hamed, 2012). The termination of the magnetic field has been reported to have occurred between the formation of Ladon and Hellas impact basins (Lillis et al., 2008) and the formation of Ladon and Promethus basins, the latter of which is dated to be older than Hellas (Robbins and Hynek, 2012). The large impact events resulted in primary impact basins, rim materials, basin deposits, and both deep-seated and shallow basement structures. As such, the giant impact events significantly influenced the geologic and hydrologic histories at regional to local scales from the time of impact until the present-day.

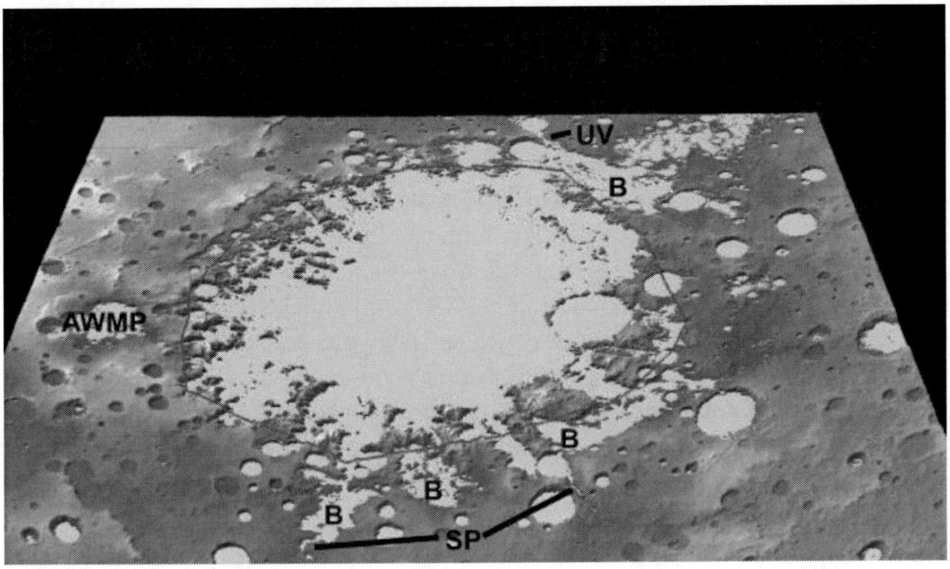

Figure 9. Based on Dohm et al. (2011b) schematic paleolake map of the Argyre basin using a maximum topographic elevation of 0 km based on MOLA topography (regions in blue).An estimated extent of the hypothesized Argyre lake based on geomorphologic and topographic analyses, as well as detailed geologic mapping is also shown (red line). In addition to the estimated extent, sapping channel systems (SP), local basins (B) which occur among the crater rim materials, and the Uzboi Vallis system (UV) correspond to the blue-highlighted region. Also shown is a small extent (near base level) of the Argyre west margin paleolake (AWMP). The volumes of the hypothesized AWMP and Argyre lakes are estimated to be 1.6 x 10^4 and 1.9x10^6 km^3, respectively, using MOLA. There is significant evidence of water-ice modification (e.g., glaciation) (e.g., Dohm et al., 2011b; Dohm and Kargel, 2012).The possibility of an interplay between lakes, ice sheets, glaciers, and ever changing conditions through time including waning water bodies.Further detailed geologic and geomorphic investigation is warranted.

In addition to destroying the magnetic signature and harvesting upper mantle and lower crustal materials (e.g., Buczkowski et al., 2010; Louzada et al., 2011), the Hellas and Argyre impact events influenced the geologic evolution at regional scales (Scott and Tanaka, 1986; Greeley and Guest, 1987; Parker and Gorsline, 1991; Crown et al., 1992; Dohm et al., 2001a; Moore and Wilhelms, 2007). For example, the Argyre event controlled the development of the Thaumasia plateau (Dohm and Tanaka, 1999; Dohm et al., 2001 a, b), the drainage pattern of the Uzboi drainage system due to its multi-ring structure (Parker and Gorsline, 1991), and the routing of glaciers from near the south pole (Kargel and Strom, 1994; Kargel, 2004). The Argyre basin was also the repository of large bodies of liquid water, ice, and sedimentary sequences in the impact-induced primary basin (Kargel 2004) and the secondary (smaller) basins which occur among and nearby the rim materials (Dohm et al., 2011b). Additionally, Hellas basin was a significant catchment of water and sediments (Greeley and Guest, 1987; Crown et al., 1992; Moore and Wilhelms, 2007). Distinct sequence stratigraphy through detailed mapping is observed in and along the margins of the Argyre basin recording significant geologic and hydrologic activity at regional to global scales (Dohm and Kargel, 2012; Dohm et al., map in preparation), including the proposed Tharsis-induced global perturbations from prevailing ice-house conditions, including hydrologic activity (glacial, periglacial, fluvial, spring-fed activity, etc.), solifluction, and climate change (Baker et al.,

1991; Dohm et al., 2007a; Dohm and Kargel, 2012). The Argyre basin contained large lakes with volumes as much as nearly 1.9 million cubic km. This is based on a Geographic Information Systems (GIS) calculation using Mars Orbiter Laser Altimeter (MOLA) topographic data at a 0 km datum (Fig. 9), sourced Uzboi Valles during the Noachian Period, and may still contain a sediment-enriched frozen lake protected from the atmosphere by a relatively thick mantle; this is indicated by numerous topographic depressions best explained through ablation (Dohm et al., geologic map in preparation).

Also, the giant impact events would form basement structures that could serve as conduits for the migration of magma to form Malea Planum and Tyrrhena and Hadriacus Montes volcanic fields and other volatiles such as water and carbon dioxide. Several Argyre-induced basement structures transect the transition zone that links the Argyre basin and the Thaumasia highland mountain range. Such structures would be significant conduits for the transfer of volatiles in the subsurface from the margin of the Thaumasia plateau and Thaumasia highland mountain range to the deep-seated Argyre catchment basin during major phases of Tharsis activity, for example.

Following the shutdown of the dynamo and inferred plate tectonism (Baker et al., 2007), such giant impact-induced structures would source Malea Planum, Tyrrhena and Hadriaca, and Syrtis Major volcanic provinces (Anderson et al., 2008). Interestingly, Argyre and highly-degraded Chryse impact basins do not comprise volcanic fields along their margins as do Hellas and Isidis. One possible explanation is that the nearby Tharsis superplume was the major heat release site for the western equatorial region of Mars, restricting the formation of volcanic fields along the margins of the two giant impact basins.

The Isidis impact basin also records a significant geologic history (Anderson et al., 2008). The basin occurs on the margin of the dichotomy boundary (Greeley et al., 1987; Tanaka et al., 2005), and is marked by a volcanic field with evidence of hydrothermal activity, Syrtis Major, on its western margin (Marzo et al., 2010; Ehlmann et al., 2010; Ehlmann et al., 2011), as well as other aqueous activity (Mustard et al., 2009). Thick sedimentary infill, diverse geomorphologic features, and basement structural control point to diverse and complex geologic and hydrologic histories, including impact-induced structures and subsequent control of the migration of subsurface volatiles (Komatsu et al., 2011) as well as the formation of water bodies ranging from lakes (Scott et al., 1995) to oceans (Fairén et al., 2003; 2009c).

Tharsis and Other Volcanic Provinces Such as Elysium Rise, Malea Planum, Hadriaca and Tyrrhena Mons, Apollinaris Mons, and Syrtis Major (Stages 7 and 8; Figs. 3 and 6; Table 1)

In the GEOMARS hypothesis, the most influential process on the evolution of Mars since the shutdown of the dynamo is the evolution of the Tharsis superplume. Though it is difficult to determine whether the incipient development of Tharsis occurred during the tail end of plate tectonism or during the giant impacts of Stage 6, especially since it has masked remanent magnetism in the ancient basement that underpins the volcanic lava flows (e.g., Lillis et al., 2009), its evolution dates back to at least the Middle Noachian epoch (Dohm et al., 2001b), or greater than 3.8 Ga according to the cratering model of Hartmann and Neukum (2001). Parts of Tharsis may be still active into the future (Dohm et al., 2008). An ancient, large promontory located to the southwest of Syria Planum, Claritas rise, with thousands of faults pointing back to the prominent feature with an associated remanent magnetic signature, may mark the superplume's incipient development or pre-Tharsis development (Dohm et al., 2009a). Occupying nearly 25% of Mars, Tharsis has been the focus of geologic and geophysical investigation (Phillips et al., 2001; Dohm et al., 2001b; Zuber, 2001; Solomon et al., 2005; Nimmo and Tanaka, 2005; Golombek and Phillips, 2010; Carr and Head, 2010). Existing hypotheses include endogenic activity (e.g., hotspot, mantle plume, and/or superplume activity) (Carr, 1974; Wise et al., 1979a; Mège and Masson, 1996a,b; Baker et al., 2007; Dohm et al., 2001b, 2007; Zhong, 2009; Hynek, et al., 2011) or impact cratering (Solomon and Head, 1982; Reese et al., 2004; Yin, 2012-in review).

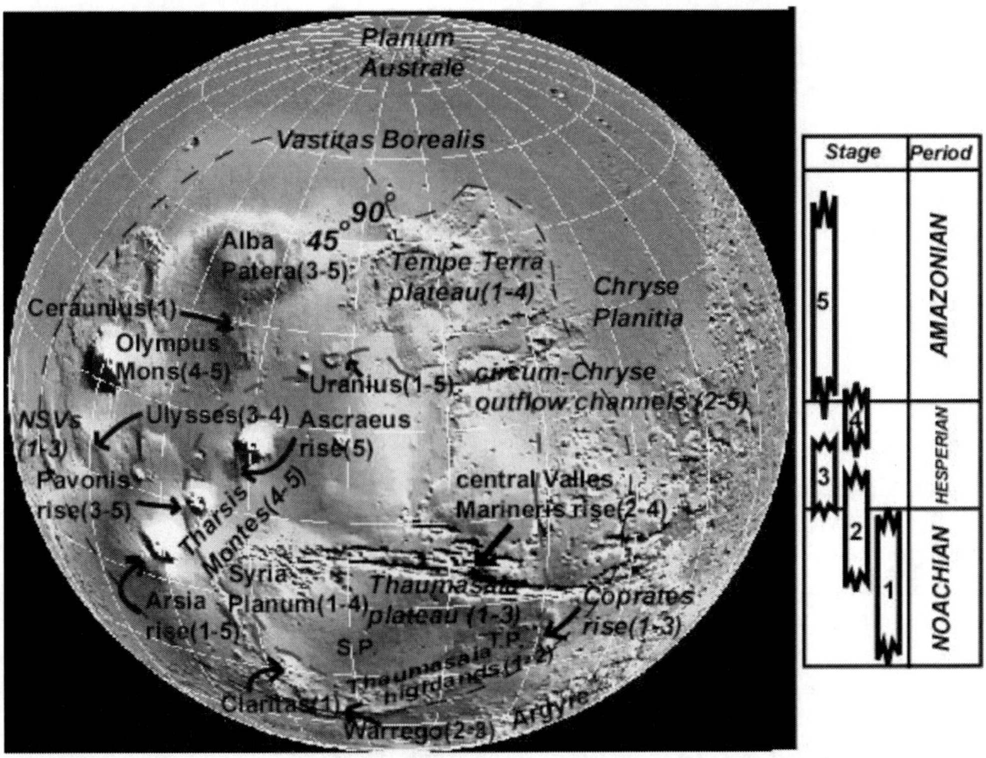

Figure 10. Adapted from Dohm et al. (2001b, 2007), MOLA shaded relief map (courtesy of the MOLA team) showing features of interest and their relative stage of Tharsis Superplume evolution, which includes northwestern slope valleys (NSVs) and circum-Chryse outflow channel systems (northwest and northeast watersheds of Tharsis, respectively), Tharsis Montes, Olympus Mons, and Alba Patera shield volcanoes, centers of magmatic-driven tectonic activity (Claritas, Ceraunius, Warrego, Syria Planum, and central Valles Marineris rise), Tempe Terra and Thaumasia igneous plateaus, ancient mountain ranges, Thaumasia Highlands and Coprates rise, Argyre impact basin, and parts of the northern plains (Chryse Planitia, Vastitas Borealis, and Planum Boreum).The five major stages of Tharsis Superplume activity (stages 1-5 after Dohm and Tanaka, 1999; Dohm et al.,2001a,b) correspond to the geologic periods of Scott and Tanaka (1986).Also shown is the estimated extent of the Tharsis Superplume (black dashed line), which encompasses a total surface area of approximately 2 X 10^7 km^2 compared to the largest terrestrial igneous plateau, Ontong-Java, which encompasses a total surface area of approximately 2 X 10^6 km^2, and the Europe-sized Tharsis Noachian drainage basin/aquifer system (blue dashed line) of Dohm et al. (2001b), which is considered to be the primary source of water for the northwestern slope valleys (NSVs) and circum-Chryse outflow channel systems.

Tharsis records at least five major stages of activity (Dohm and Tanaka, 1999; Dohm et al., 2001b, 2007; Anderson et al., 2001), with intensive early magmatic/tectonic activity that declines with time from the Noachian Period to the geologically recent past. Tharsis superplume activity included ascending magma bodies manifesting in igneous plateaus, volcanoes of diverse sizes and shapes, lava flow fields, and shield fields (e.g., Scott and Tanaka, 1986; Frey and Grant, 1990; Hodges and Moore, 1994; Dohm et al., 2001a; Hauber et al., 2009) (Fig. 10). This activity included development of local and regional centers of tectonic activity of varying age, size, and duration of formation (Anderson et al., 2001), as well as fault and rift systems of varying extent and relative age of formation, including Valles Marineris (Dohm et al., 2009b).

Valles Marineris, a vast and deep canyon system, is one of the most conspicuous features of Tharsis; its length would extend across the United States from Los Angeles to New York and its depth fitting Mount Everest which has a peak at about 8.9km above sea level. Faults of the western margin of the Thaumasia Planum tectonic province (highlighted as T.P. in Fig. 10) are overlain by lavas from the west in Sinai Planum (S.P. in Fig. 10); here there is very little to no evidence of extensional deformation. These stratigraphic and cross-cutting relations among lavas and faults delineate a distinct geologic contact extending from the central part of Valles Marineris to the Thaumasia highlands mountain range, thousands of kilometers to the south. The Early Hesperian lavas of Sinai Planum constrain incipient Late Noachian Valles Marineris-related deformation (Dohm and Tanaka, 1999; Dohm et al., 2001c, 2009b).

A variety of processes have been proposed to explain the development of Valles Marineris, including: (1) rifting (e.g., Schulz, 1991), (2) collapse of near-surface materials due to the withdrawal of underlying material such as magma (e.g., McCauley et al., 1972) or opening of tension fractures at depth (e.g., Tanaka and Golombek, 1989), (3) development of keystone grabens at the crest of a bulge (e.g., Lucchitta et al., 1992 and references therein),(4) massive dikeemplacement causing ground-ice melting and thus catastrophic formation of outflow channels(McKenzie and Nimmo, 1999), (5) rotation of the Thaumasia plateau (as a lithospheric block) (e.g., Anguita et al., 2001), or (6) right-slip or left-slip faulting related to plate tectonics, lateral extrusion, or continental-scale mega landslide emplacement (e.g., Montgomery et al., 2009; Yin, 2012 and references therein). Many of these hypotheses regarding the formation of Valles Marineris may not be mutually exclusive (see Dohm et al., 2009b and references therein).

Most of the local and regional centers of tectonic activity (Anderson et al., 2001) noted above are interpreted to be the sites of plume-related activity, including uplift, extensional and contractional tectonism, dike emplacement, volcanism, and hydrothermal activity, such as recorded for the central part of Valles Marineris, Melas Chasma (e.g., Dohm et al., 2001b, 2007, 2009b). This activity and interaction with the Tharsis basin and associated productive aquifer and sedimentary deposits, which include evaporite deposits composed of various salts, highly influenced the formation Valles Marineris (Figs. 10-12). Such activity may have resulted in clockwise Thaumasia block rotation and left-lateral movement in the canyon among other dynamic processes, as observed through detailed geologic investigation (Yin, 2012). Faults and folds within the canyon may point to primitive plate tectonism such as hypothesized for Stages 1-6 of the GEOMARS hypothesis (Yin, 2012).

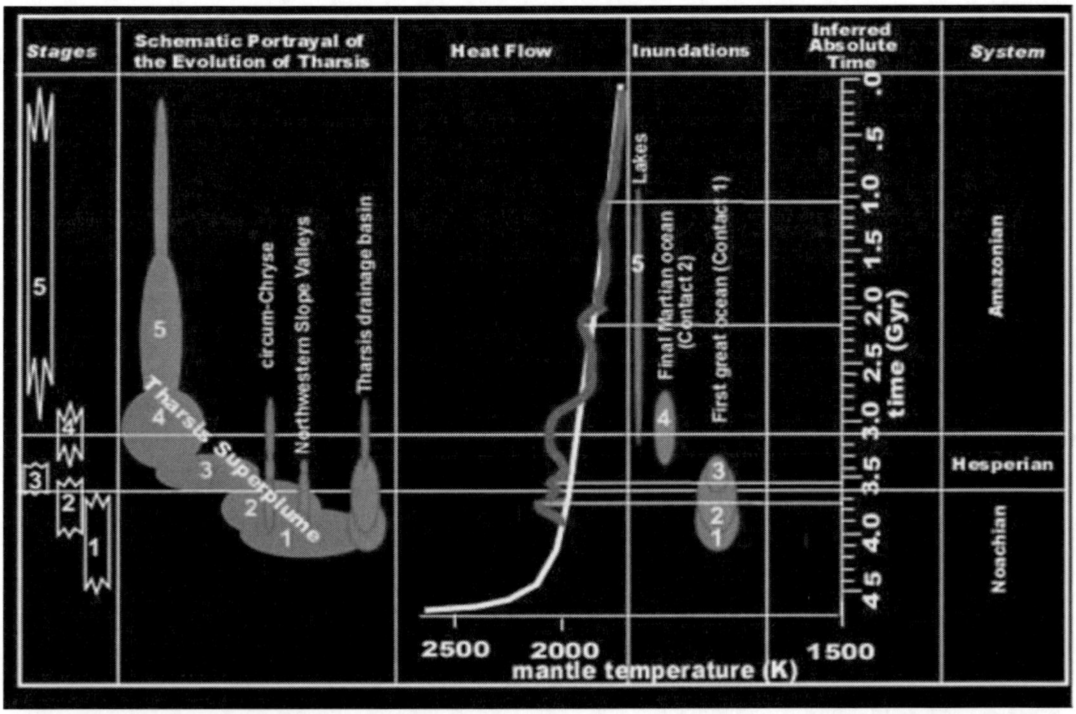

Figure 11. Adapted from Dohm et al. (2007), chart comparing the evolutional stages of geologic activity in the Tharsis Superplume (similar to Figure 1), which includes circum-Chryse, NSVs, and Tharsis drainage basin/aquifer system (Dohm et al., 2001b), with: (1)heat flow; note the maximum effective heat flow from the core to lithosphere in the Early and Middle Noachian (white line) and non-steady-state decline in subjective heat flow extending from Middle Noachian to present (red line) compared to proposed steady-state decline in mantle temperature with time (white line; Schubert et al., 1992) based on published geologic information (e.g., Dohm and Tanaka, 1999; Dohm et al., 2001a-b; Anderson et al., 2001; Fairén et al., 2003; Baker et al., 2007), (2) hypothesized Tharsis-triggered inundations in the northern plains ranging from oceans to lakes (Fairén et al., 2003), (3) inferred absolute time (Hartmann and Neukum, 2001), and (4) System information of Scott and Tanaka (1986).Sizes of solid areas are roughly proportional to the degree of exposed activity.

Such interaction among Tharsis magmatism, the Tharsis basin (Dohm et al., 2001b; Komatsu et al., 2004), and basement structures dominate the evolution of Mars during the stagnant-lid phase of Mars (Stages 7-8). Tharsis-associated, magmatic-driven activity triggered outflow channel flooding and associated ponding to form oceans (e.g., Baker et al., 1991; Fairén et al., 2003; Komatsu et al., 2004) and/or lakes (Scott et al., 1995) (Fig. 11), short-lived climatological perturbations on the order of tens of thousands of years, transient hydrological cycling. This transient environmental change included fluvial activity such as spring-fed activity along the highland-lowland boundary and into the giant impact basins, local valley network development such as within parts of Valles Marineris (Mangold et al., 2004), the formation of glaciers and ice sheets, especially in the northern and southern hemispheres (e.g., Kargel and Strom, 1994; Baker, 2001; Kargel, 2005; Fairén et al., 2012), and solifluction (Squyres, 1989) involving melting ice and rock materials such as clays along martian hill slopes. Prominent Late Noachian-Early Hesperian fluvial activity could be linked to Tharsis activity (Hynek et al., 2010; Hoke et all., 2011), specifically defined as stage 2 activity by Dohm et al. (2001b). Tharsis activity may have also resulted in ash flow and far-reaching air-fall deposits (Hynek et al., 2003). Magmatic activity related to the early growth of Tharsis, which included fissure-fed eruptions and widespread tectonism, transitioned into more concentrated activity to forge the giant shield volcanoes (Dohm and Tanaka, 1999; Dohm et al., 2001a,b; Anderson et al., 2001).

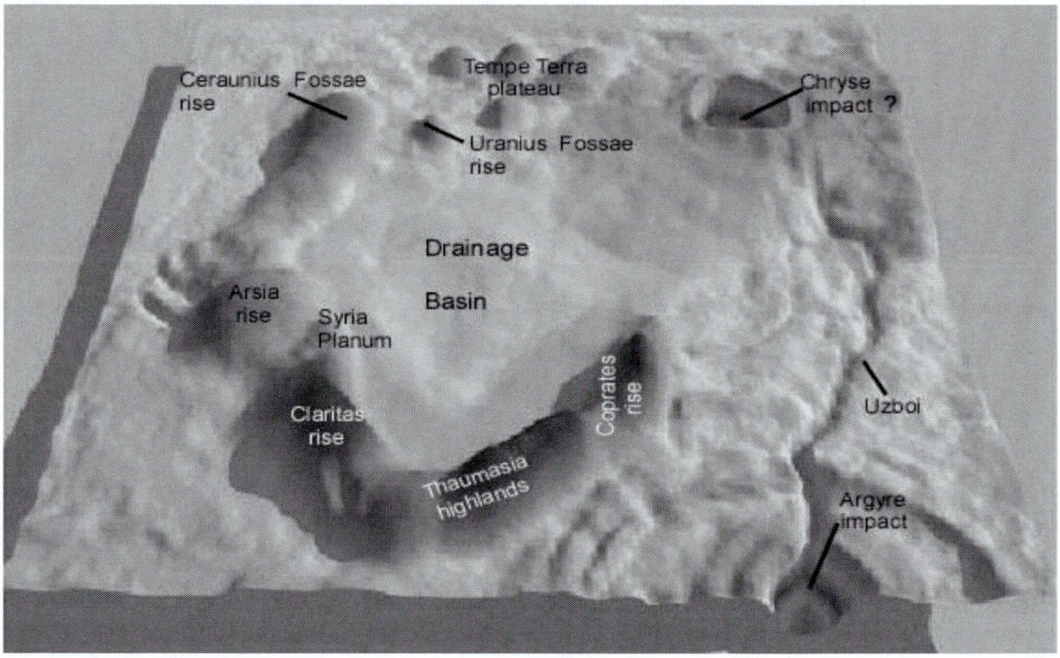

Figure 12. Adapted from Dohm et al. (2001a, 2007), 3-D portrayal of the major geologic features of the Tharsis Superplume during the Early into Middle Noachian Period (part of stage 1), including the drainage basin. Significant highlights include: (1) drainage basin, (2) distinct mountain ranges of Coprates rise, Thaumasia highlands, and multi-ringed structures of the Argyre impact basin (the Charitum and Nereidum Montes), (3) incipient Ceraunis Fossae, Claritas, and Arsia rises, Syria Planum, and Tempe plateau, (4) absence of highland-lowland boundary (further to the north outside of the scene; interpreted to be partly the result of Early into Middle Noachian plate tectonism (Dohm et al., 2002; Fairén et al., 2002; Fairén and Dohm, 2004; Baker et al., 2007), Valles Marineris, circum-Chryse outflow channel systems, and Thaumasia plateau, (5) distinct Argyre and putative Chryse impact basins, and (6) Uzboi Vallis.

Wrinkle ridge plains are also a prominent feature of parts of Tharsis. In Tharsis, wrinkle ridge plains have been attributed to flexural thrust faulting related to the growth of the volcanic province (e.g., Banerdt et al., 1992). Wrinkle ridge plains outside of Tharsis, such as at Hesperia Planum, point to additional possible mechanisms for their formation. These include basement structural control such as induced by the large impacts and subsequent geologic activity (e.g., Anderson et al.,, 2008), shrinkage due to planetary cooling (e.g., Watters, 1988), and dewatering of basin-infill materials, such as likely during the expulsion of groundwater from evaporite deposits of the Tharis basin associated with Tharsis-triggered outflow channel activity (Dohm et al., 2001b; Montgomery et al., 2009; Wezeland Baioni, 2010).

Without efficient cycling of internal heat energy since the termination of the inferred early plate tectonism, Tharsis internal heat releases were often catastrophic, perturbing prevailing ice-house conditions into a dynamic atmosphere and landscape (Baker et al., 1991, 2001, 2007). Such dynamic activity is recorded in various parts of Mars, including sequence stratigraphy identified and mapped in the Argyre basin and surroundings (Dohm and Kargel, 2012; Dohm et al., geologic map in preparation), regional deposits in the northern plains (Tanaka et al., 2005; Skinner et al., in preparation) and associated volatile-release features (Mahaney et al., 2004), and sedimentary deposits in the structurally-controlled basins, especially highlighted in the Terra Sirenum region (Anderson et al., 2012; Anderson et al., in preparation). In addition to flexure related to the load of Tharsis (Banerdt et al., 1992; Watters, 1993), partial dewatering of the basin-associated regional aquifer and salts during the development of the circum-Chryse outflow channel system may have contributed to the formation of wrinkle ridges through contractional deformation (Montgomery et al., 2009), among other structures. Detailed mapping related to the compilation of the new global geologic map of Mars (Tanaka et al., 2011) has identified wrinkle ridges which transition into graben and fractures and even volatile-release features, adding credence to such a possibility, while some could mark dike emplacement (Scott, 1981; Scott and Dohm, 1990).

What triggered the giant superplume and when was its initial stage of development with respect to the shutdown of the dynamo (e.g., Stevenson, 2001; Baker et al., 2007)? Several hypotheses have been proposed to explain the formation of the superplume, including the concentration of hydrated crustal materials into the core-mantle boundary region through plate tectonism (Baker et al., 2007), a migrating plume from the south (Zhong, 2009; Hynek et al., 2011), a mantle plume (Mege et al., 2001), and a giant impact event (Reese et al., 2004). Also, why was the prominent superplume, encompassing a total surface area of approximately 2 X 10^7 km^2 compared to the largest igneous plateau on Earth, Ontong-Java, covering a total surface area of approximately 2 X 10^6 km^2, so long-lived (i.e., for more than 3.8 Ga) (Dohm et al., 2001b; Stevenson, 2001)? Superplumes on Earth, which are influential on climate change, environmental conditions, and life, persist only for millions of years (Maruyama, 1994; Li et al., 2003; Maruyama and Santosh, 2008; Li and Zhong, 2009). If hydrated slabs of continental material were thrust into the core-mantle boundary region, then the water could have driven a long-lived, venting heat engine (Baker et al., 2007; Dohm et al., 2007), likened to a lava lamp with blobs of magma manifesting at the surface in the form of igneous plateau, volcanoes, lava flow fields, and shield fields. Large impact events such as Hellas, Isidis, and Argyre may have influenced the development of the superplume.

Other volcanic provinces also had an influence on the climate, environmental conditions, geology, hydrology, and possibly life, but to a much less extent when compared to Tharsis. These include the Elysium rise (e.g., Tanaka et al., 2005), Apollinaris Mons (Scott et al., 1993; Robinson et al., 1993; El Maarry et al., 2012), Malea Planum (Williams et al., 2010a), Tyrrhena and Hadriaca Mons (Crown et al., 1992; Williams et al., 2010), and Syrtis Major (Skok et al., 2010). All display evidence for magma-water interactions. The Elysium rise, for example, comprises a flow on its northwestern flank interpreted to be a lahar (Christiansen, 1989), distinctly observed through elemental information of the Gamma Ray Spectrometer (Boynton et al., 2004, 2007), as well as structurally-controlled outflow channels that dissect its flanks (e.g., Tanaka et al., 2005). Water releases from basement structures such as observed at Cerberus Fossae (Burr et al., 2002) could be linked to magmatic-driven activity related to Elysium or to fissure-fed eruptions not necessarily related to Elysium.

Apollinaris Mons, including its surroundings, displays valley networks at its summit, outflow-channel-like outflow features on the southern margin of flow interpreted to be basaltic (Scott et al., 1993) or mud/lahar flows (Ghail and Hutchison, 2003; Farrell and Lang, 2010; El Maarry et al., 2012). The flow may have sourced from a summit caldera lake which interacted with volcanism (El Maarry et al., 2012). In addition, other lahar-like flows on its flanks, as well as several other types of features, point to a long-lived interaction among magmatism and water such as mounds and possible vent structures on its flanks and within the surrounding plains, thus making it a prime hydrothermal target for future reconnaissance (El Maarry et al., 2012).

Syrtis Major and its surroundings, including interaction with a complex basement structural fabric related to the Isidis, Utopia, and Arabia Terra events (Anderson et al., 2008), such as Nili Fossae, also display a rich history of hydrothermal activity, especially supported through the analysis of NIR-Visible spectroscopy (Mustard et al., 2007; Mangold et al., 2007; Brown et al., 2010). The volcanic fields that occur on the southern and eastern margins of Hellas basin, Malea Planum and Tyrrhena-Hadriaca, respectively, show evidence of the interaction among magmatism, tectonism, and water. Tyrrhena and Hadriaca display lava-flow morphologies indicative of volatile-enriched magma, including rille-like valley forms, and explosive activity (Crown et al., 1992; Williams et al., 2010b), as well as distinct structurally-controlled outflow channels that debouch into the eastern part of the basin such as highlighted along the southeastern margin of the latter volcano. Hesperian Planum, a Hellas-induced basin which comprises wrinkle-ridged materials, shows evidence of volatile enrichment (Ivanov et al., 2005) and could be a source of water for fluvial activity such as associated with Reull Vallis. Malea Planum, a volcanic field that developed in a large water-enriched valley, also shows evidence of magma-water interactions. The distinct valley underlying the volcanic field highlighted by MOLA topography, was most likely carved by a large south polar region-sourcing glacier and/or ice sheet during the Noachian Period when a large ocean may have inundated the northern plains and terrain further south and above the MER Opportunity landing site (Fairén et al., 2003, 2004); on the other hand, gas release has been hypothesized for the sculpted southern basin margin (Tanaka et al. 2002). The northern flank of the volcanic field, which drapes the southern part of the Hellas basin, is dissected by valleys, also pointing to a water-enriched region.

Stage 7-8 of the GEOMARS hypothesis, including the development of superplumes and volcanic fields and their influence on climate and environmental change, explains aspects of the stagnant-lid phase of Mars' evolution, and is summarized in Table 1. The sites of magma, structure, and water interactions, many of which record significant hydrothermal activity, some extremely long-lived such as at Tharsis (billions of years), could have been highly influential on life if it existed on Mars, especially in the subsurface. Energy conditions, though, would be highly variable especially when comparing the highly dynamic Tharsis with the interface between the long-lived productive regional aquifer of Arabia Terra and Syrtis Major volcanic province (Dohm et al., 2011a).

Northern Plains (Stages 2-8; Fig. 7)

The heavily cratered basement of the northern plains is well established to be ancient, reaching as far back as the Early Noachian epoch (Frey et al., 2002; Nimmo and Tanaka, 2005). The origin of the dichotomy, one of the oldest recorded features on Mars according to Solomon et al. (2005), remains uncertain and highly controversial. Several hypotheses have been invoked to explain the origin of the martian northern lowlands, which include extensive southward erosional retreat of the heavily cratered highland plateau (Scott, 1978; Hiller, 1979), subcrustal erosion resulting from mantle convection (Wise et al., 1979b), excavation by one or more large impacts (Wilhelms and Squyres, 1984; Frey et al., 2002; Andrews-Hanna et al., 2008), overturn of unstable cumulates in an initial magma ocean (Hess and Parmentier, 2001), degree-1 mantle convection induced by a layered viscosity structure (Roberts and Zhong, 2006), or flexure of the southern highlands in response to vertical loading (Watters, 2003). Alternatively, the landscape of the northern lowlands may have been the result of sea-floor spreading, linked to plate tectonism (Sleep, 1994). Fairén and Dohm (2004) highlighted the potential for plate tectonism, while considering other factors.

The highland-lowland dichotomy has degraded southward over time, through fluvial modification such as spring-fed activity (e.g., Tanaka et al., 2005), glacial activity (e.g., Kargel et al., 1995), solifluction (Squyres, 1989) of clay-enriched materials, and slumping. Thus, the mélange of materials that partly in filled the northern plains is comprised partly of highly degraded cratered highland materials. In addition, the far-reaching and diverse geologic and hydrologic activities recorded in the northern plains (e.g., Dohm et al., 2009c) include impact cratering, magmatic- and impact-driven tectonism and related marsquakes, periglacial and glacial modification, putative marine and paleolake activity, aeolian, fluvial, debris-flow, volcanic resurfacing, subterranean gas release such as sedimentary (mud) volcanism (Mahaney et al., 2004; Oehler and Allen, 2010;Komatsu et al., 2011), possible burial of ice bodies from a more extensive north polar ice cap and/or frozen water bodies ranging from lakes to oceans, and possible impact-generated hydrothermal lakes (El Maarry et al., 2010).

There is increasing evidence for catastrophic volatile release related to Tharsis activity and to a lesser extent Elysium and the Tharsis/Elysium corridor region that links the two, as detailed below, and associated ponding in the northern plains to form bodies of water ranging from oceans to lakes (e.g., Dohm et al., 2009c). In addition to the controversial shoreline demarcations (Parker et al., 1989, 1993), geomorphic evidence such as deltas at certain topographic zonal boundaries (e.g., Di Achille and Hynek, 2010), giant polygons similar to those found on the ocean floors of Earth (McGowan, 2011), and impact craters and associated features reminiscent of impacts into Earth oceans (Ormö et al., 2004), point to oceans which occupied the northern plains of Mars. In addition, Mars Odyssey Gamma Ray Spectrometer-based elemental information showing distinctions between the regions below and above the putative shorelines (Dohm et al.,. 2009c), long-wavelength topography consistent with deformation caused by true polar wandering (Perron et al., 2007), and Mars Advanced Radar for Subsurface and Ionosphere Sounding (MARSIS) data which indicates that regionally occurring fine-grained sediments such as marine deposits and/or ice rather than lava flows best explain the radar signatures (Mouginot et al., 2012) also corroborate such a potential.

Mappable regionally occurring geologic units are likely related to the circum-Chryse outflow channel activity (Tanaka et al., 2005; Skinner et al., in preparation), driven by late-stage Tharsis activity (Stages 4-5 of Tharsis; see Dohm and Tanaka, 1999; Dohm et al., 2001b; Anderson et al., 2001). The rapid emplacement of Tharsis-induced, sediment-charged floodwaters would entrap volatiles that could be released in the form of vent structures such as mud volcanoes (e.g., Mahaney et al., 2004; Skinner and Tanaka, 2007; Skinner and Mazzini, 2009; Oehler and Allen, 2010; Komatsu et al., 2011). Such features should be considered as prime targets for future exobiologic reconnaissance, as they are energy-water habitats (e.g., Dohm et al., 2011a; ElMaarry et al., 2012).

Tharsis-Elysium Corridor Province (Stage 8; Figs. 8 and 13)

A youthful region of Mars, the Tharsis/Elysium corridor (Dohm et al., 2008), uniquely highlighted through a new global impact crater data base (Robbins and Hynek, 2012), records long-lived hydrologic activity, tectonism, and volcanism including fissure-fed eruptions and shield-volcano field development at least within tens of millions of years (Fig. 13). For example, pristine lava flows of Tharsis Montes and Olympus Mons display few if any superposed impact craters (Hartmann et al., 1999; Neukum et al., 2004), as do structurally-controlled releases of water and other possible volatiles from Cerberus Fossae (e.g., Burr et al., 2002; Berman and Hartmann, 2002; Mitchell and Wilson, 2003) and structurally-controlled channels that dissect Amazonian lavas of Tharsis Montes and Medusae Fossae Formation materials possibly marking extensive ignimbrites (Malin, 1979; Mouginis-Mark, 1990; Scott et al., 1993, 1995; Dohm et al., 2004). In addition, anastomosing dark slope streaks in Mangala Valles and the northwestern slope valleys region (NSVs), which may be indicative of fluvial activity (Ferris et al., 2002; Miyamoto, 2004) occur in regions which have recorded magmatic, tectonic, and hydrologic activity since the Noachian Period (Dohm et al., 2001a,b, 2004). Tectonic features such as fractures, faults, graben, and structurally-controlled pit crater chains cut stratigraphically young rock materials, including the flanks and aureole deposits of the Tharsis Montes shield volcanoes, plains-forming materials in the region that straddles Tharsis and Elysium, and Elysium rise-forming materials (Scott and Zimbelman, 1995; Scott et al., 1998; Anderson et al., 2001; Wyrick et al., 2004; Márquez et al., 2004; Vaucher et al., 2009). Also in the Tharsis/Elysium corridor, MO Gamma Ray Spectrometer (GRS)-based patterns of elevated hydrogen (H) and chlorine (Cl) recorded in rock materials up to 0.33 m depth are distinct particularly in the NSVs region and other regions along the highland-lowland boundary west of the Tharsis Montes shield volcanoes (Boynton et al., 2002, 2007; Feldman et al., 2002; Keller et al., 2007). The broad Tharsis/Elysium corridor region records magmatic, tectonic, and hydrologic activity extending from the Noachian to the Late Amazonian (Scott and Tanaka, 1986; Greeley and Guest, 1987), indicating possible hydrothermal environments for parts of the region (Dohm et al., 2004).

Geologically recent activity in the Tharsis/Elysium corridor region is particularly highlighted in the Amazonis and Elysium basins and their link, Marte Vallis (hereafter referred to as AME). A youthful shield-volcano field with at least seven gentle sloping shield volcanoes and fissure-fed vent structures have been identified, mapped, and characterized in the western part of the Elysium basin (Vaucher et al., 2009). In addition, the lava flows on and near the margins of the shield volcanoes display crenulated lava-flow margins, possibly marking lava-water-ice interactions. Both the shield volcanoes and pristine lavas located within AME, which are marked by a paucity of relatively small (~few-km-diameter) superposed impact craters and modified by faults and fractures and valleys, point to geological and hydrological activity on Mars in recent geologic time. For example, there are only two superposed impact craters with pristine eject a blankets on the floor of Marte Vallis. Marte Vallis has been interpreted to be a spillway (Scott et al., 1995) that connects the

Elysium and Amazonis basin or tsunami sourced by an impact into an ocean (Mahaney et al., 2011).

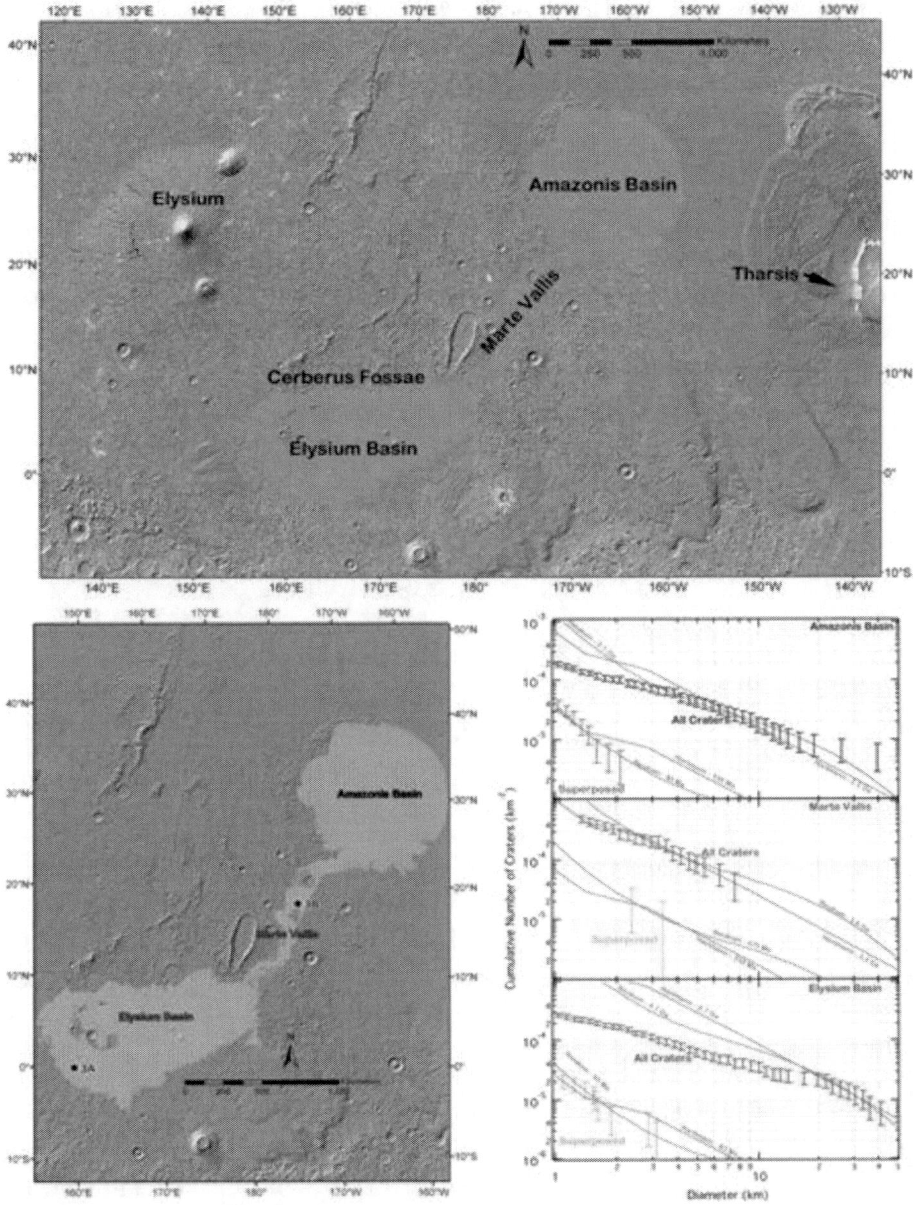

Figure 13. Adapted from Dohm et al. (2012), (Top) large part of the Tharsis-Elysium corridor (Dohm et al. (2008) with the new global crater database (Robbins, 2011; Robbins and Hynek, 2011) overlaid on MOLA shaded relief (yellow dots are mapped impact craters, red dots mark superposed craters on AME).Mapped region colors correspond to regions identified in bottom left and right. (Bottom left) Mapping regions separated into Amazonis Basin, Marte Vallis, and Elysium Basin (AME). (Bottom right) Same-scale cumulative size-frequency distributions showing all crater counts (darker colors, higher density data) and superposed crater counts (lighter colors, lower density data).Grey lines indicate best-fit isochrons from the two main systems in use (Hartmann, 2005; Neukum et al., 2001) and are labeled with ages.

In addition to the youthful magma, water, and tectonic interactions presented here, Icelandic-like activity (Burr et al., 2002), possible ice enrichment and related aqueous phenomena (Page, 2007; Page et al., 2009), and GRS-based elevated chlorine (Boynton et al., 2004, 2007; Keller et al., 2006; Dohm et al., 2009c) collectively make AME a significant target for future reconnaissance, including testing the hypotheses of whether Mars is geologically, hydrologically, and biologically active. In particular, the following could be addressed through international reconnaissance missions to AME, which would include instrument suites with optimal geologic, geochemical, geophysical (including seismic), environmental, and biological capabilities, including whether Mars: (1) is geologically and hydrologically active, (2) contains salty groundwater and magma at relatively shallow depths, (3) has sustained elevated heat flow, (4) records seismic activity, and (5) comprises fossilized and/or extant life. Certainly, AME and its various landforms, which include geologically young drainage networks and a shield volcano with collapse and vent-release features, can be readily and feasibly assessed through landing and roving agents within current engineering constraints. This is a golden age of Mars exploration and discovery, and AME as a prime target in the not-too-distant future could answer many long-standing questions about the red planet.

DISCUSSION/SUMMARY

The GEOMARS hypothesis distinctly expressed through the geologic provinces of Mars, as windows through time, will be tested through future international orbital and in situ investigations. Through the various aspects of the geologic provinces, prime targets can be identified, mapped, and characterized to address remaining queries and anomalies. These include determining: (1) what are the primary rocks that compose the ancient cratered highlands province, including the ancient mountain ranges, Thaumasia highlands and Coprates rise (e.g., are some of the ancient promontory-forming rock outcrops such as Claritas rise composed of granites/granodiorites?), the basin-forming materials such as observed in the Basin and Range-like topography of Terra Sirenum, and the highly degraded promontories among the structurally controlled basins that could be andesitic in composition; (2) whether there is groundwater, including extent and composition; (3) the extent and conditions of subterranean environments including caves; (4) whether there is magma at depth, elevated heat flow, and seismic activity in areas thought to be still active, including AME and the calderas of the prominent Tharsis volcanoes, and whether future endogenic-driven activity could result in yet another phase of environmental and global climate change; (5) the extent of the extremely dynamic stages of Mars evolution, as highlighted through stages 1-5 of the GEOMARS hypothesis; and (6) whether there is extant or fossilized life in potential life-containing habits as summarized by Dohm et al. (2011a), such as in the salts of the ancient structurally controlled basins as observed in the Atacama Desert, in the ice, in subsurface environment such as caves, in paleosols, in groundwater, including brines, in ancient and possible active hydrothermal environments, in basin deposits, including those of fluvial, lacustrine, marine, glacial, and periglacial origin, in the vent structures such as those identified, mapped, and characterized in the northern plains and on and near the flanks of Apollinaris Mons, in weathering rinds, and in desert varnish.

This is a golden age of paradigm-changing, multidisciplinary and international reconnaissance and discovery. Through such future effort, the above queries can be optimally addressed. The paradigm that Mars is a dry, dead, and mineralogically simple planet is no longer plausible. Other prevailing paradigms such as a one-plate planet through its entire history and a lifeless planet will be tested through such efforts. As of this writing, the Mars Science Laboratory is on its way to Mars, having the unique opportunity to test the ocean and life hypotheses. The mound-containing Gale Crater and surroundings could record the more ancient, higher-standing ocean during the Noachian and possibly in the Early Hesperian (shoreline 1 as portrayed by Fairén et al., 2003), a hiatus in magmatic-driven hydrologic activity consistent with an observed unconformity (Milliken et al., 2010), and yet another phase of Tharsis-driven flooding and ocean formation (Shoreline 2), as well as glaciation and fluvial activity that modified the mound and surrounding materials (e.g., Fairén et al., 2012). In addition, the deep crustal materials of the ancient cratered highlands possibly harvested by the impact event, as well as materials which record long-term aqueous activity, and thus habitat and potential life information, make Gale Crater a prime target.

References

Acuña, M.H., Connerney, J.E.P., Ness, N.F., Lin, R.P., Mitchell, D., Carlson, C.W., Mcfadden, J., Anderson, K.A., Reme, H., Mazelle, C., Vignes, D., Wasilewski, P., and Cloutier, P. (1999) Global distribution of crustal magnetization discovered by the Mars Global Surveyor MAG/ER experiment. *Science*284, 790–793, doi: 10.1126/science.284.5415.790.

Acuña, M.H., Connerney, J.E.P., Wasilewski, P., Lin, R.P., Mitchell, D., Anderson, K.A., Carl-son, C.W., McFadden, J., Réme, H., Mazelle, C., Vignes, D., Bauer, S.J., Cloutier, P., and Ness, N.F. (2001) Magnetic field of Mars: Summary of results from the aerobraking and mapping orbits. *Journal of Geophysical Research*106, 23,403–23,417, doi:10.1029/2000JE001404.

Anderson, R.C., Dohm, J.M., Golombek, M.P., Haldemann, A., Franklin, B.J., Tanaka, K.L., Li-as, J., and Peer, B. (2001) Significant centers of tectonic activity through time for the western hemisphere of Mars. *Journal of Geophysical Research*106, 20,563–20,585, doi:10.1029/2000JE001278.

Anderson, R.C., et al. (2008) Centers of tectonic activity in the eastern hemisphere of Mars. *Icarus*, doi:10.1016/j.icarus.2007.12.027.

Anderson, R.C., Dohm, J.M., Robbins, S., Hynek, B., Andrews-Hanna, J. (2012) Terra Sirenum: window into pre-Tharsis phases of Mars evolution. *Lunar Planetary Science Conference, XXXXIII,* Abstract #2803 (CD-ROM).

Andrews-Hanna, J.C., Zuber, M.T., and Banerdt, W.B. (2009) The Borealis basin and the origin of the martian crustal dichotomy. *Nature* 453, 1212-1215.

Andrews-Hanna, J.C. (2009) Planetary science: A mega-landslide on Mars. *Nature Geoscince: News and Views* 2, 248-249.

Anguita, F., Farelo, A.,-F., López, V., Mas, C., Muñoz-Espadas, M.–J., Márquez, Á., Ruiz, J. (2001) Tharsis dome, Mars: new evidence for Noachian-Hesperian thick-skin and Amzonian thin-skin tectonics. *J. Geophys. Res.* 106, 7577-7589.

Baker, V.R., Maruyama, S., and Dohm, J.M. (2007) Tharsis superplume and the geological evolution of early Mars. In Yuen, D.A., Maruyama, S., Karato, S.-I., and Windley, B.F., eds., Super-plumes: *Beyond plate tectonics:* Springer, p. 507–523.

Banerdt, W.B., Golombek, M.P., and Tanaka, K.L. (1992) *Stress and tectonics on Mars* (book chapter). Mars, H.H. Kieffer, B.M. Jakosky, (eds.), University of Arizona Press, p. 249-297.

Berman, D.C. and Hartmann, W.K. (2002) Recent fluvial, volcanic, and tectonic activity on the Cerberus plains of Mars. *Icarus* 159, 1-17.

Bandfield, J. L., Hamilton, V.E., Christensen, P.R. (2000) A Global View of martian Surface Compositions from MGS-TES. *Science*, 287, 1626-1630.

Bandfield, J.L., Hamilton, V.E., Christensen, P.R., and McSween, H.Y. Jr. (2004) Identification of quartzofeldspathic materials on Mars. *Journal of Geophysical Research* 109, DOI:10.1029/2004JE002290.

Boynton, W.V., et al. (2004) The Mars Odyssey Gamma-Ray Spectrometer Instrument Suite. *Space Science Reviews*110, 37–83.

Boynton, W.V., et al. (2007) Concentration of H, Si, Cl, K, Fe, and Th in the low- and mid-latitude regions of Mars. *Journal of Geophysical Research* 112, E12S99, doi: 10.1029/2007JE002887.

Brown A.G., et al., (2010) Hydrothermal formation of Clay-Carbonate alteration assemblages in the Nili Fossae region of Mars. *Earth and Planetary Science Letters, Volume 297*, Issue 1-2, p. 174-182.

Buczkowski, D.L., Murchie, S., Clark, R., Seelos, F., Malaret, E., and Hash, C. (2010) Investigations of an Argyre basin ring structure using Mars Reconnaissance Orbiter/Compact Reconnaissance Imaging Spectrometer for Mars. *Journal of Geophysical Research 115*, E12011, doi:10.1029/2009JE003508.

Burr, D.M., Grier, J.A., McEwen, A.S., and Keszthelyi, L.P. (2002) Repeated aqueous flooding from the Cerberus Fossae: Evidence for very recently extant, deep groundwater on Mars 159, 53-73.

Carr, M.H. (1974) Tectonism and volcanism of the Tharsis Region of Mars. *J. Geophys. Res.* 79, 3943–3949.

Carr, M.H., and Head, J.W.III (2010) Geologic History of Mars. *Earth and Planetary Science Letter* 294, 185-203.

Christiansen, E.H., (1989), Lahars in the Elysium region of Mars. *Geology* 17, 203-206.

Connerney, J.E.P., Acuña, M.H., Wasilewski, P.J., Kletetschka, G., Ness, N.F., Rème, H., Lin, R.P., and Mitchell, D.L. (1999) The global magnetic field of Mars and implications for crustal evolution. *Science* 284, 790–793.

Connerney, J.E.P., Acuña, M.H., Ness, N.F., Kletetschka, G., Mitchell, D.L., Lin, R.P., and Reme, H. (2005) Tectonic implications of Mars crustal magnetism. *Science* 102, 14970–14975.

Craddock, R.A., Greeley, Ronald, and Christensen, P.R. (1990) Evidence of an ancient impact basin in Daedalia Planum, *Mars: Journal of Geophysical Research*95, 10,729-10,741.

Crown, D.A., Price, K.H., and Greeley, R. (1992) Geologic evolution of the east rim of the Hellas basin, Mars: *Icarus*100, 1–25, doi:10.1016/0019-1035(92)90014-X.

Davila, A.F., Gross, C., Marzo, G.A., Fairén, A.G., Kneissl, T., McKay, C.P., and Dohm, J.M. (2011) A large sedimentary basin in the Terra Sirenum region of the southern highlands of Mars 212, 579-589.

Di Achille, G., and Hynek, B.M. (2010) Ancient ocean on Mars supported by global distribution of deltas and valleys. *Nature Geoscience*3, p. 459–463, doi:10.1038/ngeo891.

Dohm, J.M., and Tanaka, K.L. (1999) Geology of the Thaumasia region, Mars: plateau development, valley, and magmatic evolution. *Planetary and Space Science*47, 411–431, doi:10.1016/S0032-0633(98)00141-X.

Dohm, J.M., and Kargel, J.S. (2012) Geologic mapping investigation of the Argyre and surrounding regions of Mars. *Lunar Planetary Science Conference, XXXXIII*, Abstract #1639 (CD-ROM).

Dohm, J.M., Tanaka, K.L., and Hare, T.M. (2001a) Geologic map of the Thaumasia region of Mars. *U.S. Geological Survey Map* I-2650 at 1:5,000,000 scale.

Dohm, J.M., Ferris, J.C., Baker, V.R., Anderson, R.C., Hare, T.M., Strom, R.G., Barlow, N.G., Tanaka, K.L., Klemaszewski, J.E., and Scott, D.H. (2001b) Ancient drainage basin of the Tharsis region, Mars: Potential source for outflow channel systems and putative oceans or paleolakes. *Journal of Geophysical Research* 106, 32,942–32,958, doi:10.1029/2000JE001468.

Dohm, J.M., Maruyama, S., Baker, V.R., Anderson, R.C., Ferris, J.C., and Hare, T.M. (2002) Plate tectonism on early Mars: Diverse geological and geophysical evidence *Planetary Science Conference, XXXIII,* Abstract #1639 (CD-ROM).

Dohm, J.M., Ferris, J.C., Barlow, N.G., Baker, V.R., Mahaney, W.C., Anderson, R.C., and Hare, T.M. (2004) The Northwestern Slope Valleys (NSVs) region, Mars: A prime candidate site for the future exploration of Mars. *Planetary and Space Science*52, 189–198, doi:10.1016/j.pss.2003.08.013.

Dohm, J.M., Kerry, K., Keller, J.M., Baker, V.R., Maruyama, S., and Anderson, R.C., Ferris, J.C., and Hare, T.M. (2005) Mars geological province designations for the interpretation of GRS data. *Lunar Planetary Science Conference, XXXVI*, #1567 (abstract) [CD-ROM], 2005.

Dohm, J.M., Maruyama, S., Baker, V.R., and Anderson, R.C. (2007a) Traits and evolution of the Tharsis superplume, Mars. In Yuen, D.A., Maruyama, S., Karato, S.-I., and Windley, B.F., eds., *Super-plumes: Beyond plate tectonics:* Springer, p. 523-537.

Dohm, J.M., et al. (2007b) Possible ancient giant basin and related water enrichment in the Arabia Terra province, Mars, doi: 10.1016/j.icarus.2007.03.006.

Dohm, J.M., Anderson, R.C., Barlow, N.G., et al., 2008, Recent geological and hydrological activity on Mars: The Tharsis/Elysium Corridor: *Planetary and Space Science* 56, 985–1013, doi:10.1016/j.pss.2008.01.001.

Dohm and 13 other authors (2009a) Claritas rise: Pre-Tharsis magmatism. *Journal of Volcanology and Geothermal Research* 185, 139-156.

Dohm, J.M., Williams, J.-P., Anderson, R.C., et al., 2009b, New evidence for a magmatic influence on the origin of Valles Marineris, Mars: *Journal of Volcanology and Geothermal Research,* v. 185, p. 12–27, doi:10.1016/j.jvolgeores.2008.11.029.

Dohm, J.M., Baker, V.R., Boynton, W.V., et al., 2009c, GRS evidence and the possibility of ancient oceans on Mars: *Special: Planetary and Space Science,* v. 57, p. 664–684, doi:10.1016/j.pss.2008.10.008.

Dohm, J.M., and 39 other authors (2011a) An inventory of potentially habitable environments on Mars: geological and biological perspectives. In Garry, W.B., and Bleacher, J.E., eds., *Analogs for Planetary Exploration: Geological* Society of America Special Paper 483, 317–347, doi:10.1130/2011.2483(21).

Dohm, J.M., Ferris, J.C., Baker, V.R., Komatsu, G., Buczkowski, D.L., El Maarry, M.R., Hare, T.M., Mahaney, W.C., Kim, K.J., Davila, A.F., and Fairén, A.G. (2011b) Did a large Argyre lake source the Uzboi Vallis drainage system?: Post-Viking-era geologic mapping investigation. *Lunar Planetary Science Conference, XXXIII,* Abstract #2255 (CD-ROM).

Dohm, J.M., Robbins, S.J., and Hynek, B.M. (2012) Recent geological and hydrological activity in Amazonis and Elysium basins and their link, Marte Valles (AME): prime target for future reconnaissance. *Lunar Planetary Science Conference, XXXXIII,* #1948 (abstract) [CD-ROM].

Ehlmann, B.L., Mustard, J.F., and Murchie, S.L. (2010) Geologic setting of serpentine-bearing rocks on Mars. *Res. Letters,* doi:10.1029/2010GL042596.

Ehlmann, B.L., Mustard, J.F., Clark, R.N., Swayze, G.A., Murchie, S.L. (2011) Evidence for low-grade met morphism, diagenesis, and hydrothermal alteration on Mars from phyllosilicate mineral asse blages. *Clays and Clay Minerals,* 59(4), 359-377.

El Maarry, M.R., Markiewicz, W.J., Mellon, M.T., Goetz, W., Dohm, J.M., and Pack, A. (2010) Crater floor polygons: Desiccation patterns of ancient lakes on Mars? *J. Geophys. Res.* 115, E10006, doi:10.1029/2010JE003609.

El Maarry, M.R., et al. (2012) Searching for evidence of hydrothermal activity at Apollinaris Mons, Mars 217, 297-314.

Fairén, A.G., and Dohm, J.M. (2004) Age and origin of the lowlands of Mars: *Icarus*, v. 168, p. 277–284, doi:10.1016/j.icarus.2003.11.025.

Fairén, A.G., Ruiz, J., and Anguita, F. (2002) An origin for the linear magnetic anomalies on Mars through accretion of terranes: implications for dynamo timing. *Icarus*160, 220–223, doi:10.1006/icar.2002.6942.

Fairén, A.G., Dohm, J.M., Baker, V.R., de Pablo, M.A., Ruiz, J., Ferris, J.C., and Anderson, R.C. (2003) Episodic flood inundations of the northern plains of Mars. *Icarus* 165, 53–67, doi:10.1016/S0019-1035(03)00144-1.

Fairén, A.G., Fernández-Remolar, D., Dohm, J.M., Baker, V.R., and Amils, R. (2004) Inhibition of carbonate synthesis in acidic oceans on early Mars. *Nature* 431, 423–426, doi:10.1038/nature02911.

Fairén, A.G., Dohm, J.M., Uceda, E.R., Rodríguez, A., Baker, V.R., Fernández-Remolar, D., Schulze-Makuch, D., Amils, R. (2005). Prime candidate sites for astrobiological exploration through the hydrogeological history of Mars. Planet. *Space Sci.* 53, 1355-1375.

Fairén, A.G., Davila, A., Uceda, E.R., Dohm, J.M., Baker, V.R., Mckay, C.P., and Stokes, C.R. (2012) Glacial paleomorphologies in Gale Crater, Mars. *Lunar Planetary Science Conference, XXXIII*, Abstract #2182 (CD-ROM).

Farrell, A.K., Lang, N.P., 2011. Distribution of explosive and effusive volcanic deposits at Apollinaris Patera, Mars. *Lunar Planet. Sci.* 41. Abstract #2072.

Feldman, W.C., Boynton, W.V., Tokar, R.L., Prettyman, T.H., Gasnault, O., Squyres, S.W., Elphic, R.C., Lawrence, D.J., Lawson, S.L., Maurice, S., McKinney, G.W., Moore, K.R.,

Reedy, R.C. (2002) Global distribution of neutrons from Mars: Results from Mars Odyssey. *Science* 297, 75-78.

Ferris, J.C., Dohm, J.M., Baker, V.R., and Maddock, T. (2002) Dark slope streaks on Mars: Are aqueous processes involved?. *Geophysical Research Letters*29, doi:10.1029/2002GL014936.

Frey, H., and Grant T.D. (1990) Resurfacing history of Tempe Terra and surroundings. *Journal of Geophysical Research* 95, 14,249-14,263, 1990.

Frey, H.V., Roark, J H., Shockey, K.M., Frey, E.L., Sakimoto, S.E.H., 2002. Ancient lowlands on Mars. *Geophys. Res. Lett.* 29, 10.1029/2001GL013832.

Ghail R.C., and Hutchison J.E. (2003) An alluvial fan at Apollinaris, Mars. *Lunar Planetary Science Conference, XXXIIII,* abstract#1775 (CD-ROM).

Golombek, M.P., and Phillips, R.J. (2010) *Mars tectonics. In Planetary Tectonics,* edited by T.R. Watters and R.A. Schultz, Cambridge University Press, New York, 183-232.

Greeley, R., and Guest, J.E. (1987) Geologic map of the eastern equatorial region of Mars. *USGS Misc. Inv. Ser. Map I-1802B* (1:15,000,000).

Hartmann, W.K. (2005) martian cratering 8: isochron refinement and the chronology of Mars. *Icarus* 174, 294-320.

Hartmann, W.K., and Neukum, G. (2001) Cratering chronology and the evolution of Mars *Science Reviews96*, 165–194, doi:10.1023/A:1011945222010.

Hartmann, W.K., Malin, M.C., McEwen, A., Carr, M., Soderblom, L., Thomas, P., Danielson, E., James, P., and Veverka, J. (1999) Recent volcanism on Mars from crater counts. *Nature* 397, 586-589.

Hauber, E., Bleacher, J., Gwinner, K., Williams, D., and Greeley, R. (2009) The topography and morphology of low shields and associated landforms of plains volcanism in the Tharsis region of Mars. *Journal of Volcanology and Geothermal Research* 185, 69-95.

Hess, P.C., and Parmentier, E.M. (2001) Implications of magma ocean cumulate overturn on Mars. *Lunar Planetary Science Conference XXXII,* 1319 (abstract).

Hiller, K. (1979) Geologic map of the Amenthes quadrangle of Mars. *USGS Misc. Inv. Ser. Map I*-1110 (1:5,000,000).

Hodges, C.A., and Moore, H.J. (1994) Atlas of Volcanic Landforms on Mars: *U.S. Geological Survey Professional Paper* 1534, 194 p.

Hoke, M.R.T., Hynek, B.M., and Tucker, G.E. (2011) Formation timescales of large martia valley networks. *Earth and Planetary Science Letters* 312, 1-12.

Hynek, B.M., Phillips, R.J., Arvidson, R.E. (2003) Explosive volcanism in the Tharsis region: global evidence in the martian geologic record. Journal *of Geophysical Research* 108, doi:10.1029/2003JE002062.

Hynek, B.M., Beach, M., and Hoke, M.R.T. (2010) Updated global map of martian valley networks and implications for climate and hydrologic processes. *Journal of Geophysical Research* 115, doi: 10.1029/2009JE003548.

Hynek, B.M., Robbins, S.J., Šmárek, O., and Zhong, S. (2011) Geological Evidence for a Migrating Tharsis Plume. *Earth and Planetary Science Letters* 310, 327-333.doi: 10.1016/j.epsl.2011.08.020.

Ivanov, M.A., Korteniemi, J., Kostama, V.-P., Aittola, M., Raitala, J., Glamoclija, M., Marinangeli, L., and Neukum, G. (2005) Major episodes of the hydrologic history in the region of Hesperia Planum, Mars. *J. Geophys. Res.,* 110, E12S21, doi:10.1029/2005JE002420.

Karasozen, E., Andrews-Hanna, Dohm, J.M., Anderson, R.C. (2012) The formation mechanism of the south Tharsis ridge belt, Mars. *Lunar Planetary Science Conference, XXXXIII*, Abstract #2592 (CD-ROM).

Kargel, J.S., 2004, *Mars: A Warmer Wetter Planet*: Praxis-Springer, 557 p.

Kargel, J. S. et al. (1995) Evidence of continental glaciation in the martian northern plains. *J. Geophys. Res*. 100: 5351-5368.

Kargel, J.S., and Strom, R.G. (1994) Ancient glaciation on Mars. *Geology* 20, 3–7, doi:10.1130/0091-7613(1992)020<0003:AGOM>2.3.CO;2.

Kaula, W.M. (1975) The seven ages of a planet. *Icarus* 26, 1-15.

Keller, J.M., et al. (2006) Equatorial and midlatitude distribution of chlorine measured by Mars Odyssey GRS. *Journal of Geophysical Research*111, p. E03S08, doi:10.1029/2006JE002679.

Komatsu, G, Dohm, J.M., and Hare, T.M. (2004) Hydrogeologic processes of large-scale tectono-magmatic complexes in Mongolia-southern Siberia and on Mars. *Geology* 32, 325-328.

Komatsu, G., et al. (2011) Roles of methane and carbon dioxide in geological processes on Mars. *Planet. Space Sci.* 59, 169-181.

Li, Z.X., and Zhong, S., 2009, Supercontinent-superplume coupling, true polar wander and plume mobility: plate dominance in whole-mantle tectonics: *Physics of the Earth and Planetary Interiors*, v. 176, p. 143–156, doi:10.1016/j.pepi.2009.05.004.

Li, Z.X., Li, X.H., Kinny, P.D., Wang, J., Zhang, S., and Zhou, H., 2003, Geochronology of Neoproterozoic syn-rift magmatism in the Yangtze Craton, South China and correlations with other continents: evidence for a mantle superplume that broke up Rodinia: *Precambrian Research*, v. 122, p. 85–109, doi:10.1016/S0301-9268(02)00208-5.

Lillis, R.J. et al. (2008) An improved crustal magnetic field map of Mars from electron reflectometry: Highland volcano magnetic history and the end of the martian dynamo. *Icarus* 194, 575-596. doi: 10.1016/j.icarus.2007.09.032.

Lillis, R.J., Dufek, J., Bleacher, J.E., and Magna, M. (2009) Demagnetization of crust by magmatic intrusion near the Arsia Mons volcano: magnetic and thermal implications for the development of the Tharsis province, Mars. *Journal of Volcanology and Geothermal Research* 185, 123-138.

Louzada, K., Stewart, S.T., Weiss, B.P., Gattacceca, J., Lillis, R.J., and Halekas, J.S. (2011) Impact demagnetization of the martian crust: current knowledge and future directions. *Earth and Planetary Science Letters* 305, 257-269.

Lucchitta, B.K., McEwen, A.S., Clow, G.D., Geissler, P.E., Singer, R.B., Schultz, R.A., and Squyres, S. W. (1992) *The canyon system on Mars. in Mars*, pp. 453-492, Univ. of Ariz. Press, Tucson, 1992.

Mahaney, W.C., Milner, M.W., Netoff, D., Malloch, D., Dohm, J.M., and Baker, V.R., Miyamoto, Hideaki, Hare, T.M., and Komatsu, G. (2004) Ancient wet Aeolian environments on Earth: clues to presence of fossil/live microorganisms on Mars. *Icarus*171, 39–53.

Mahaney, W.C., et al. (2010) Tsunamis on Mars: Earth analogues of projected martian sediment. *Planet. Space Sci.* 58, 1823-1831.

Malin, M.C. (1979) Mars: evidence of indurated deposits of fine materials. (abstract)*NASA Conf. Pub.*2072, 54.

Mangold, N., Quantin, C., Ansan, V., Delacourt, C., and Allemand, P. (2004) Evidence for precipitation on Mars from dendritic valleys in the Valles Marineris area. *Science* 305, 78-81.

Mangold, N., Poulet, F., Mustard, J.F., Bibring, J.P., Gondet, B., Langevin, Y., Ansanm V., Masson, P., Fassett, C., Head, J.W., Hoffman, H., and Nuekum, G. (2007*). Journal of Geophysical Research* 112, 1-25.

Márquez, A., Fernández, C., Anguita, F., Farelo, A., Anguita, J., de la Casa, M-A. (2004) New evidence for a volcanically, tectonically, and climatically active Mars 172, 573-581.

Maruyama, S. (1994) Plume tectonics. *Journal of Geological Society of Japan*100, 24–49.

Maruyama, S., and Santosh, M. (2008) *Models on Snowball Earth and Cambrian Explosion: A synopsis. Gondwana Research* 14, 22–32, doi:10.1016/j.gr.2008.01.004.

Marzo, G.A., et al. (2010) Evidence for Hesperian impact-induced hydrothermalism on Mars. *Icarus* 208, 667-683.

McCauley, J.F., Carr, M.H., Cutts, J.A., Hartmann, W.K., Masursky, H., Milton, D.J., Sharp, R.P., Wilhelms, D.E. (1972) Preliminary Mariner 9 report on the geology of Mars. *Icarus* 17, 289–327.

McGowan, E.M. (2011) The Utopia/Isidis overlap: possible conduit for mud volcanism on Mars. *Icarus* 212, 622-628.

McKenzie, D., Nimmo, F. (1999) The generation of martian floods by the melting of ground ice above dikes. *Nature* 397, 231-233.

Mège, D., and Masson, P. (1996a) Stress models for Tharsis formation. Mars, Planet. *Space Sci.*, 44, 1471–1497.

Mège, D., and Masson P. (1996b) A plume tectonics model for the Tharsis Province, Mars. Planet. *Space Sci.,* 44, 1499– 1546.

Mège, D., and Ernst, R.E. (2001) Contractional effects of mantle plumes on Earth, Mars, and Venus, in Ernst, R.E., and Buchan, K.L., eds., Mantle plumes: Their identification through time, *Geological Society of America Special Paper* 352, 103–140.

Milliken, R.E., Grotzinger, J.P., and Thomson, B.J. (2010) Paleoclimate of Mars as captured by the stratigraphic record in Gale Crater. *Geophysical Research Letters* 31, doi:10.1029/2009GL041870.

Mitchell, K.L., and Wilson, L. (2003) Mars: a geologically active planet. *Astronomy and Geophysics* 44, 4.16-4.20.

Miyamoto, H., Dohm, J.M., Beyer, R.A., and Baker, V.R. (2004) Fluid dynamical implications of anastomosing slope streaks on Mars. *Journal of Geophysical Research* 109, E06008, doi:10.1029/2003JE002234.

Montgomery, D.R., Som, S.M., Jackson, M.P.A., Schreiber, B.C., Gillespie, A.R., and Adams, J.B. (2009) Continental-scale salt tectonics on Mars and the origin of Valles Marineris and associated outflow channels. *Geology* 37, 691-694.

Moore, J.M., and Wilhelms, D.E. (2007) Geologic map of part of western Hellas Planitia, *Mars. USGS map* SIM-2953.

Mouginis-Mark, P.J., 1990, Recent water release in the Tharsis region of Mars: *Icarus*, v. 84, p. 362–373, doi:10.1016/0019-1035(90)90044-A.

Mouginot, J., Pommerol, A., Beck, P., Kofman, W., Clifford, S.M. (2012) Dielectric map of the martian northern hemisphere and the nature of plain filling materials. *Geophysical Research Letters* 39, DOI: 10.1029/2011GL050286

Mustard, J.F., et al. (2007) Mineralogy of the Nili Fossae region with OMEGA/Mars Express data: 1. Ancient impact melt in the Isidis Basin and implications for the transition from the Noachian to Hesperian. *Journal of Geophysical Research* 112, 1-14.

Mustard, J.F., Ehlmann, B.L., Murchie, S.L., Poulet, F., Mangold, N., Head, J.W., Bibring, J., -P., Roach, L.H. (2009) Composition, morphology, and stratigraphy of Noachian crust around the Isidis basin. Journal *of Geophysical Research* 114, 10.1029/2009JE003349.

Neukum, G., Ivanov, B.A., and Hartmann, W.K. (2001) Cratering records in the inner solar system in relation to the lunar reference system. In *Chronology and Evolution of Mars* (R. Kallenbach, J. Geiss, and W. K. Hartmann, eds.), Kluwer Academic Publishers, 55-86.

Neukum, G., Jaumann, R., Hoffmann, H., Hauber, E., Head, J. W., Basilevsky, A. T., Ivanov, B. A., Werner, S. C., van Gasselt, S., Murray, J. B., McCord T., and The HRSC Co-Investigator Team (2004) Recent and episodic volcanic and glacial activity on Mars revealed by the High Resolution Stereo Camera. *Nature* 432, 971-979.

Nimmo, F., and Tanaka, K.L. (2005) Early crustal evolution of Mars. Annu. Rev. Earth Planet. Sci. 33, 133-161.

Oehler, D.Z., and Allen, C.C. (2010) Evidence for pervasive mud volcanism in Acidalia Planitia, Mars. *Icarus*208, 636–657, doi:10.1016/j.icarus.2010.03.031.

Ormö, J., Dohm, J.M., Ferris, J.C., Malvitte, A.L., Fairen, A.G. (2004) Marine-target craters on Mars? An assessment study. *Meteoritics and Planetary Science*39, 333–346b.

Page, D.P. (2007) Recent low-latitude freeze–thaw on Mars. *Icarus* 189, 83–117, doi:10.1016/j.icarus.2007.01.005.

Page, D.P., Balme, M.R., and Grady, M.M. (2009) Dating martian climate change: *Icarus*, v. 203, p. 376–389, doi:10.1016/j.icarus.2009.05.012.

Parker, T.J., and Gorsline, D.S. (1991) Where is the source for Uzboi Vallis, Mars? *Lunar Planet Sci.* XXII, 1033-1034.

Parker, T.J., Saunders, R.S., Schneeberger, D.M. (1989) Transitional morphology in the west Deuteronilus Mensae region of Mars: Implications for modification of the lowland/upland boundary. *Icarus* 82, 111-145.

Parker, T.J., Gorsline, D.S., Saunders, R.S., Pieri, D.C., Schneeberger, D.M. (1993) Coastal geomorphology of the martian northern plains. *J. Geophys. Res.* 98, 11061-11078.

Perron, J.T., Mitrovica, J.X., Manga, M., Matsuyama, I., and Richards, M.A. (2007) Evidence for an ancient martian ocean in the topography of deformed shorelines. *Nature* 447, 840-843.

Phillips, R.J., Zuber, M.T., Solomon, S.C., Golombek, M.P., Jakosky, B.M., Banerdt, W.B., Smith, D.E., Williams, R.M.E., Hynek, B.M., Aharonson , O., and Hauck, S.A. (2001) Ancient geodynamics and global-scale hydrology on Mars. *Science* 291, 2587–2591, doi: 10.1126/science.1058701.

Reese, C.C., Solomatov, V.S., Baumgardner, J.R., Stegman, D.R., and Vezolainen, A.V. (2004) Magmatic evolution of impact induced martian mantle plumes and the origin of Tharsis. *J. Geophys. Res.* 109, E08009, doi:10.1029/2003JE002222.

Robbins, S.J., and Hynek, B.M. (2012) A new global database of Mars impact craters ≥1 km: 1. Database creation, properties, and parameters. *J. Geophys. Res.*, (in press), doi: 10.1029/2011JE003966.

Robbins, S.J., and Hynek, B.M. (2012) Impact History of Large Bollides at Mars: Implications for the Late-Heavy Bombardment and Isochron Uncertainties. *Lunar Planetary Science Conference, XXXXIII*, Abstract #1649 (CD-ROM).

Roberts, J. H., and S. Zhong (2006), Degree-1 convection in the martian mantle and the origin of the hemispheric dichotomy, *J. Geophys. Res.*, 111, E06013, doi:10.1029/2005JE002668.

Roberts, J.H., and Arkani-Hamed, J. (2012) Impact-induced mantle dynamics on Mars 218, 278-289.

Roberts, J.H., Lillis, R.J., and Manga, M.(2009) Giant impacts on early Mars and the cessation of the martian dynamo. *J. Geophys. Res.* 114, E04009, doi: 10.1029/2008JE003287.

Robinson, M.S., Mouginis-Mark, P.J., Zimbelman, J.R., Wu, S.S.C., Ablin, K.K., and Howing-ton-Kraus, A.E., 1993, Chronology, eruption duration, and atmospheric contribution of the Martian volcano, Apollinaris patera: *Icarus*, v. 104, p. 301–323, doi:10.1006/icar.1993.1103.

Ruff, S. W., et al. (2011), Characteristics, distribution, origin, and significance of opaline silica observed by the Spirit rover in Gusev crater, Mars.*J. Geophys. Res.* 116, E00F23, doi:10.1029/2010JE003767.

Schubert, G., Solomon, S.C., Turcotte, D.L., Drake, M.J., and Sleep, N.H. Origin and thermal evolution of Mars. In *Mars*, University of Arizona Press, Tucson, pp. 147-183, 1992.

Schultz, R.A. (1991) Structural development of Coprates Chasma and western Ophir Planum, central Valles Marineris rift, Mars. *J. Geophys. Res.* 96, 22,777–22,792, doi:10.1029/91JE02556.

Scott, D. H (1978) Mars highlands-lowlands: Viking contribution to Mariner relative age studies. *Icarus* 34, 479-485.

Scott, D.H. (1981) New evidence – old problems: wrinkle ridge origin (abstract), in *MEVTV Workshop on Tectonic features on Mars* (eds. T.R. Watters and M.P. Golombek). Houston: Lunar and Planetary Institute, LPI Tech. Rpt. 98-06, pp. 52-4.

Scott, D.H., and Carr, M.H. (1978) Geologic map of Mars. *U.S. Geological Survey Miscellaneous Investigations Series Map I-1083*, scale 1:25,000,000.

Scott, D.H., and Tanaka, K.L. (1986) Geologic map of the western equatorial region of Mars. *U.S. Geological Survey Miscellaneous Investigations Series Map I-1802-A*, scale 1:15,000,000.

Scott, D.H., and Dohm, J.M. (1990) Chronology and global distribution of fault and ridge systems on Mars. In *Lunar and Planetary Science Conference, 20th,* Houston, March 13-17, 1989, Proceedings: Houston, Lunar and Planetary Institute, p. 487-501.

Scott, D.H., and Zimbelman, J.R. (1995) Geologic map of Arsia Mons volcano, Mars. *USGS Misc. Inv. Ser. Map I-2480* (1:1,000,000).

Scott, D.H., Dohm, J.M., and Applebee, D.J., 1993, Geologic map of science study area 8, Apollinaris Patera region of Mars: *U.S. Geological Survey Miscellaneous Investigations Series Map I-2351*, scale 1:500,000.

Scott, D.H., Dohm, J.M., and Rice, J.W., Jr. (1995) Map of Mars showing channels and possible paleolake basins. *U.S. Geological Survey Miscellaneous Investigations Series Map I-2461*, scale 1:30,000,000.

Scott, D.H., Dohm, J.M., and Zimbelman, J.R. (1998) Geologic map of Pavonis Mons volcano, Mars. *U.S. Geological Survey Miscellaneous Investigations Series Map I-2561*, scale 1:1,000,000.

Selvans, M.M., Plaut, J.J., Aharonson, O., and Safaeinili, A. (2010) Internal structure of Planum Boreum, from Mars advanced radar for subsurface and ionospheric sounding data. Journal of Geophysical Research 115, doi:10.1029/2009JE003537.

Skinner, J.A., Mazzini, A. (2009) Martian mud volcanism: terrestrial analogs and implications for formational scenarios. *Marine and Petroleum Geology* 26, 1866–1878.

Skinner, J.A., Tanaka, K.L. (2007) Evidence for and implications of sedimentary diapirism and mud volcanism in the southern Utopia highland-lowland boundary plain, Mars. *Icarus* 186, 41–59.

Skok, J.R., Mustard, J.F., Ehlmann, B.L., Milliken, B.L., and Murchie, S.L. (2010) Silica deposits in the Nili Patera caldera on the Syrtis Major volcanic complex on Mars. *Nature Geoscience* 3, DOI:10.1038/NGEO990.

Sleep, N. H. (1994) Martian plate tectonics. *J. Geophys. Res.* 99, 5639-5655.

Sleep, N.H. (2000) Evolution of the mode of convection within terrestrial planets. *Journal of Geophysical Research* 105, 17,563-17,578.

Solomon, S.C., and Head, J.W. (1982) Evolution of the Tharsis Province of Mars: The importance of heterogeneous lithospheric thickness and volcanic construction. *J. Geophys. Res.* 87, 9755-9774.

Solomon, S.C., Aharonson, O., Aurnou, J.M., Banerdt, W.B., Carr, M.H., Dombard, A.J., Frey, H.V., Golombek, M.P., Hauck, S.A.II, Head, J.W.III, Jakosky, B.M., Johnson, C.L. McGovern, P.J., Neumann, G.A., Phillips, R.J., Smith, D.E., and Zuber, M.T. (2005) New Perspectives on Ancient Mars. *Science* 307, 1214-1220.

Squyres, S.W. (1989) Urey Prize Lecture: Water on Mars: *Icarus* 79, 229-288.

Stevenson, D.J. (2001) Mars' core and magnetism. *Nature* 412, 214-219.

Tanaka, K.L. (1986) The stratigraphy of Mars. Proc. 17th Lunar Planet. Sci. Conf., Part 1, *J. Geophys. Res.,* 91, E139-E158, 1986.

Tanaka, K. L., Golombek, M. P. (1989) Martian tension fractures and the formation of graben and collapse features at Valles Marineris. *Proc. Lunar Planet. Sci. Conf.* 19, 383–396.

Tanaka, K.L., and Skinner, J.A., Jr., Hare, T.M. (2005) Geologic map of the northern plains of Mars: *U.S. Geological Survey Scientific Investigations Map SIM-2888*, scale 1:15,000,000.

Tanaka, K.L., Kargel, J.S., Mackinnon, D.J., Hare, T.M., And Hoffman, N. (2002) Catastrophic erosion of Hellas basin rim on Mars induced by magmatic intrusion into volatile-rich rocks. *Geophysical Research Letters* 29, doi:10.1029/2001GL013885.

Tanaka, K.L., et al. (2011) Completing the new global geologic map of Mars. *Lunar Planet. Sci. Conf. [CD-ROM], XXXII*, abstract 2265.

Vaucher, J., Baratoux, D., Toplis, M.J., Pinet, P., Mangold, N., and Kurita, K. (2009) The morphologies of volcanic landforms at central Elysium Planitia: evidence for recent and fluid lavas on Mars. *Icarus* 200, 39-51.

Watters, T.R. (1988) Wrinkle ridge assemblages on the terrestrial planets. *Journal of Geophysical Research* 93, 10236-10254.

Watters, T.R. (1993) Compressional tectonism on Mars: *Journal of Geophysical Research*98, 17,049-17,060.

Watters, T. R. (2003) Lithospheric flexure and the origin of the dichotomy boundary on Mars 31, 271– 274.

Wezel, F.C., and Baioni, D. (2010) Evidence for subaqueously resedimented sulphate evaporates on Mars. *Planetary and Space Science* 58, 1500-1505.

Wilhelms, D.E. and Squyres, S.W. (1984) The martian hemispheric dichotomy may be due to a giant impact. *Nature* 309, 138–140.

Williams, D.A., Greeley, R., Manfredi,, L., Fergason, R.L., Combe, J.-P., Poulet, F., Pinet, P., Rosnmberg, C., Clenet, H., McCord, T.B., Raitala, J., Neukum, G., and the HRSC Co-Investigator Team (2010a) Surface-compositional properties of the Malea Planum region of the Circum-Hellas volcanic province, Mars. *Earth and Planetary Science Letters* 294, 451-465.

Williams, D.A., Greeley, R., Zuschneid, W., Werner, S.C., Neukum, G., Crown, D.A., Gregg, T.K.P., Gwinner, K., and Raiala, J. (2010b) Hadriaca Patera: insights into its volcanic history from Mars Express High Resolution Stereo Camera. *Journal of Geophysical Research* 112, doi:10.1029/2007JE002924 .

Williams, J. P., Nimmo, F., Moore, W.B., and Paige, D.A. (2008), The formation of Tharsis on Mars: What the line-of-sight gravity is telling us. *Journal of Geophysical Research* 113, E10011, doi:10.1029/2007JE003050.

Wise, D.U., Golombek, M.P., and McGill, G.E. (1979a) Tharsis province of Mars: Geologic sequence, geometry, and a deformation mechanism. *Icarus* 38, 456–472.

Wise, D.U., Golombek, M.P., McGill, G.E. (1979b) Tectonic evolution of Mars. *J. Geophys. Res*. 84, 7934–7939.

Wyrick, D., Ferrill, D.A., Morris, A.P., Colton, S.L., and Sims, D.W. (2004) Distribution, morphology, and origins of martian pit crater chains. *Journal of Geophysical Research* 109, p. E06005, doi:10.1029/2004JE002240.

Yin, A. (2011) Structural analysis of the southern Valles Marineris trough zones and implications for large-scale left-slip faulting on Mars. *Lunar Planetary Science Conference, XXXXII*, Abstract #1529 (CD-ROM).

Yin, A. (2012) Structural analysis of the Valles Marineris fault zone: possible evidence for large-scale strike-slip faulting on Mars. *Lithosphere*, doi: 10.1130/L192.1

Yin, A. (2012—in review) A Slab Rollback Model for the Origin of the Tharsis Rise on Mars: Implications for Initiation of Local Plate Subduction and Final Unification of a Kinematically Linked Global Plate Tectonic Network on Earth. *Lithosphere—in review*.

Zhong, S. (2009) Migration of Tharsis volcanism on Mars caused by differential rotation of the lithosphere. *Nature Geoscience* 2, 19-22.

Zuber, M.T. (2001) The crust and mantle of Mars. *Nature*, 412, 220–227.

Zuber, M.T., Solomon, S.C., Phillips, R.J., Smith, D.E., Tyler, G.L., Aharonson, O., Balmino, G., Banerdt, W.B., Head, J.W., Johnson, C.L., Lemoine, F.G., McGovern, P. J., Neumann, G.A., Rowlands, D.D., and Zhong, S. (2000) Internal structure and early thermal evolution of Mars from Mars Global Surveyor topography and gravity. *Science* 287, 1788-1793.

In: Mars
Editor: Alberto G. Fairén

ISBN: 978-1-62618-102-1

Chapter 2

EVIDENCE FOR ICEBERG RAFTING AND GROUNDING ON THE COLD ANCIENT OCEANS OF MARS

Alberto G. Fairén*

Space Science and Astrobiology Division,
NASA Ames Research Center, Moffett Field, CA, US

ABSTRACT

The smooth topography of the Martian northern lowlands has been classically equated to an ancient ocean basin. The High-Resolution Imaging Science Experiment (HiRISE) onboard the Mars Reconnaissance Orbiter (MRO) is providing unprecedented images of the Martian surface at scales of 25 to 32 cm per pixel. The analysis of this high-resolution imaging is revealing the presence of meter-sized boulders distributed forming clusters over the northern plains, along with a variety of mounds, and also scour marks ranging from plough marks to chains of craters. Here I show that the distribution of boulders, mounds and scours on the northern plains of Mars is related to iceberg rafting and grounding. The association of boulders, mounds, crater chains and scour marks on the northern plains of Mars can be related to the dual processes of ice keel scouring and ice rafting of both glacial and non-glacial detritus by a floating ice canopy and icebergs. These processes are well documented on Earth and result in characteristic morphologies on the ocean floor, which are analogous to features observed in the Martian basins. Therefore, the presence of icebergs floating in primeval cold oceans explains the distribution of boulders and the occurrence of mounds and scours on the northern plains of Mars.

* Space Science and Astrobiology Division, NASA Ames Research Center, 94035 Moffett Field, CA, US, alberto.g.fairen@nasa.gov

Introduction

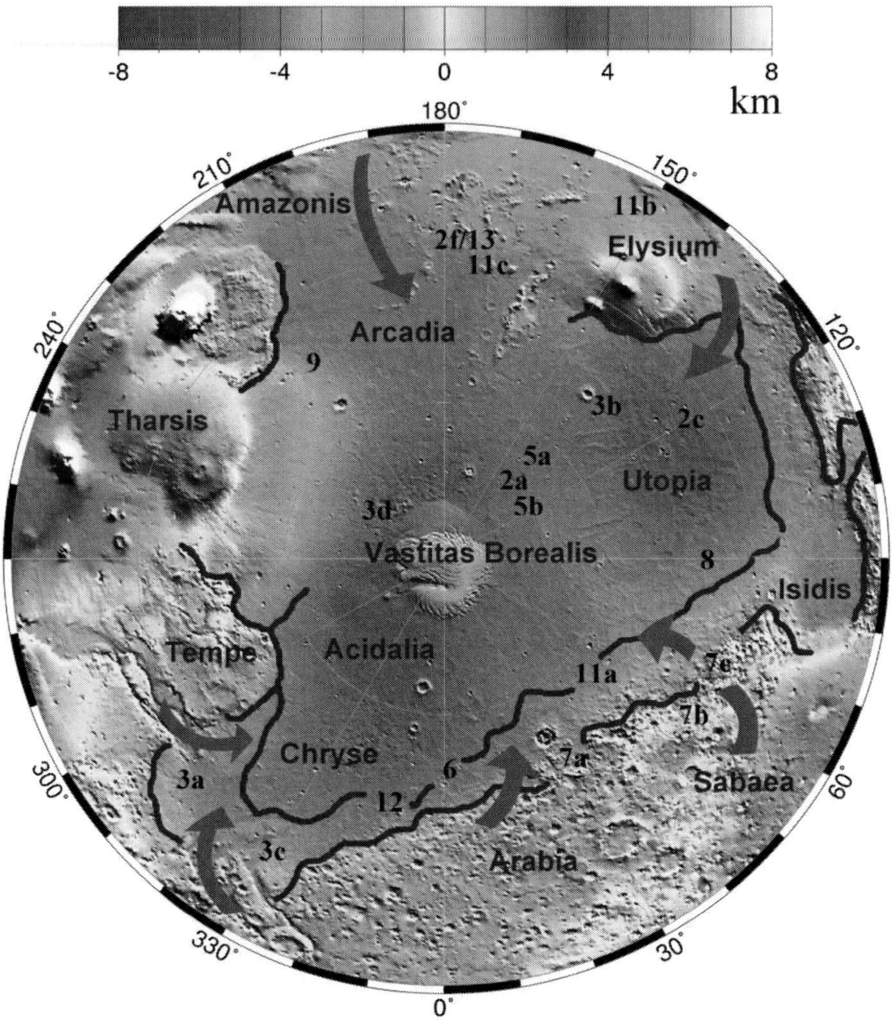

Figure 15a. Topographic labeled map of the northern lowlands. The blue lines represent the mapped boundaries of the putative ancient Martian oceans (Fairén et al., 2003). Red arrows indicate possible sources of icebergs from glaciers streaming through valleys in the highlands. Background is a Mars Orbiter Laser Altimeter (MOLA) Lambert equal-area projection of pole-to-equator topography in the northern hemisphere. Background map credit: MOLA Science Team.

The northern plains of Mars have been previously interpreted as the floor of an ocean that once covered the Martian northern hemisphere (Parker et al., 1993; Clifford and Parker, 2001; Carr and Head, 2003; Fairén et al., 2003; see Figure 15a). This hypothesis is based on the unusual topography of the northern lowlands, which appeared in early orbital images as persistently flat and showing few evident topographic features, closely resembling that of the ocean basins on Earth (Parker et al., 1993; Head et al., 1999; Withers and Neuman, 2001; Carr and Head, 2003). The High-Resolution Imaging Science Experiment (HiRISE) onboard the Mars Reconnaissance Orbiter (MRO) is providing unprecedented images of the Martian surface at scales of 25 to 32 cm per pixel (McEwen et al., 2009). This high-resolution imaging allow for more detailed analyses of geomorphologic traits on the surface, particularly of features that may have slipped by in previous imaging datasets. I have identified in HiRISE images the presence of three differentiated geomorphologies throughout the northern lowlands of Mars (Figure 15a) and the Hellas basin (Figure 15b), which are informative of the presence of icebergs floating in ancient oceans and/or seas. These morphologies are: (i) boulders ranging in size from 0.5 m to ~2 m in diameter (McEwen et al., 2007), distributed forming clusters with sizes from several hundred meters to 1-2 km; (ii) the previously described flat-topped and conical circular fractured mounds (Dundas and McEwen, 2009); and (iii) scattered scour marks, including plough marks several km long and some meters deep, and chains of crater-like structures several hundred meters wide and 1 to 5 km long. Here I describe these features and assert their origin related to iceberg rafting, scouring and grounding in comparison with well documented similar processes occurring on glacial oceans on Earth.

Figure 15b. MOLA color topography superposed on a shaded relief map of the Hellas Basin. Map credit: MOLA Science Team.

ICEBERG-DERIVED MORPHOLOGIES

1. Boulders

Dark clusters of boulders appear concentrated in different parts of the lowlands, and also within the Hellas and Argyre basins. Clusters have sizes ranging from several hundred meters to 1-2 km, and are irregularly shaped, from almost circular to highly elongated or totally amorphous. At large scale, boulder concentrations are revealed by their slightly darker tonality with respect to the surrounding terrain. At smaller scale, the clusters are characterized by a high density of relatively large boulders. The surrounding terrains appear slightly brighter and have a lower density of boulders, which are also smaller in size (Figure 16a). Grounding of icebergs on the ocean floor would explain the occurrence of clusters in the lowlands of Mars. Iceberg-rafted rock fragments were eroded by glaciers in the highlands (see Chapter 2 by A. Davila) and transported towards the lowlands during times when the lowlands and the main martian basins were covered by large bodies of standing waters. On Earth, continental glaciers intensively erode the terrain and transport large rocks on their surface, and within the ice flowing. Icebergs form as a result of glacier melt and break up, and are released into the ocean, where they can be transported thousands of km before they disappear (Woodworth-Lynas et al., 1991; Bass and Woodworth-Lynas, 1991; Figure 16b). When an iceberg anchors on the ocean floor due to a sudden decrease in sea level or to local higher topography, it discharges its rock cargo in a relatively constrained area, resulting in localized clusters of rock fragments (Fetch and Tallman, 1978; Driemanis, 1979).

Figure 16a. Dark regions with a high concentration of boulders on the lowlands, located on the northern rim of Utopia Planitia, and all situated over local topographic rises (HiRISE image TRA_000848_2890). Scale bar = 1 km. HiRISE image credit: NASA/JPL/University of Arizona.

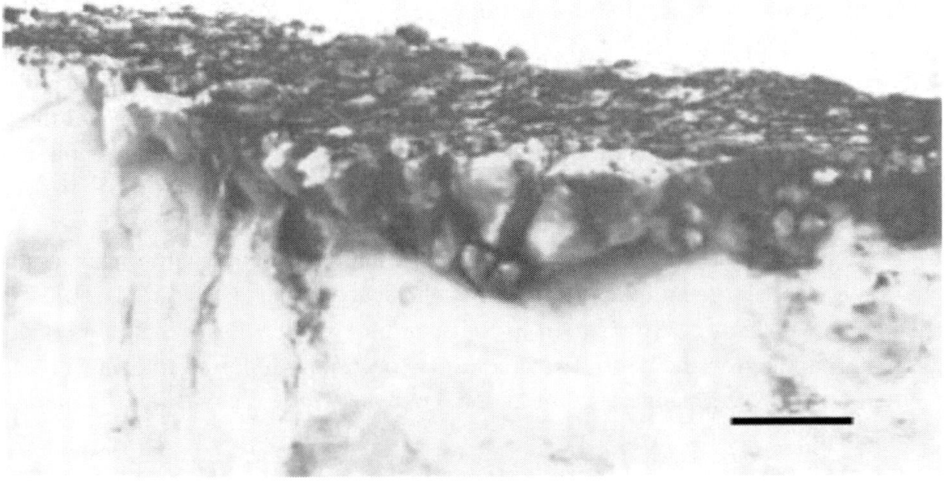

Figure 16b. Photograph of an overturned iceberg. Sediment grain size ranges from large boulders approximately 0.5 m diameter to cobbles and probably pebbles and fines. Scale bar = 1 m.

2. Mounds

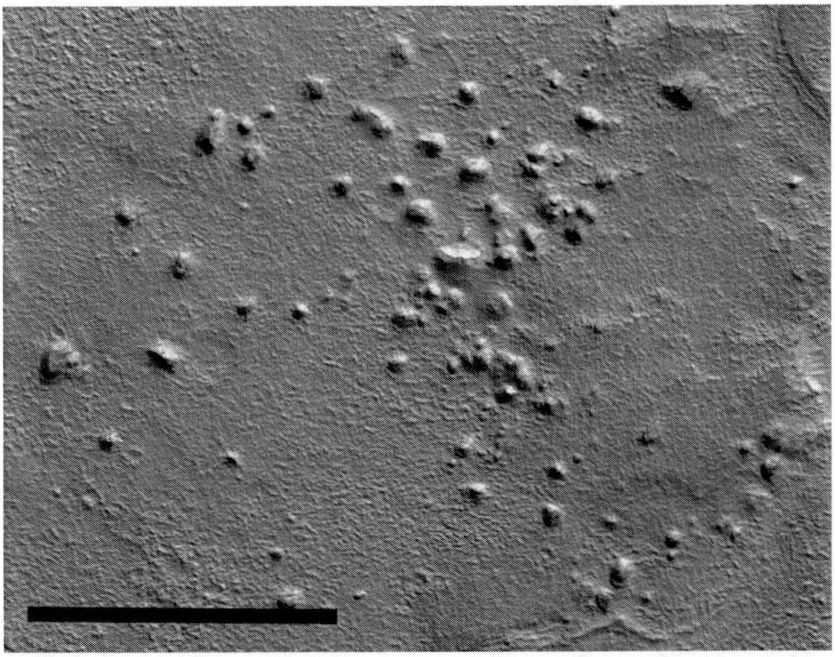

Figure 16c. Cones in Tinjar Valles (HiRISE image ESP_014371_2185). Scale bar = 1 km. HiRISE image credit: NASA/JPL/University of Arizona.

The presence of a number of circular fractured mounds in different places on the lowlands of Mars, previously analyzed by means of HiRISE imagery (Dundas and McEwen, 2009), is also confirmed here. Mounds on Mars appear isolated or in groups. Two types of mounds have been described (Dundas and McEwen, 2009): Type 1 is flat-topped, slightly domed or depressed, or even collapsed in rare occasions; and Type 2 is conical, smaller in diameter but of comparable height (Figure 16c). Both are purported to have an undefined erosional origin (Dundas and McEwen, 2009). Here I propose that the origin of the Type 1 mounds is due to the movement of an iceberg passing over an open depression. As the iceberg advances, dragged material from the leading edge is bulldozed into the depression, and then overridden by the passing iceberg, forming planar surfaces. This phenomenon has been observed in the Labrador continental shelf (Woodworth-Lynas et al., 1991). Type 2 mounds seem to be related to the breakup and overturning of boulder-laden, floating icebergs and the consequent release of large amounts of debris to the seabed assisted by gravitational sorting. This phenomenon, which is different from that forming clusters of boulders as it does not imply actual grounding on the bottom sediment, has been long reported to occur in terrestrial glacial environments (Ovenshine, 1970; Anderson et al., 1980; Thomas and Connell, 1985; Syvitski et al., 1996).

3. Scour Marks

HiRISE images also reveal scoured lineations which can be observed in the northern plains, including near the dichotomy boundary, and in the Hellas Basin (Figure 16d). Scours are typically 1-5 km in length, meters to tens of meters wide, and some meters deep. I propose that scours were carved in ocean-bottom sediments when the keels of icebergs touched the oceanic floor and shoved wet and loose material to the sides. Similar morphologies in the Hellas Basin (Ormö and Komatsu, 2003), as well as within the Kasei Valles system and parts of Echus Chasma (Woodworth-Lynas and Guigné, 2003) have previously been related to iceberg scouring on Mars. Terrestrial icebergs form scours as they drag their keel on the ocean floor (Figure 16e), usually when they drift into areas with water depths less than the iceberg's keel depth. Therefore, scouring of the sea floor by icebergs takes place where keels reach through the full depth of the water column to impinge on the sediments below (Thomas and Connell, 1985; Woodworth-Lynas et al., 1991; Bass and Woodworth-Lynas, 1991; O'Brien et al., 1997; Anderson, 1999; Hill et al., 2008). Typical scours are km long, tens of meters wide and 0.5 to 5 meters deep (Woodworth-Lynas et al., 1991; Ormö and Komatsu, 2003), consistent with the sizes of the scour marks identified on Mars.

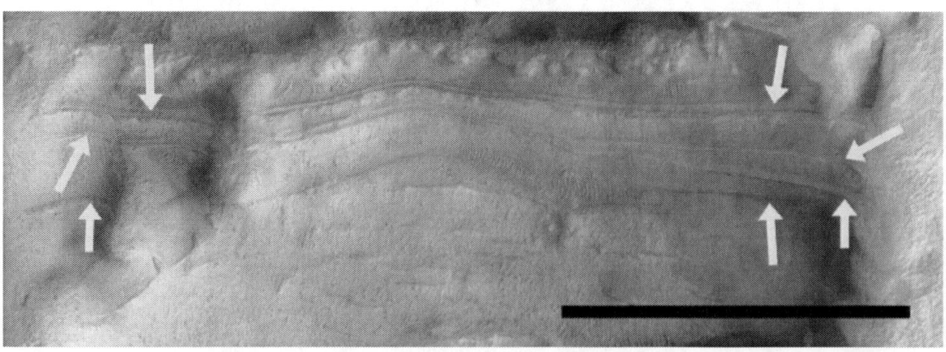

Figure 16d: Scour marks (yellow arrows) in Hellas Basin (HiRISE image PSP_009548_1420). Scale bar = 1 km. HiRISE image credit: NASA/JPL/University of Arizona.

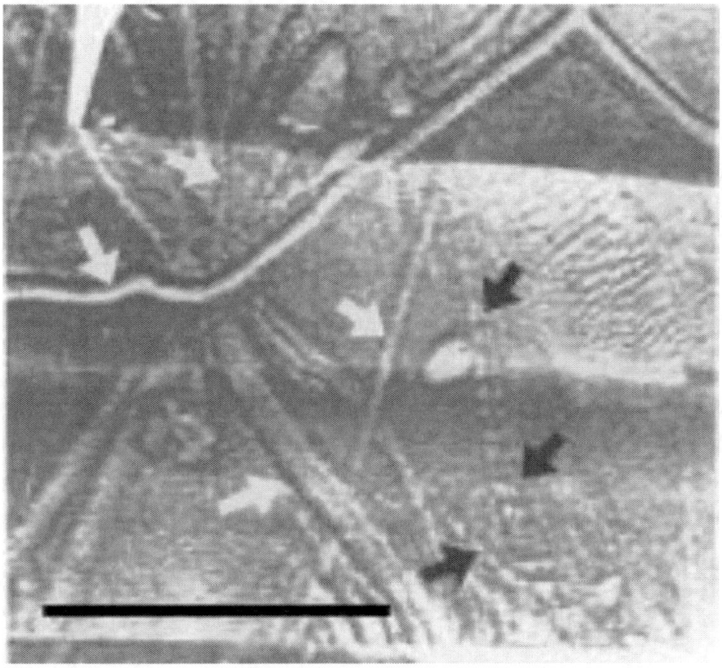

Figure 16e. Scour marks (yellow arrows) and chains of craters (red arrows) in Saglek Bank, Labrador Shelf. Scale bar = 500m.

4. Chains of Craters

Chains of crater-like structures several hundred meters wide and 1 to 5 km long are also revealed by high-resolution images of the Martian lowlands (Figure 16f). These structures are relatively shallow when compared to neighboring isolated craters, and have straight contours and a strong anisotropic orientation. Some of the crater-chains contain up to 15-20 circular or elongated depressions with diameters ranging between 50 and 100 m. All crater-like structures within a chain have a similar size and shape, and are interconnected or superimposed. Occasionally chains are composed of circular individual units separated some tens of meters from their neighbors. Here I suggest that the chains of craters are created by iceberg rafting and scouring. A similar process has been proposed to have occurred in Athabasca Vallis (Gaidos and Marion, 2003). On the Earth, iceberg crater marks and pits form on the seabed as response to impact and loading by rolling and grounding icebergs (Barrie et al., 1986; Bass and Woodworth-Lynas, 1991). The vertical motion of the iceberg due to wave action induces the excavation of craters up to several tens of meters in diameter (Bass and Woodworth-Lynas, 1991). Iceberg-generated craters form sometimes in linear associations or in chains, with neighboring craters being interconnected, superimposed or separated approximately 20-30 meters (Figure 16e). These chains form as icebergs ground and roll due to wave motions and ocean currents (Bass and Woodworth-Lynas, 1991). Documented iceberg-generated crater chains on Earth show that crater shape and size vary between different chains, but are similar among craters in the same chain (Bass and Woodworth-Lynas, 1991).

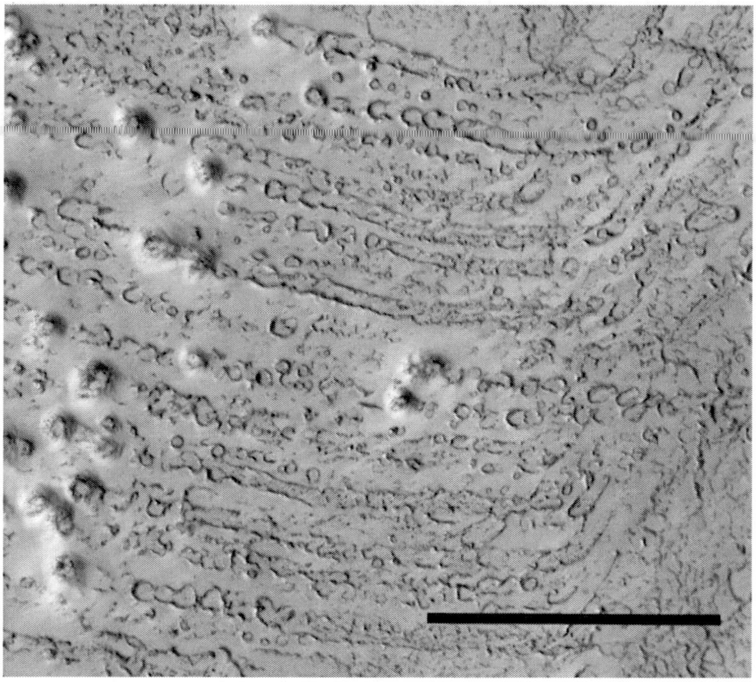

Figure 16f. Chains of craters in HiRISE image PSP_006959_2060 in the Phlegra Dorsa region. Scale bar = 1 km. HiRISE image credit: NASA/JPL/University of Arizona.

Discussion

Different alternative hypotheses for the formation of each one of the morphologies described here have been considered and analyzed but, after careful considerations, they were disregarded as they raised more problems than they help to solve: some not interconnected hypotheses may separately offer a possible explanation to one or another individual feature of those presented here, albeit never thoroughly but facing serious problems difficult to solve; and no alternative hypothesis to iceberg-related processes can explain together all the features shown in this investigation.

1. Boulders

Boulders over the plains of Mars have been proposed to be related to periglacial and glacial processes and to impact gardening (McEwen et al., 2007). On Earth, the marine record of dropstones from icebergs has been used to document ice rafting (Heinrich, 1988). I tested my hypothesis that boulder accumulation on the lowlands of Mars is related to iceberg activity, to demonstrate that it offers the more solid explanation to the boulder features, as follows.

1a. Shape of the Boulders

The debris forming boulder accumulations over the lowlands of Mars consists of mainly angular boulders. Angular boulders are in fact dominant all over the plains, as shown in direct Viking, Pathfinder and Phoenix observations (Figure 17). This distribution is consistent with iceberg deposition, as observed on Earth. Terrestrial debris transported by glaciers is derived either supraglacially from mountain rocky elements not covered by ice and from valley sides, or from basal erosion of the subglacial bed. As a result, supraglacial or englacial debris is passively transported inside the glacier, and often tends to be angular (Figure 18); while ice-rafted debris transported through the basal tractional zone is commonly subjected to large forces generating rounded boulders which often have glacial striae (Boulton, 1978). Rounded basal debris is preferentially released to the ocean while the iceberg is still pinned to the iced shore margins, and therefore tends to dominate the sedimentary record near the shore; on the contrary, angular supraglacial and englacial debris is transported by the icebergs more far away from shore (Syvitski et al., 1996).

Figure 17. Angular and flat-topped boulders in landscapes captured by (a) Viking 1 in one of the lowlands basins, Chryse Planitia; (b) Viking 2, in Utopia Planitia, a great basin on the lowlands; (c) Pathfinder, in the outlet of a giant outflow channel, Ares Vallis; and (d) Phoenix, in the plains close to the North Pole. Landscapes covered with boulders with this preferential morphology have been observed all over the lowlands since the Viking missions landed on Mars in the '70s. Such dominant morphology has remained unexplained to date, as water- and/or ice-related transport would have shaped mainly rounded boulders by potent erosion over the lowlands of Mars. The hypothesis of englacial debris transported by icebergs and gardening the bottom of a northern ocean offers the first self-consistent explanation to a four decades long dilemma. (Images NASA/JPL/Caltech).

Figure 18. Debris-covered iceberg calved from the terminus of Alaska's Sheridan Glacier in 2009. (American Geological Institute/Bruce Molnia, US Geological Survey).

1b. Distance of Deposition

Although the amount of ice-rafted debris released by a glacier is variable, in terrestrial glaciomarine settings the abundance of boulder accumulations generally decreases as the distance from the shore increases. This is so because Earth oceans are tempered, and icebergs melt rapidly near the shore. But martian oceans have been suggested to be very cold (Fairén et al., 2009, 2011, 2012; Fairén, 2010), and therefore iceberg melting would have been a slower, more sequential process on Mars, releasing their boulders cargo more uniformly during a longer period of time than terrestrial icebergs, including far distances from the shore. The fact that massive iceberg sediments dating back from glacial periods on Earth are found thousands of kilometers away from the shore both in the North (Bond et al., 1992) and the South (Williams et al., 2010) oceans, strongly supports this interpretation.

1c. Relationship with Craters

In some locations, clusters of boulders appear concentrated around circular structures (Figure 19), which have been proposed to be impact-derived craters exhuming subsurface boulders (McEwen et al., 2007). But such very small craters would have been erased by the periglacial ice (as noted by Dundas and McEwen, 2009), the same ice invoked (McEwen et al., 2007) as the generator of periglacial processes distributing the boulders. Therefore, the simultaneous advocation for preserved very small craters and periglacial processes is complicated. In addition, it is challenging to find any kind of relationship between impact craters and boulders in locations with extremely high concentration of boulders which also show pervasive crater gardening (Figure 20). Conversely, if the boulders were transported into the lowlands inside icebergs, it is expected that icebergs anchored in local topographic rises, such as crater rims, which would have assisted to the formation of the circular bouldered structures detected in HiRISE imagery. In this sense, HiRISE images of clusters of boulders shown here in Figures 16a, 19a and 19b are all located on the northern rim of Utopia Basin, as mapped by Thomson and Head (2001).

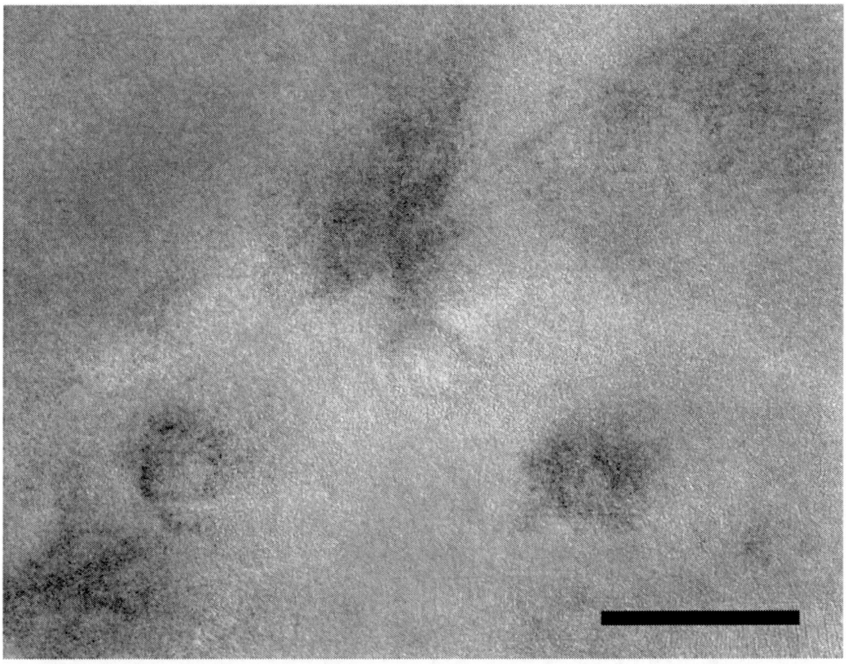

Figure 19a. Portion of the image TRA_000841_2460, showing clusters of boulders northern Utopia Planitia. In all HiRISE images in this document scale bars are 1 km, and North is up.

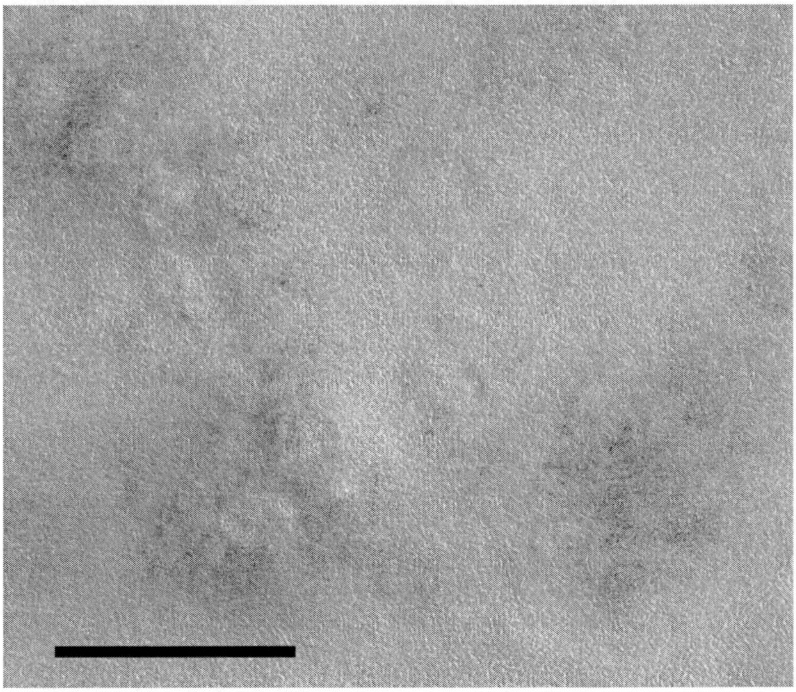

Figure 19b. Clusters of boulders in the northern rim of Utopia Planitia in the image PSP_001448_2470.

2. Mounds

2a. Types 1 and 2

The circular mounds on Mars have been generally regarded as erosional features (Dundas and McEwen, 2009), after disregarding a possible pingo origin (Dundas and McEwen, 2009) (pingos are perennial features of some periglacial landscapes, formed by small hills with cores of ice, generated by injection and freezing of pressurized water; morphologically are small, elongate to circular, ice-cored mounds). Alternatively, mounds have been proposed to be analogous to sub-aqueous mud volcanoes, similar to those identified in terrestrial marine basins (Oehler and Allen, 2012). The interpretation of Martian mounds as iceberg-related features, advanced here, is based in their distribution and composition, as follows.

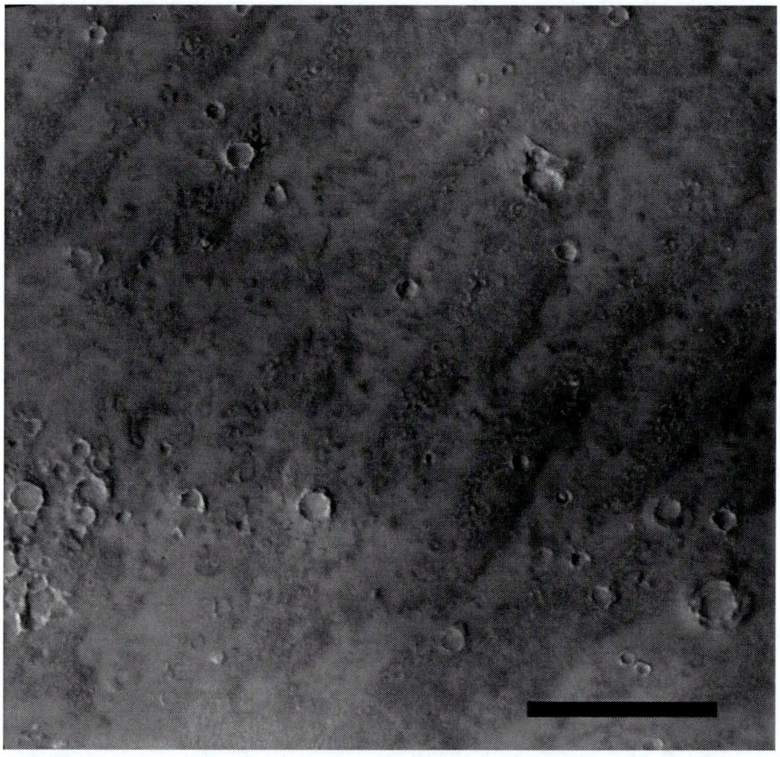

Figure 20. "Thumbprint terrain" near the highlands/lowlands dichotomy boundary northern Arabia Terra/Terra Sabaea in the HiRISE image ESP_011949_2255. These features are distributed across many parts of the Northern Plains from 30°N to 50°N (Kargel and Strom, 1992). High resolution imagery allows the identification of bouldered ridges as the agents shaping the "thumbprints". At large scale, boulders are distributed forming very dark ridges of an unusually high concentration of rocks alternating with clear strips almost devoid of boulders. Ridges are typically some hundreds of meters wide and up to 3 to 5 km long. The distribution of boulders forming ridges is likely driven by periglacial processes similar to those creating de Geer moraines on Earth (Lindén and Möller, 2005; Larsen et al., 2006; Todd et al., 2007; Golledge and Phillips, 2008), as suggested by Kargel and Strom (1992). Moraines formed beneath the grounded part of an ice sheet that extended into a lake or sea, and the ridges mark the underwater winter positions of the glacier margins in a retreating ice sheet. The final morphology shows ridges capped by a layer of boulders. The ridges are intensively cratered by subsequent impacts, and the craters do not show any kind of relationship with the presence or absence of boulders.

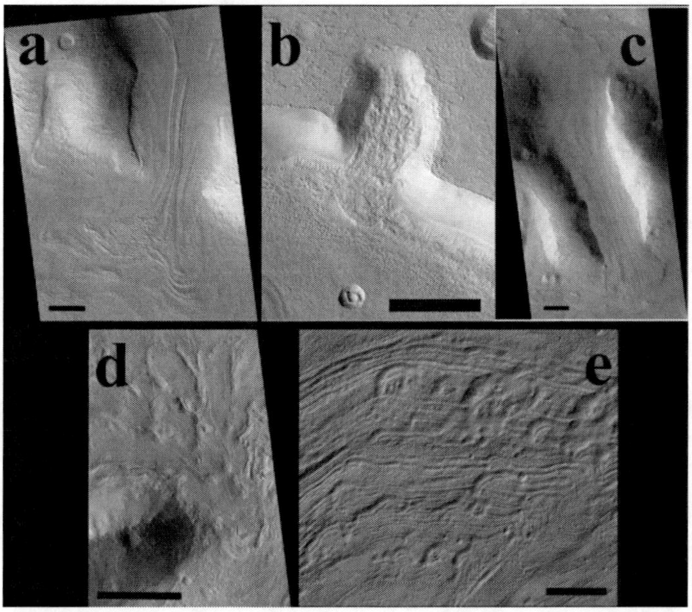

Figure 21. Some examples of fretted terrains and lobate landforms. (a) to (c) in the highlands/lowlands dichotomy boundary Northern Arabia Terra/Terra Sabaea, the most common location of fretted terrains on Mars: (a) portion of HiRISE image PSP_006806_2215 in Deuteronilus Mensae; (b) cutout of PSP_006950_2150 in Coloe Fossae; and (c) PSP_007451_2225 Northern Terra Sabaea. (d) and (e) in Hellas Planitia: (d) at the Head of Dao and Niger Valles (ESP_011656_1465); and (e) in Reull Vallis (PSP_007120_1380).

Distribution: Mounds on Earth and Mars occur isolated or in groups. On Mars, they are mostly distributed between 30 and 40°N in Utopia Planitia (Dundas and McEwen, 2009) and Chryse and Acidalia planitiaes (Oehler and Allen, 2012), just north of the greater concentration of outflow channels and fretted terrains, purported to be ancient glacial developments (Lucchitta, 2001; Dickson et al., 2008; Fairén et al., 2011) (Figure 21). Additional mounds appear in other places on the lowlands, but always near the hemispheric dichotomy boundary, and associated to valleys and the rims East of Hellas Planitia and other craters on the southern highlands (Dundas and McEwen, 2009), but in all instances strikingly close to proposed martian paleoglacier sites (Kargel and Strom, 1992; Lucchitta, 2001; Kite and Hindmarsh, 2007; Davila, chapter 2 in this volume). This distribution appears to be reflecting both concentrated iceberg activity near the shore forming Type 1 mounds; and rapid initial breakup of recently released icebergs from their forming glaciers, discharging significant amounts of debris while icebergs were still pinned to the immediate shore, forming Type 2 mounds. Type 2 mounds are also reported to occur in some craters in the southern highlands (Dundas and McEwen, 2009). The close relationship of mounds with the ancient sea shores strongly suggests an origin related to processes derived from the transport of materials from the highlands into the lowlands, such as iceberg activity. Alternative hypotheses for the origin of the mounds (Dundas and McEwen, 2009; Oehler and Allen, 2012) need to explain this relationship.

Composition: On the Earth's seafloor, mounds are composed of coarser materials, generally sorted gravel, than the surrounding sediments, and they are generally conical (Thomas and Conell, 1985). Terrestrial iceberg-derived subaqueous mounds are different from mud volcanoes in marine basins, particularly because the abundant presence of gravel in the composition of the mounds. Mounds on Mars are equally covered with small boulders. Therefore, describing the mounds in the lowlands of Mars as ancient subaqueous mud volcanoes (Oehler and Allen, 2012) appears complicated.

Finally, Martian mounds are fractured. The fractures in the Martian mounds, not present in their terrestrial counterparts, may be informative of partial dehydration due to the disappearance of the ocean, followed by cycles of more and less intense freezing resulting in episodes of dilation and contraction that have ultimately cracked the mounds.

2b. Types 3 and 4

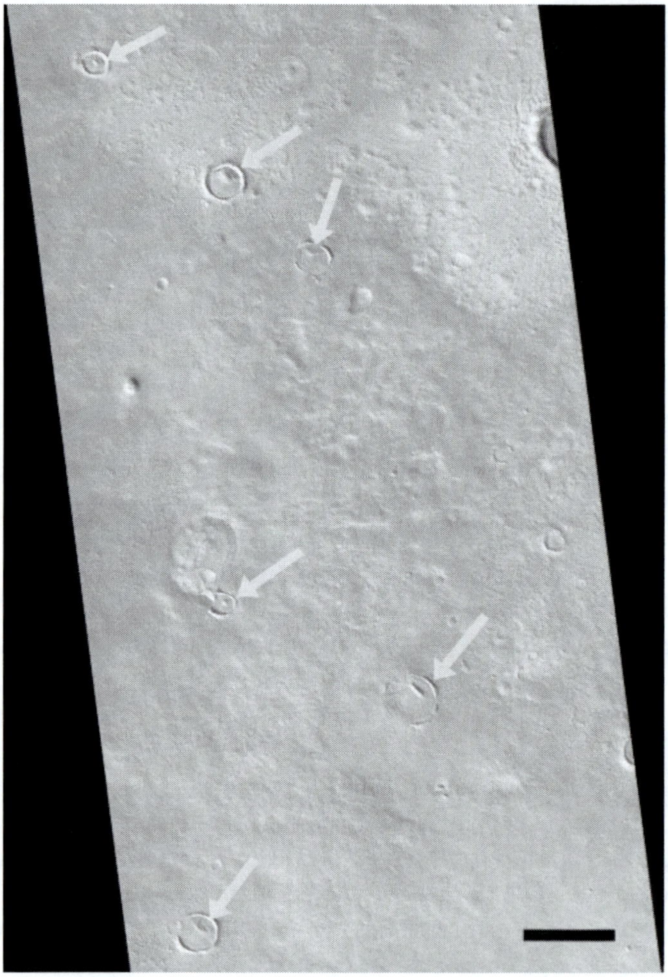

Figure 22. Field of craters partially filled with sediments in Nilosyrtis Mensae. Yellow arrows indicate sediment accumulations inside the craters, mostly in the same location in all craters. Image HiRISE PSP_006250_2200.

Interestingly enough, another two types of mounds have been described on the lowlands, also characterized as enigmatic erosional features (Dundas and McEwen, 2009). The formation of Type 3 and 4 mounds cannot confidently be attributed to iceberg-related processes. Type 3 is formed by crater-filling mounds that occur within putative small impact craters. At large scale, mounds can be appreciated inside fields of craters with a marked tendency to occur in the same location inside all craters, indicating a marked directionality in the deposition process (Figure 22). At small scale, they are covered with boulders, and therefore a wind-related origin can be disregarded. These two facts seems to indicate that the mounds are sedimentary deposits accumulated by means of aquatic currents, acting in the ancient ocean simultaneously or before the time proposed here for the widespread boulder gardening of the surface of the lowlands. Finally, Type 4 consists in flat mounds which have been interpreted as inverted-relief craters (Dundas and McEwen, 2009) (Figure 23a). This hypothesis is difficult to sustain, as there are no other inverted features (such as rivers or channels which would have carried the now hardened sediments) spatially related to the mounds. Here I suggest that these features are in fact related to glacial processes, similar to those described in Iceland (Chapman et al., 2001) and the Antarctic (Lawver et al., 2007) (Figure 23b). Other possible mode of formation could be sub-ice hyaloclastic volcanism, but this alternative is not very strong as the corresponding terrestrial planar mounds lack an anomalous magnetic signature (Lawver et al., 2007). The origin of Type 4 mounds is still in discussion for the terrestrial occurrences, so it is difficult to advance any hypothesis for their Martian possible counterparts as yet.

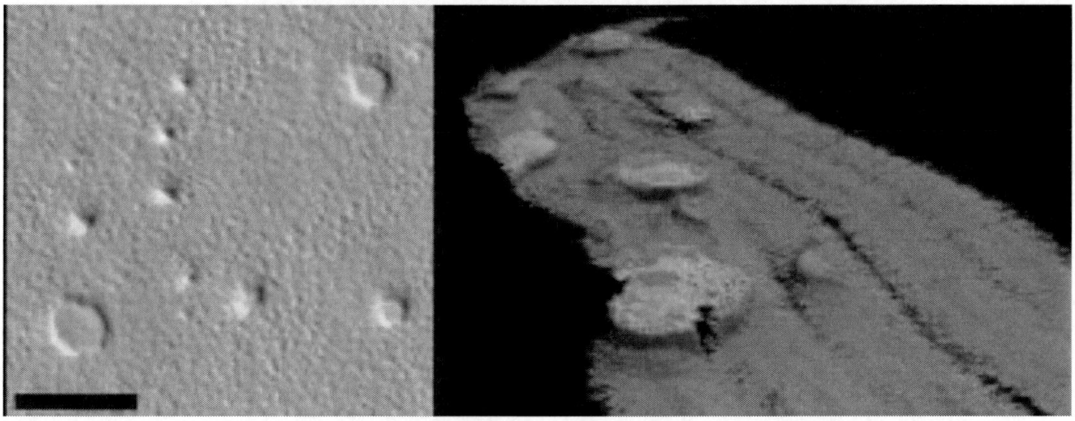

Figure 23. (Left) Roughly circular features with nearly planar tops of varying diameter in Arcadia Planitia, as shown in HiRISE image PSP_006852_2200. Scale bar = 200 m. (Right) Roughly circular features with nearly planar tops in the Victoria Land Basin, Antarctica, analyzed by multibeam bathymetry data. Features have a diameter of up to 4 km and rise up to 120 m above the surrounding seafloor. Figure modified from Lawver et al. (2007).

3. Scour Marks

Ormö and Komatsu (2003) made a comprehensive description of the different processes acting on a glacial Mars and creating linear and curvilinear features. I have followed here their descriptions to make an accurate distinction between aeolian processes creating dust devils, periglacial and fluvial traits, and glacial and iceberg features.

Ice scouring of lake and sea-floor substrates by the keels of drifting ice masses is a common geological process in modern northern lakes and continental shelves (Eyles and

Meulendyk, 2008). Figure 24 shows terrestrial scour marks detected on the seabed on the Chatham Rise, east of New Zealand (Carter et al., 2002). Modern icebergs impact the seabed up to 350 m depth in terrestrial polar oceans, even though the geological record informs of icebergs contacting the bottom of the ocean at water depths in excess of about 500 m (Dowdeswell and Bamber, 2007), with extreme values of 750 m on the Labrador margin (Piper, 2005), 820 m in the Iceland-Faeroe Ridge (Kuijpers and Werner, 2007) and almost 1000 m offshore West Greenland (Kuijpers et al., 2007) during the Late Pleistocene due to massive deglaciation. The period of seabed interaction may last from a few minutes to several months (Woodworth-Lynas et al., 1991). Scour marks and crater chains are iceberg features more fragile than the mounds or boulders, and therefore they stand for a shorter time on Earth's oceanic floor. Scours dating from Pleistocene or pre-Pleistocene times are rare (Eden and Eyles, 2001), but it is true that low sedimentation rates favor the preservation of scours on terrestrial settings (Syvitski et al., 2001). Albeit this fragility, some patterns are still identifiable on the surface of Mars' plains near 3 billion years after their formation. Figure 25 shows some additional examples of possible ice scouring in different localities on the martian basins.

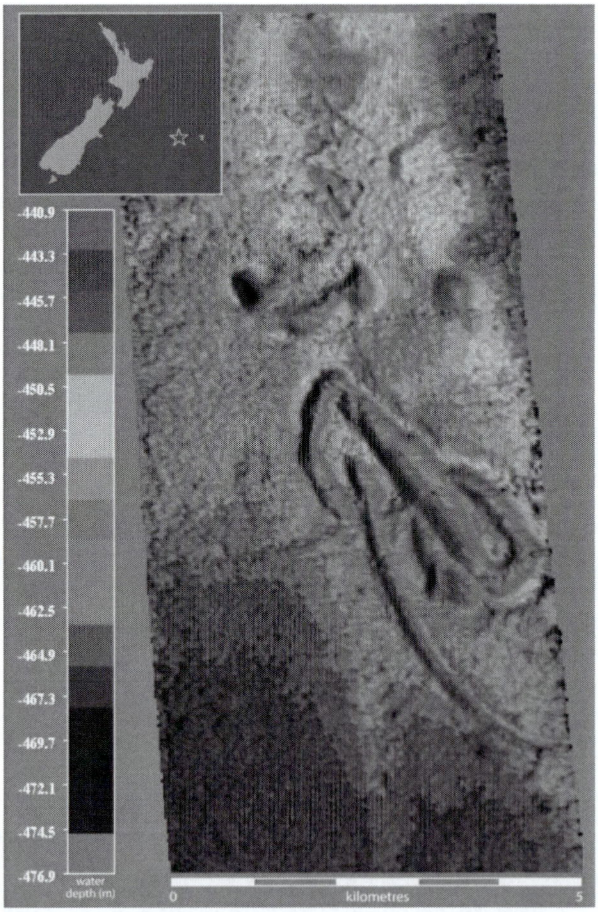

Figure 24. Scour marks on the seabed on the Chatham Rise. (Image courtesy of NIWA, New Zealand's National Institute of Water and Atmospheric Research).

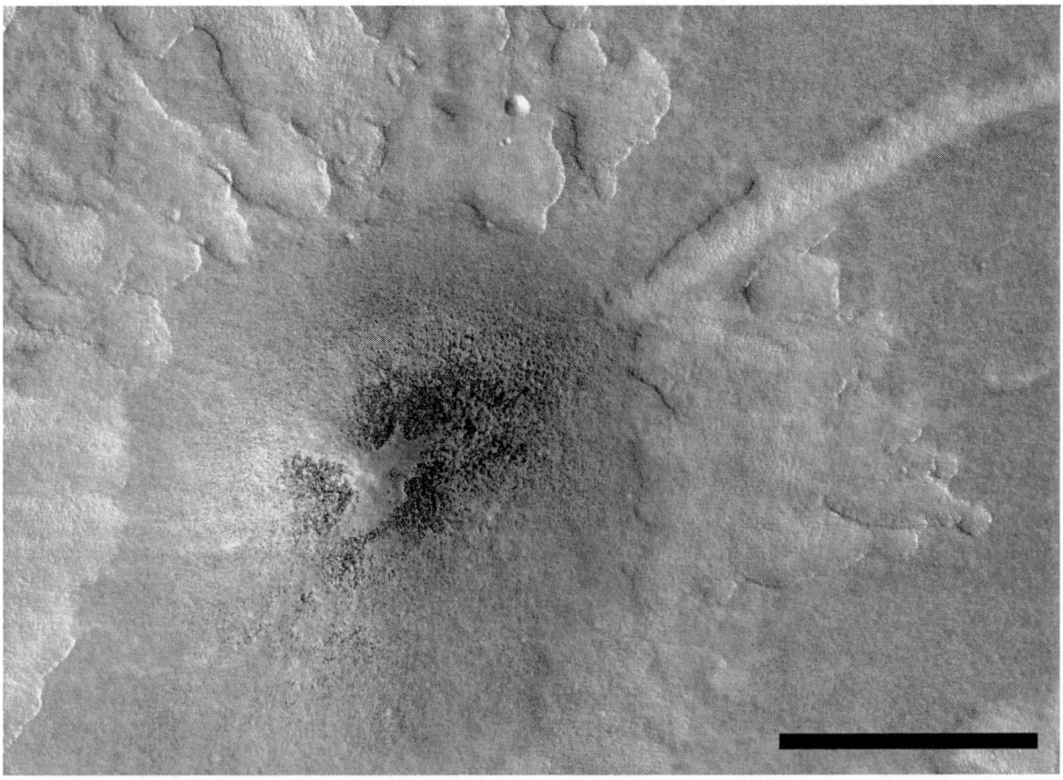

Figure 25a. Bouldered knob and linear feature northern Arabia Terra (PSP_008796_2325).

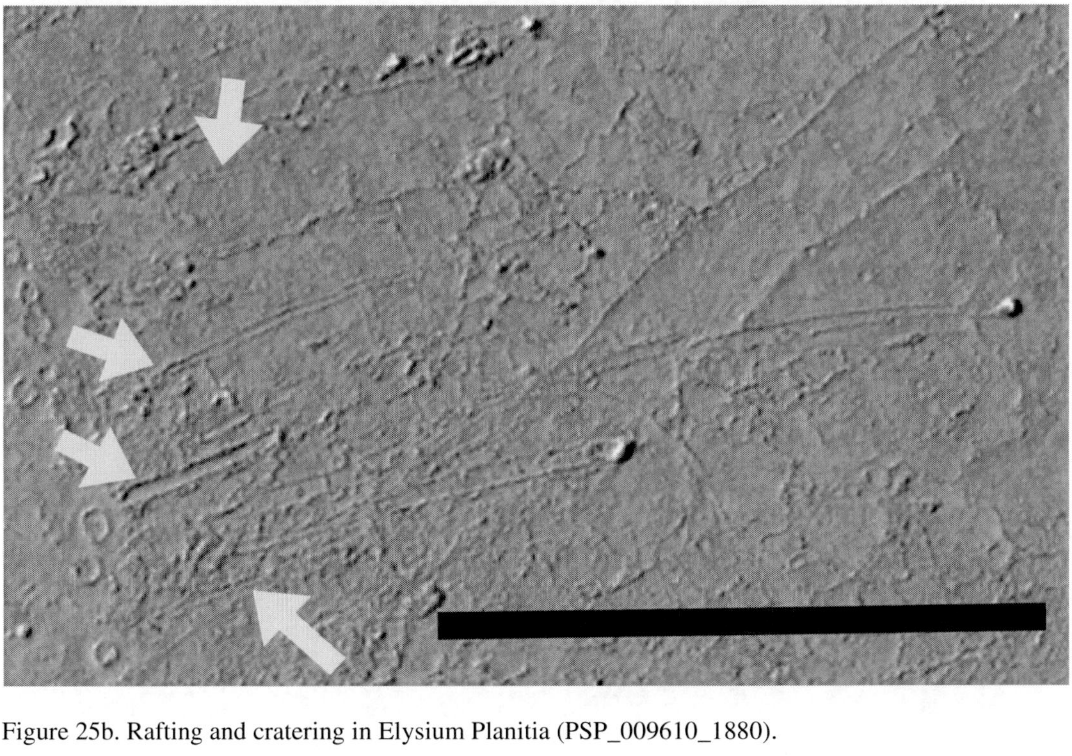

Figure 25b. Rafting and cratering in Elysium Planitia (PSP_009610_1880).

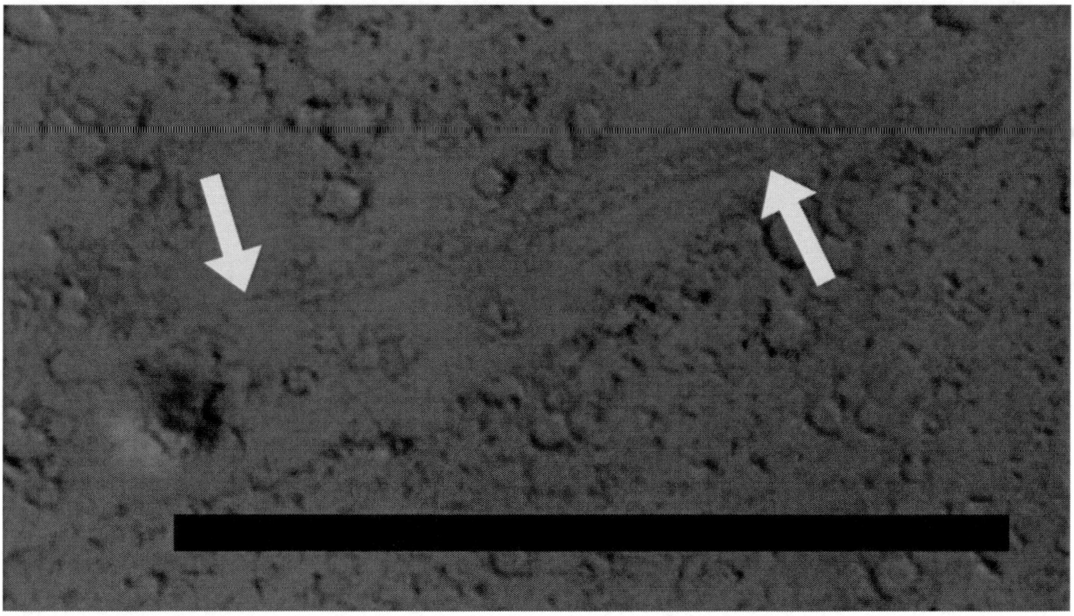

Figure 25c. Linear scour or chain of craters in the Tartarus Colles Region (ESP_012655_2070).

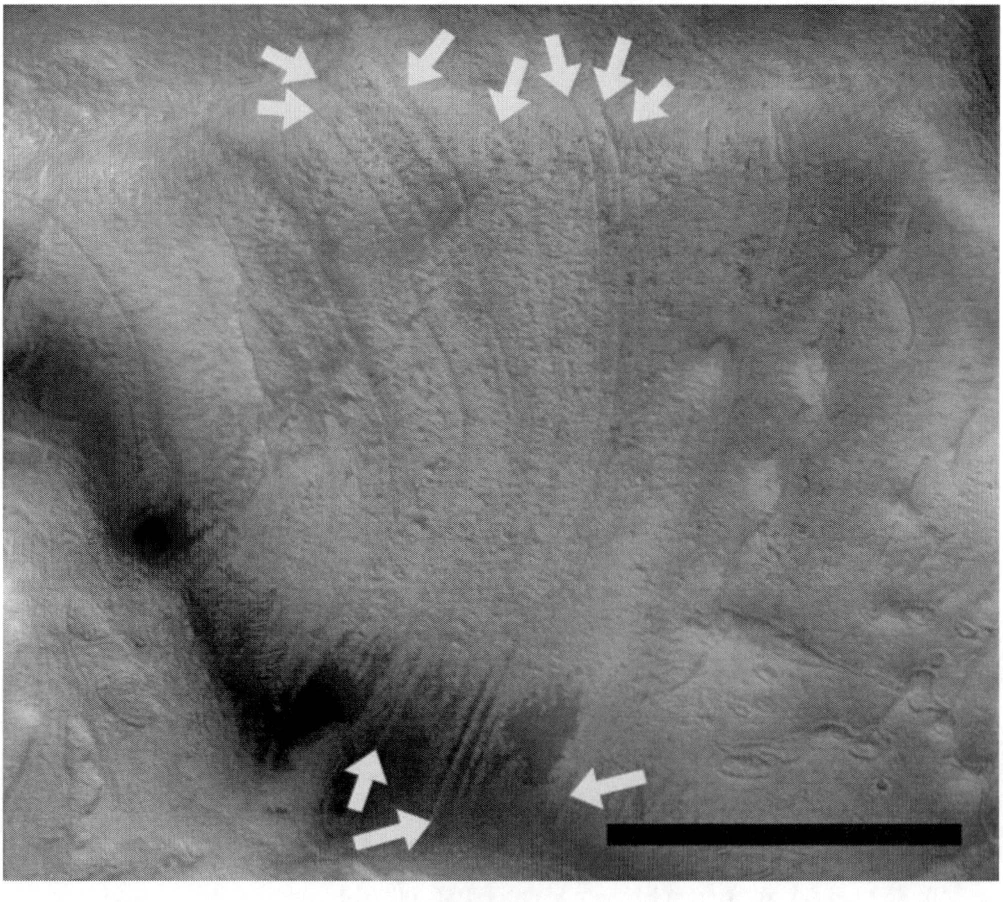

Figure 25d. Scour marks in Hellas Basin (PSP_009548_1420).

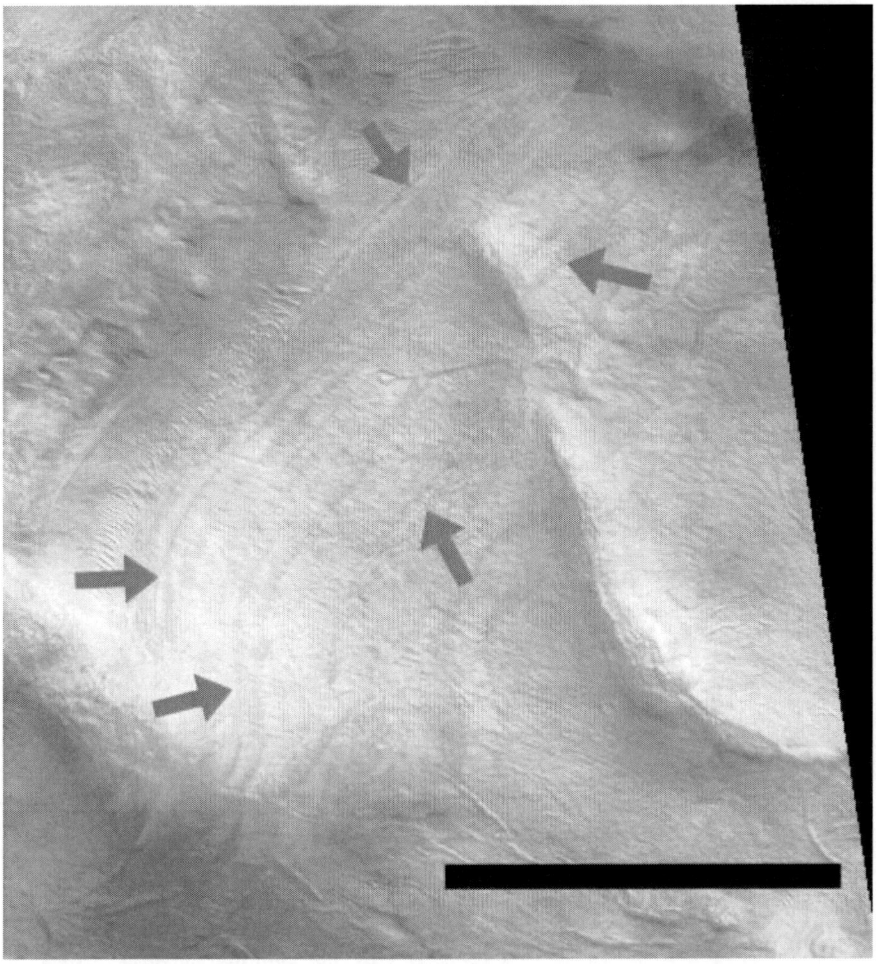

Figure 25e. Scour marks in Hellas Basin (PSP_009548_1420).

4. Crater Chains

Although meter-scale surface features could not be confidently resolved in Viking or MOC data, craters were previously attributed to volcanism (Frey and Jarosewich,1982) and/or pingo structures more or less collapsed (Soare et al., 2005; De Pablo and Komatsu, 2009). Here I show that different chains of craters have been shaped by alternative processes. The unprecedented meter-scale resolution of MRO datasets offers for the first time a way to differentiate between small impact craters, volcanic-like features (Figure 26), chains of hydrovolcanoes (Keszthelyi et al., 2009), fractured mounds with an unknown erosional origin (Dundas and McEwen, 2009) and craters related to iceberg rafting and grounding (Longva and Thoresen, 1991). No evidence for pingos on the lowlands is found after an extensive search (Dundas and McEwen, 2009). Volcanoes have a well defined tridimensional cone shape, and chains of volcanoes must be related to some tectonic feature, such as faults, and triggered by seismic shaking (Keszthelyi et al., 2009). On the contrary, chains of iceberg craters lack a well defined vertical structure and are not associated with tectonic features.

Chains use to be formed by craters all open to the same side, and with higher rims in the opposed side, as expected in a process of shoving wet and loose material by an iceberg's keel. Also, some are covered by boulders, and in occasions chains terminate in bouldered mounds. Crater chains are also common on Earth (Bass and Woodworth-Lynas, 1986, 1991; Longba and Bakkejord, 1990).

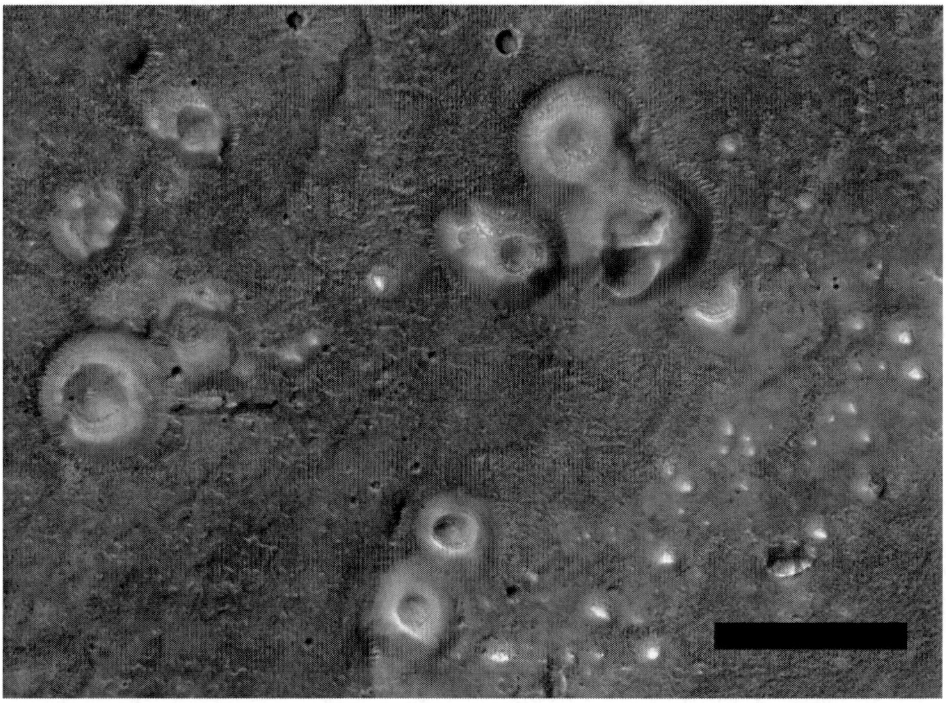

Figure 26. Mounds between 20 and 500 m in diameter, with central crater-like depressions, in Southern Acidalia Planitia. Similar mounds are common all over the lowlands. Cutout of HiRISE image PSP_008548_2205.

I want to particularly deeply discuss the interpretation offered here for the structures in Figure 16f, as chains of craters formed by iceberg grounding. These features have been previously interpreted as chains of rootles cones formed by successive explosions as the crust was translating over a fixed steam source (Keszthelyi et al., 2009). I here dispute this interpretation basing in:

1) Keszthelyi et al. (2009) analyze just a cutout of HiRISE image PSP_006959_2060, which shows only one cluster of chains of craters. When the complete image of the region is observed (Figure 27a), they appear multiple and differentiated clusters, each one formed by craters with different depths and more or less interconnected (from almost scours to differentiated craters), indicating different icebergs and/or different times of rafting. In the complete image are also clearly visible some scour marks independent from the chains of craters, giving strong support to the iceberg hypothesis.
2) The purported description as rootless cones faces the challenge of explaining how the Martian crust gained mobility by the time of formation of the chains, as such tectonic

activity is incompatible with decades of investigation on the thermal evolution of Mars (e.g., Schubert et al., 1992). Also, a hypothesis involving a translating crust fails to explain the existence of different directions in the chains, some of them opposite with respect to others.

3) The chains of craters shown in Figure 16f are not associated with tectonic features.
4) Some chains are imprinted over different more ancient chains, and these previous chains are still clearly identifiable. Cones resulting from volcanic explosions are not expected to be superimposed over ancient cones, but more likely they would deform and destroy ancient cones. Particularly, rootles cones formed by successive explosions as the crust was translating over a fixed steam source would have deformed the preexisting cones from the bottom up, while the more recent features identified in the Phlegra Dorsa region are modifying preexisting chains of cones from the top.
5) The main chains all begin as simple lines of bouldered strips, which progressively turn into scour marks, and finally chains of craters. The bouldered strips are intrinsically related to the chains of craters, and they are not explicable as volcanic features. On the contrary, they are clear diagnostic of the path followed by the iceberg or icebergs as they were moved by the oceanic currents and/or the wind (Smith and Banke, 1983).
6) Crater shape and size are exactly the same in all craters in a given chain, a morphology which is very hard to explain as a result of explosive crustal volcanism, but is actually expected after iceberg grounding.
7) The craters forming the chains are all open to one side, and all have higher rims in the opposite side. This morphology, present in all the craters in all the chains, is very difficult to explain as a result of steam explosions. On the contrary, it is easily attributable to the direction of movement of the iceberg or icebergs rafting the oceanic floor, likely caused by the oceanic and/or wind currents dominating in that area.
8) Crater rims are covered by boulders, in a concentration strikingly contrasting with the virtual absence of boulders in the surface between chains. It is challenging to find reports of a great amount of boulders extruded by mud volcanism.
9) The main chains terminate in mounds covered with boulders. No association is evident between a chain of some tens of rootless volcanic cones and one single mound. More difficult is to find such association when the phenomenon is observed in all chains of three clusters of five or more parallel chains each one. On the contrary, these morphologies are easily attributable to iceberg rafting and final grounding, creating long chains of craters and final bouldered mounds where the iceberg was longer stranded.
10) The clusters are located near the dichotomy boundary, northern Marte Vallis and Mangala Valles, and northwest the Northwestern Slope Valleys (Figure 27b), a system of giant outflow channels which has been proposed as the site of significant glacier development (Kite and Hindmarch, 2007). These glaciers were likely the source of repeated iceberg release into the northern plains ocean near the analyzed location.
11) As expected if the craters were formed by iceberg rafting, the clusters of chains are imprinted on a local topographic rise, as observed in MOLA data (Figure 27c).

12) In sum, I argue that each group/cluster of multiple, sub-parallel, curved chains of craters in this HiRISE observation was made by a single ice mass with multiple keels, and drifted by oceanic currents. A terrestrial analogue is shown in Figure 28.

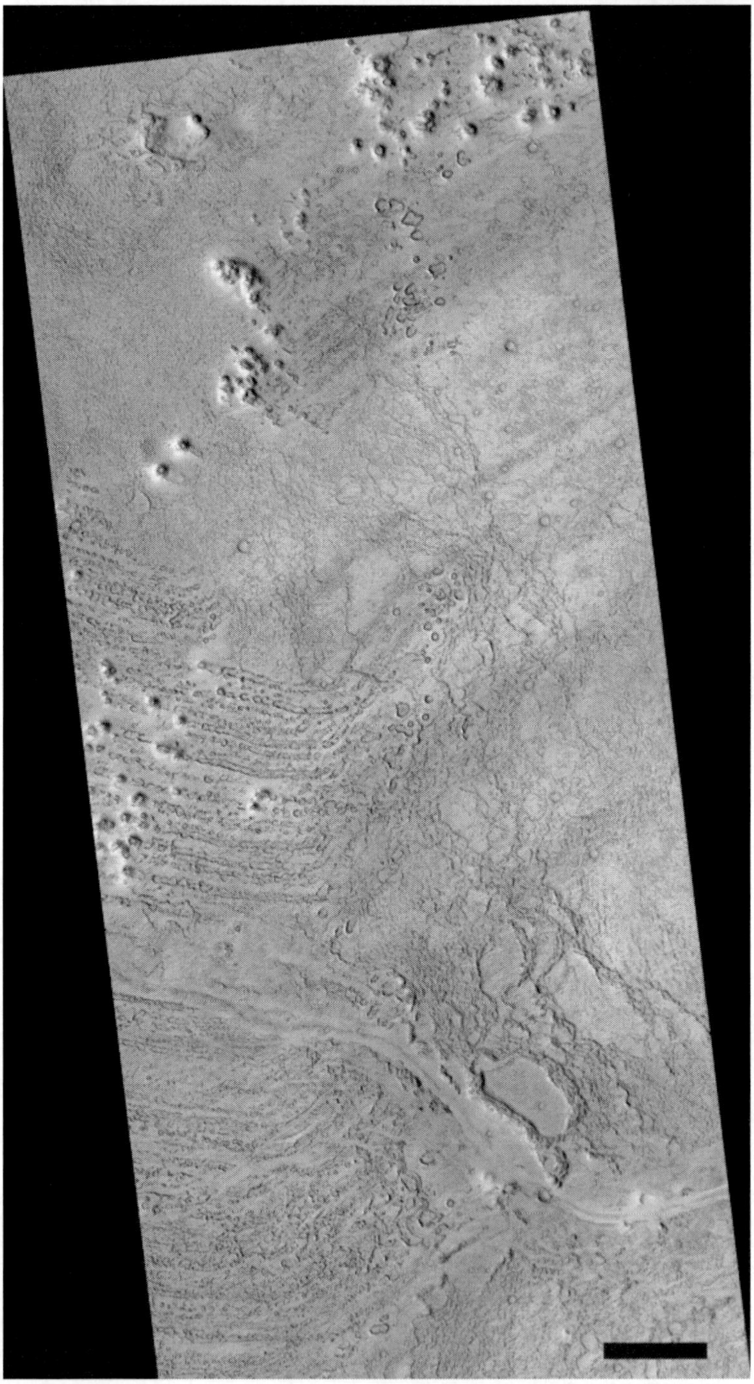

Figure 27a. Complete coverage of the chains of craters shown in HiRISE image PSP_006959_2060 in the Phlegra Dorsa region.

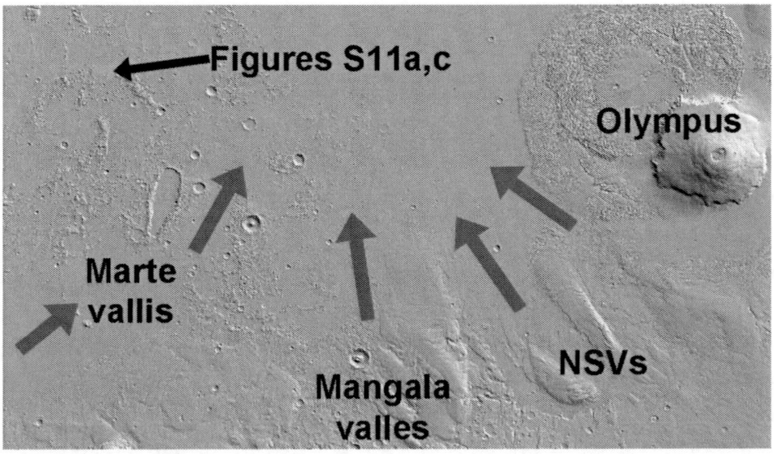

Figure 27b. MOLA shaded relief map of the location of HiRISE image PSP_006959_2060, as shown in Fig S11 a and c. Red arrows mark the possible source of icebergs toward the location in the image.

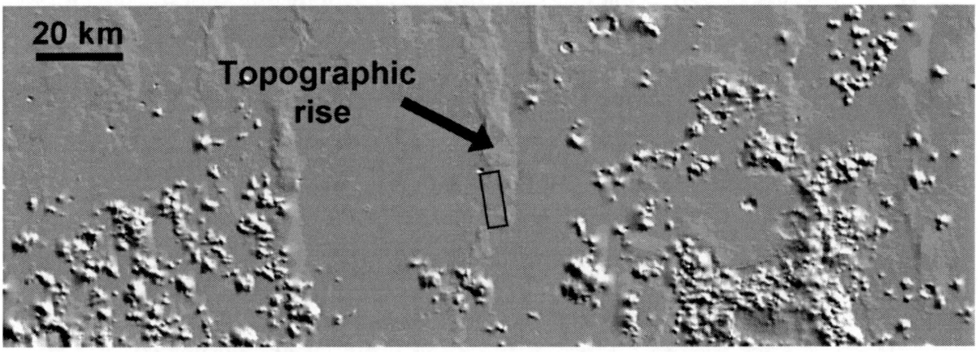

Figure 27c. MOLA shaded relief map showing the location of HiRISE image PSP_006959_2060 in Fig. S11a (black box).

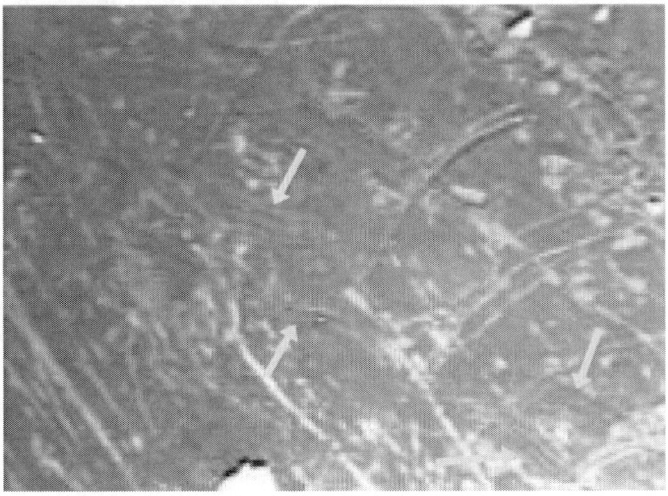

Figure 28. Multi-keeled scour marks (yellow arrows) in the St. Lawrence estuary, Quebec. Aerial photo, image width approximately 750 m. (Figure modified from Woodworth-Lynas and Guigné, 2003).

THE AGE OF ICEBERG ACTIVITY

The precise dating of the iceberg activity is challenging, because at least 3 Gy of periglacial processes over the lowlands of Mars have obscured most evidences for the existence of icebergs in the early oceans. Evidence for the development of periglacial features over different regions on the lowlands during the Amazonian, time after the disappearance of the oceans, is well recognized (Kargel and Strom, 1992; Head et al., 2003; McEwen et al., 2007; Mellon et al., 2009; Dundas and McEwen 2009; Keszthelyi et al., 2009). But periglacial traits must not be confused with iceberg-related features (Gilbert, 1989): cold Noachian and/or Hesperian seas/oceans harboring icebergs are perfectly compatible with Amazonian periglacial processes over the lowlands, and in fact the latest is the more consistent final outcome of the former, considering the global climatic evolution of Mars (e.g., Fairén et al., 2003). Given the well constrained severe limitations that suffered both the Martian atmosphere and hydrosphere during the Amazonian, I suggest that the glaciomarine geological environment forming the iceberg-related features described here occurred during the Late Hesperian. This age estimation is also consistent with the time of development of kilometer-scale polygonal terrains in the northern plains, suggested to have formed during or after the Late Hesperian, as deep-water marine deposits in a lowlands ocean (Oehler and Allen, 2012; Moscardelli et al., 2012). In any case, the evidences for iceberg activity presented here should have been formed at the time when the last major standing body of water was covering the northern lowlands of Mars.

It is remarkable that billions of years of enhanced degradation rates all over the lowlands (Carr and Head, 2003), mostly triggered by pervasive periglacial and eolian erosion, have still left some scarce and dispersed, but readily identifiable, evidences for iceberg grounding and scouring in the primeval martian oceans. This is more significant when these same geomorphologies are not easy to discern even on the Earth (Woodworth-Lynas et al., 1991). In fact, on the geologically active Earth, the older iceberg-related erosional features date from less than 300 million years (Eyles et al., 1997), one order of magnitude less time than that proposed in this investigation for iceberg activity on Mars. The same as on Mars, such ancient landforms are very scarce on Earth, and the most common ice-related features on the seabed are post-Pleistocene (<3 million years, see e.g. Eden and Eyles, 2001), with the older identification of ice-rafted debris dating from 7 million years (Williams et al., 2010). But the strong evidences presented here still remain as the fossil confirmation of the action of icebergs in the distant past.

As expected, most of the features found here to be related to iceberg rafting and grounding are located in elevated areas of the northern plains, whether near the dichotomy boundary or on local topographic highs (see Figure 15). These areas had shallow water depths, less than the iceberg's keel depth, and therefore keels reached through the full depth of the water column to impinge on the sediments below (see, for example, Figure 27c). Also, these areas have been more isolated from subsequent aqueous episodes, including periglacial processes, because the elevation prevented them to be covered again by water/ice/sediments which would have erased the previous iceberg erosive marks.

GLACIERS AND ICEBERGS

The interpretations offered here imply the occurrence of continental glaciers forming in the highlands and streaming northward towards the lowlands, and towards the Hellas and Argyre Basins. Glacier activity is well recognized and documented to have occurred on Mars at least since the Late Noachian/Early Hesperian (Irwin et al., 2004; Davila, chapter 2 in this volume) through the Hesperian (Kargel and Strom, 1992) and into the Late Amazonian (Neukum et al., 2004; Head et al., 2005), including very large-scale regional glaciation during the Amazonian (Head et al., 2006). It has been recently suggested the presence of wide highland extensions covered with glaciers rimming the topographic dichotomy at the time when oceans existed on Mars, during the Noachian (Fairén et al., 2011, 2012; Davila, chapter 2 in this volume). Also, the outflow channels terminating in Chryse, Amazonis, Isidis and Elysium Planitiaes, and flowing into the Hellas and Argyre impact Basins, all have been interpreted as erosive features of ice streaming within ancient ice sheets, creating glaciers formed in the highlands and dying in the lowlands (Kargel and Strom, 1992; Lucchitta, 2001; Forget et al., 2006; Kite and Hindmarsh, 2007). And some icebergs could have derived from the northern polar ice sheet. Large armadas of icebergs formed at the glaciers termini would then have been released into extremely cooled northern seas or oceans, and these enormous amounts of drifting ice would have distributed the boulders forming clusters, created the mounds and carved ploughs and chains of craters all over the Martian lowlands (Figure 29).

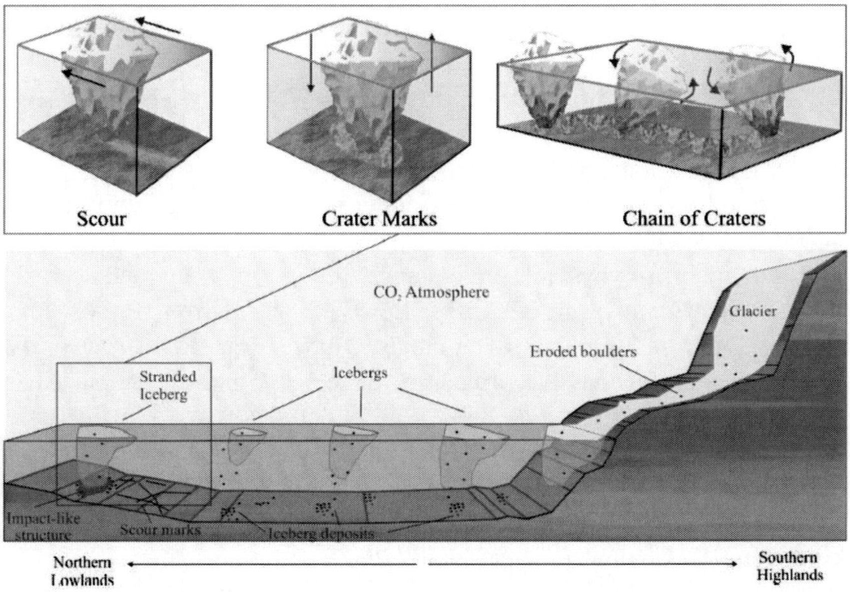

Figure 29. Schematic diagram of the formation of icebergs in the Martian highlands and of iceberg-related deposits and morphologies in a putative lowlands ocean or seas covering the major basins (Credit: A. F. Davila). Top: Iceberg grounding results in scour marks when the iceberg keel impinges on the ocean-floor sediments; mounds and crater marks form with the vertical motion of a grounded iceberg due to wave motion; chains of craters are formed by a grounded and rolling iceberg due to wave motion and ocean currents. Bottom: Icebergs form where massive glaciers covered the highlands and terminate in the northern lowlands' ocean. Their northbound migration, combined with topographic rises and/or decreases in the sea level, results in the iceberg-related ocean-floor morphologies detailed in the text.

CONCLUSION

The hypothesis of iceberg grounding and rafting in a northern ocean is the most parsimonious explanation for the origin of all the morphologies considered here, as aptly places the observations of the analyzed features under a single shared origin. This interpretation implies the presence of cold and large bodies of liquid water on the lowlands and the main basins of the martian topography. The size of the water bodies may have ranged from several local seas to a single hemispheric ocean, and they may have been continuous in time or episodic. Given the well constrained (Clifford and Parker, 2001) severe limitations that suffered both the Martian atmosphere and hydrosphere during the Amazonian, this glaciomarine episode must have occurred during or before the Late Hesperian. These interpretations are in agreement with the idea of a "cold and wet early Mars" (Fairén et al., 2009, 2011, 2012; Fairén, 2010), characterized by a primeval global hydrosphere governed by ice-semicovered oceans and/or seas, glaciers, icebergs and massive polar caps, similar to the Polar Regions on Earth.

REFERENCES

Anderson, J. B. *Antarctic marine geology* (Cambridge University Press, 1999).

Anderson, J. B., E. W. Domack, D. D. Kutz, Observations of sediment laden icebergs in Antarctic waters: implications to glacial erosion and transportation. *J. Glaciol.* 25, 387-397 (1980).

Barrie, J. V., Lewis, C. F. M., Parrott D. R. and Collins, W. T. Submersible observations and origin of an iceberg pit on the Grand Banks of Newfoundland. *Curr. Res. part A.* Geol. Surv. Can. Pap. 86-1A, pp. 251 258 (1986).

Bass, D. W. and Woodworth-Lynas, C. Iceberg crater marks on the sea floor, Labrador Shelf. *Marine Geology* 79, 243-260 (1991).

Bass, D. W., C. Woodworth-Lynas, Iceberg crater chains and scour up- and downslope. In *Ice Scour and Seabed Engineering.* C. F. M. Lewis, D. R. Parrott, P. G. Simpkin, J. T. Buckley, Eds. (Environ. Stud. Revolving Funds, Ottawa, Rep. 049, 1986), pp.122-128.

Bond, G., et al. Evidence for massive discharges of icebergs into the North Atlantic ocean during the last glacial period. *Nature* 360, 245 - 249 (1992).

Boulton, G. S. Boulder shapes and grain-size distributions of debris as indicators of transport paths through a glacier and till genesis. *Sedimentology* 25, 773-799 (1978).

Carr, M. H. and Head, J. W. Oceans on Mars: An assessment of the observational evidence and possible fate. *J. Geophys. Res.* 108, 5042 (2003).

Carter, L., H. L. Neil, L. Northcote. Late Quaternary ice-rafting events in the SW Pacific Ocean, off eastern New Zealand. *Marine Geology* 191, Pages 19-35 (2002).

Chapman, M. G., et al. Study of volcano/ice interactions gains momentum. *Eos, Transactions, American Geophysical Union* 82, 234-235 (2001).

Clifford, S. M. and Parker, T. J. The evolution of the Martian hydrosphere: Implications for the fate of a primordial ocean and the current state of the northern plains. *Icarus* 154, 40-79 (2001).

De Pablo, M. A., Komatsu, G. Possible pingo fields in the Utopia basin, Mars: Geological and climatical implications. *Icarus* 199, 49-74 (2009).

Dickson, J. L., Head, J. W., Marchant, D. R. Late Amazonian glaciation at the dichotomy boundary on Mars: Evidence for glacial thickness maxima and multiple glacial phases. *Geology* 36, 411–414 (2008).

Dowdeswell, J. A., Bamber, J. L. Keel depths of modern Antarctic icebergs and implications for sea-floor scouring in the geological record. *Marine Geol.* 243, 120-131 (2007).

Driemanis, A. The problem of waterlain till. In *Moraines and Varves* (ed. Schlucher, C.), pp. 167-177 (A. A. Balkema, 1979).

Dundas C. M., A. S. McEwen. An assessment of evidence for pingos on Mars using HiRISE. *Icarus* 205, 244-258 (2009).

Eden, D. J., Eyles, N. Description and numerical model of Pleistocene iceberg scours and ice-keel turbated facies at Toronto, Canada. *Sedimentology* 48, 1079-1102 (2001).

Eyles, N., C. H. Eyles, V. A. Gostin. Iceberg rafting and scouring in the Early Permian Shoalhaven Group of New South Wales, Australia: Evidence of Heinrich-like events? *Palaeogeography, Palaeoclimatology, Palaeoecology* 136, 1-17 (1997).

Eyles, N., Meulendyk, T., Ground-penetrating radar study of Pleistocene ice scours on a glaciolacustrine sequence boundary. *Boreas* 37, 226-233 (2008).

Fairén, A. G. A cold and wet Mars. *Icarus*, 208, 165-175 (2010).

Fairén, A. G., Dohm, J. M., Baker, V. R., de Pablo, M. A., Ruiz, J., Ferris, J., Anderson, R. M. (2003). Episodic flood inundations of the northern plains of Mars. *Icarus*, 165, 53-67.

Fairén, A. G., Davila, A. F., Gago-Duport, L., Amils, R., McKay, C. Stability against freezing of aqueous solutions on early Mars. *Nature* 459, 401-404 (2009).

Fairén, A. G., A. F. Davila, L. Gago-Duport, J. D. Haqq-Misra, C. Gil, C. P. McKay and J. F. Kasting (2011). Cold glacial oceans would have inhibited phyllosilicate sedimentation on early Mars. *Nature Geoscience*, 4(10), 667-670.

Fairén, A. G., Davila, A. F., Schulze-Makuch, D., Rodríguez, A. P., McKay, C. P. Glacial paleoenvironments on Mars revealed by the paucity of hydrated silicates in the Noachian crust of the northern lowlands. *Planetary and Space Science*, in press (2012).

Fecht, K. R. and Tallman, A. M. Bergmounds along the western margin of the channeled scablands, south central Washington. *Abs. w. prog., Geol. Soc. America* 10, p. 400 (1978).

Forget, F., Haberle, R. M., Montmessin, F., Levrard, B. and Head, J. W. Formation of glaciers on Mars by atmospheric precipitation at high obliquity. *Science* 20, 368-371 (2006).

Frey, H., M. Jarosewich. Subkilometer martian volcanoes: properties and possible terrestrial analogues. *J. Geophys. Res.* 87, 9687-9879 (1982).

Gaidos, E. and Marion, G. Geological and geochemical legacy of a cold early Mars. *J. Geophys. Res.* 108(E6), 5055, doi:10.1029/2002JE002000 (2003).

Gilbert, R. A distinction between ice-pushed and ice-lifted landforms on lacustrine and marine coasts. *Earth Surface Processes and Landforms* 15, 15-24 (1989).

Golledge, N., Phillips, E. Sedimentology and architecture of De Geer moraines in the western Scottish Highlands, and implications for grounding-line glacier dynamics. *Sedimentary Geology* 208, 1-14 (2008).

Head, J.W. et al. Possible ancient oceans on Mars: Evidence from Mars Orbiter Laser Altimeter data. *Science* 286, 2134-2137 (1999).

Head, J. W., J. F. Mustard, M. A. Kreslavsky, R. E. Milliken, D. R. Marchant. Recent ice ages on Mars. *Nature* 426, 797-802 (2003).

Head, J. W. et al. Tropical to mid-latitude snow and ice accumulation, flow and glaciation on Mars. Nature 434, 346 351 (2005).

Head, J. W., Nahm, A. L., Marchant, D. R. and Neukum, G. Modification of the dichotomy boundary on Mars by Amazonian mid-latitude regional glaciations. *Geophys. Res. Lett.* 33, L08S03, doi:10.1029/2005GL024360 (2006).

Heinrich, H., Origin and consequences of cyclic ice rafting in the northeast Atlantic Ocean during the past 130,000 years. *Quat. Res.* 29, 142–152 (1988).

Hill, J. C., Gayes, P. T., Driscoll, N. W., Johnstone, E. A. and Sedberry, G. R. Iceberg scours along the southern U.S. Atlantic margin. *Geology* 36, 447-450 (2008).

Irwin, R. P., Watters, T. R., Howard, A. D. and Zimbelman, J. R. Sedimentary resurfacing and fretted terrain development along the crustal dichotomy boundary, Aeolis Mensae, Mars. *J. Geophys. Res.* 109, E09011, doi:10.1029/2004JE002248 (2004).

Kargel, J. S. and Strom, R. G. Ancient glaciation on Mars. *Geology* 20, 3-7 (1992).

Keszthelyi, L. P., Jaeger, W. L., Dundas, C. M., Martinez-Alonso, S., McEwen, A. S. Early HiRISE observations of hydrovolcanic interaction on Mars. *Icarus* 205, 211-229 (2009).

Kite E. S., Hindmarsh, R. C. A. Did ice streams shape the largest channels on Mars? *Geophys. Res. Lett.* 34, L19202, doi:10.1029/2007GL030530 (2007).

Kuijpers, A., et al. Giant iceberg plow marks at more than 1 km water depth offshore West Greenland. *Marine Geol.* 246, 60-64 (2007).

Kuijpers, A., F. Werner. Extremely deep-draft iceberg scouring in the glacial North Atlantic Ocean. *Geo-Mar. Lett.* 27, 383–389 (2007).

Larsen, E., O. Longva, B. A. Follestad. Formation of De Geer moraines and implications for deglaciation dynamics. *J. Quarter. Sci.* 6, 263 – 277 (2006).

Lawver, L. A., et al. Neotectonic and other features of the Victoria Land basin, Antarctica, interpreted from multibeam bathymetry data. In *A keystone in a changing World*, Online Proc. 10th ISAES, Ed. A. K. Cooper, C. R. Raymond, et al., USGS Open-File Rept. 2007-1047, Extended Abstract 017, 4 p., doi:10.3133/of007-1047.ea017 (2007).

Lindén M., P. Möller. Marginal formation of De Geer moraines and their implications to the dynamics of grounding-line recession. *J. Quarter. Sci.* 20, 113–133 (2005).

Longva, O., Bakkejord, K. J., Iceberg deformation and erosion in soft sediments, Southeast Norway. *Marine Geology* 92, 87-104 (1990).

Longva, O., Thoresen, M. K., Ice scours, iceberg gravity craters and current erosion marks from a gigantic Preboreal flood in southeastern Norway. *Boreas* 20, 47–62 (1991).

Lucchitta, B. K. Antarctic ice streams and outflow channels on Mars. *Geophys. Res. Lett.* 28(3), 403-406 (2001).

McEwen, A. S. et al. A closer look at water-related geologic activity on Mars. *Science* 317, 1706-1709 (2007).

McEwen, A. S. et al., The High Resolution Imaging Science Experiment (HiRISE) during MRO's Primary Science Phase (PSP). *Icarus* 250, 2-37 (2009).

Mellon, M. T., et al. The periglacial landscape at the Phoenix landing site. *J. Geophys. Res.* 114, E00E06, doi:10.1029/2009JE003418 (2009).

Moscardelli, L., T. Dooley, D. Dunlap, M. Jackson, L. Wood. Deep-water polygonal fault systems as terrestrial analogs for large-scale Martian polygonal terrains. GSA Today, 22(8), 4-9 (2012).

Neukum, G. et al. Recent and episodic volcanic and glacial activity on Mars revealed by the High Resolution Stereo Camera. *Nature* 432, 971–979 (2004).

O'Brien, P. E., Leitchenkov, G. and Harris, P. T. Iceberg plough marks, subglacial bedforms and grounding zone moraines in Prydz Bay, Antarctica. In *Glaciated continental margins: An atlas of acoustic images* (ed. Davies, T.A. et al.), pp. 228–231, (Chapman and Hall, 1997).

Oehler, D.Z., and C. C. Allen. Giant polygons and mounds in the lowlands of Mars: Signatures of an ancient ocean? *Astrobiology*, 12(6), 601-615 (2012).

Ormö, J. and Komatsu, G. Mars orbiter camera observation of linear and curvilinear features in the Hellas basin: Indications for multiple processes of formation. *J. Geophys. Res.* 108 doi:10.1029/2002JE001980 (2003).

Ovenshine, A. T. Observations on iceberg rafting in Glacier Bay, Alaska, and the identification of ice-rafted deposits. *Bull. Geol. Soc. America* 81, 891-894 (1970).

Parker, T. J., Gorsline, D. S., Saunders, R. S., D. C. Pieri and Schneeberger, D. M. Coastal geomorphology of the Martian northern plains. *J. Geophys. Res.* 98, 11061–11078 (1993).

Piper D. J. W. Late Cenozoic evolution of the continental margin of eastern Canada. *Nor. J. Geol.* 85, 305–318 (2005).

Schubert, G., Solomon, S. C., Turcotte, D. L., Drake, M. J., Sleep, N. H. Origin and thermal evolution of Mars. In *Mars* Eds. Kieffer, H.H., Jakosky, B.M., Snyder, C.W., Matthews, M.S. (Univ. of Arizona Press, Tucson, 1992) pp. 147–183.

Smith, S. D., E. G. Banke, The influence of winds, currents and towing forces on the drift of icebergs. *Cold Reg. Sci. Technol.* 6, 241–255 (1983).

Soare, R. J., D. M. Burr, J. M. Wan Bun Tseung Possible pingos and a periglacial landscape in northwest Utopia Planitia. *Icarus* 174, 373-382 (2005).

Syvitski, J. P. M. et al. Icebergs and the sea floor of the East Greenland (Kangerlussuaq) continental margin. *Arct. Antarct. Alp. Res.* 33, 52–61 (2001).

Syvitski, J. P. M., J. T. Andrews, J. A. Dowdeswell. Sediment deposition in an iceberg-dominated glacimarine environment, East Greenland: basin fill implications. *Global Planet. Change* 12, 251-270 (1996).

Thomas, G. S. P. and Connell, R. J. Iceberg drop, dump, and grounding structures from Pleistocene glacio-lacustrine sediments, Scotland. *J. Sedim. Petrol.* 55, 243-249 (1985).

Thomson, B. J., J. W. Head, III. Utopia Basin, Mars: Characterization of topography and morphology and assessment of the origin and evolution of basin internal structure. *J. Geophys. Res.* 106, 23209-23230 (2001).

Todd, B. J., P. C. Valentine, O. Longva, J. Shaw. Glacial landforms on German Bank, Scotian Shelf: evidence for Late Wisconsinan ice-sheet dynamics and implications for the formation of De Geer moraines. *Boreas* 36, 148-169 (2007).

Williams, T., et al. Evidence for iceberg armadas from East Antarctica in the Southern Ocean during the late Miocene and early Pliocene. *Earth Planet. Sci. Lett.* 290, 351-361 (2010).

Withers, P. and Neumann, G. A. Enigmatic northern plains of Mars. *Nature* 410, 651 (2001).

Woodworth-Lynas, C. et al. The physical processes of seabed disturbance during iceberg grounding and scouring. *Cont. Shelf Res.* 11, 939–951 (1991).

Woodworth-Lynas, C. M. T., Josenhans, H. W., Barrie, J. V., Lewis, C. F. M. and Parrott, D. R. The physical processes of seabed disturbance during iceberg grounding and scouring. *Cont. Shelf Res*. 11, 939–951 (1991).

Woodworth-Lynas, C., J. Y. Guigné, Ice keel scour marks on Mars: Evidence for floating and grounding ice floes in Kasei Valles. *Oceanography* 16, 90-97 (2003).

In: Mars
Editor: Alberto G. Fairén
ISBN: 978-1-62618-102-1

Chapter 3

GLACIAL ORIGIN OF FRETTED TERRAINS ON MARS

Alfonso F. Davila*
SETI Institute, Mountain View, CA, US

ABSTRACT

Fretted terrains on Mars comprise large, flat-topped mesas flanked by steep cliffs, and separated by deep and sinuous valleys and troughs. These landforms occupy large portions of the topographic dichotomy boundary between the northern lowlands and the southern highlands, and formed towards the end of the Noachian era, a key period in the evolution of Mars when the planet transitioned from wet to dry conditions. However, the exact origin of Martian fretted terrains is unknown. I propose that landforms in Martian fretted terrains formed by the cyclic advance and retreat of warm-based glaciers, fed by equatorial and tropical ice sheets or polar caps. This interpretation is supported by geomorphologic comparisons with terrestrial counterparts in Greenland and Norway, and by the identification of glacial and periglacial features related to the fretting process in ancient areas of fretted terrains, as well as in the northern plains. Estimated erosion rates required to carve the fretted terrains are also comparable to that of Arctic warm-based glaciers and ice sheets. A vast northern body of water was required as the source of moisture feeding the dichotomy ice-masses. This model for the origin of Martian fretted terrains implies widespread and pervasive globally cold conditions throughout the history of Mars.

INTRODUCTION

There is now ample evidence of liquid water activity on early Mars, largely during the Noachian Period, ranging from 4.5 to 3.7 Gyr ago (Hartmann and Neukum 2001). Questions still remain, however, regarding the climate conditions as well as the nature and intensity of the hydrological cycle during this period. Climate models for early Mars predict globally cold

* SETI Institute, 189 N. Bernardo Avenue, Suite 100 Mountain View, CA 94043, adavila@seti.org

conditions contemporarily with the main period of aqueous activity during the Noachian. This is in apparent contradiction with the widespread formation of fluvial valleys (Carr 1995), deltas (DiAchille et al. 2010), and sedimentary deposits of secondary minerals (Poulet et al. 2005, Bibring et al. 2006; Mustard et al. 2008) during this period. Models of a cold and wet early Mars have been suggested to solve this paradox (Fairén et al. 2009, Fairén 2010). According to these models, a large portion of the planet's water inventory was trapped in the form of ice already towards the end of the Noachian period, and liquid water activity was largely the result of ice-melt during warm spells, or the formation of briny solutions with low eutectic points. An optimal analogue to a cold and wet early Mars might be regions of Earth's glaciated high Arctic, where mean annual air temperatures approximate -10 to -20ºC, and yet peak summer air temperatures rise above freezing and allow for fluvial and sedimentary processes driven by ice melt. Thus, a cold and wet early Mars ought to have left a noticeable geomorphologic imprint on the surface of the planet, largely in the form of glacial erosional and depositional landforms and associated glacio-fluvial features, similar to glacial environments on Earth (cf. Evans 2007). Here, I demonstrate that such evidence is distinctly manifested in the so-called fretted terrains.

THE MARTIAN FRETTED TERRAINS

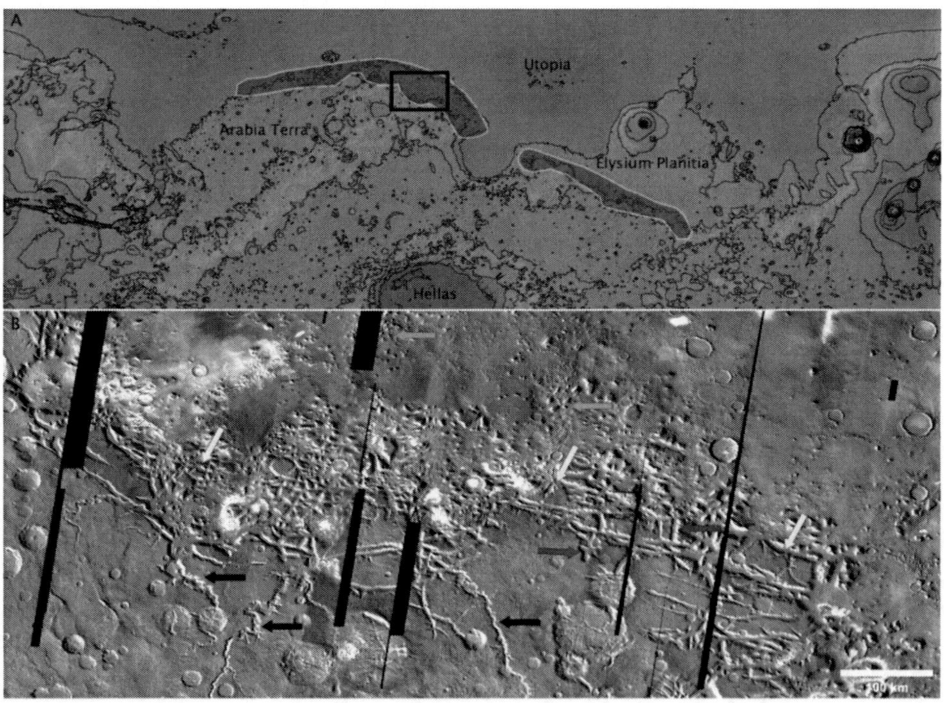

Figure 30. Distribution and morphology of fretted terrains on Mars. A. Topographic map of the Martian dichotomy boundary with regions of fretted terrains (blue areas). The black box is shown in B. and contains large mesas close to the dichotomy boundary (yellow arrows); knobby terrains nearer the northern plains (green arrows); U-shaped valleys separating, and sometimes partially dissecting, the mesas and knobby terrain (blue arrows); and meandering valleys with few tributaries that dissect rock materials of the cratered highlands (black arrows). Images courtesy of NASA/JPL/ASU.

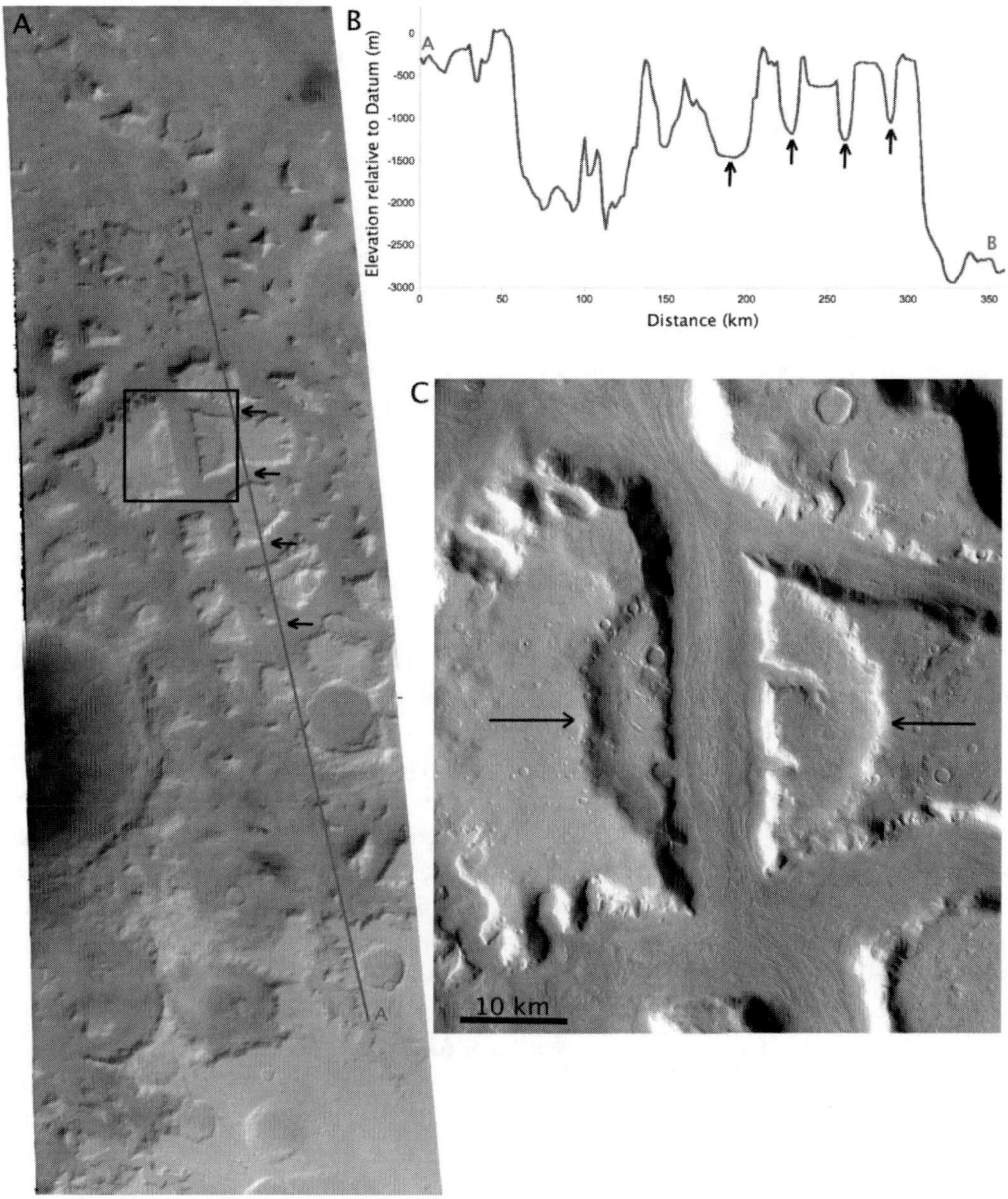

Figure 31. Geomorphology and topographic profile of fretted terrains. A. Digital Elevation Model of fretted terrains in Arabia Terra. The topographic profile along the red line is shown in B. and reveals the U-shaped morphology of the valleys. C. Adjacent mesas separated by channels share portions of an ancient impact crater (black arrows). Images courtesy of NASA/ESA/JPL/ASU.

Martian fretted terrains occur along the topographic dichotomy boundary, which separates the ancient southern cratered highlands from the younger, relatively smooth northern lowlands (Figure 30a), within latitudes ranging from 20°S to 45°N. Fretted terrain landforms range from small, isolated buttes and knobby terrain, to large, flat-topped mesas and plateaus, with major axes several hundred kilometres in length, and flanked by abrupt escarpments up to 1–2 km high (Sharp 1973; Squyres 1978) (Fig. 30b). Whereas large mesas are more frequent close to the ancient cratered highlands, isolated buttes and knobby terrain are more prevalent towards the northern plains. Mesas and plateaus are separated and/or dissected by flat-bottomed and steep-walled (U-shape) valleys, often linear to sub-linear, with most displaying low sinuosity, and depths reaching several hundred meters, and widths nearing tens of kilometres. The valleys are mostly linear, with few or no tributaries, and intersect in irregular patterns forming networks, occasionally transecting highly degraded impact craters (Fig. 31). Narrower, more sinuous valleys up to several hundred kilometres long, and with several short tributaries penetrate into the cratered highlands from areas of fretted terrains, and often originate in large impact craters. Fluvial activity (Carr 1995), ice modification (Squyres 1978; Lucchita 1984; Head et al. 2006), and structural processes (McGuill and Dimitriou, 1990) have been proposed as mechanisms responsible for the formation of the fretted terrains. However, in the absence of an established terrestrial analogue, the precise origin of these landforms is still poorly understood (Carr, 1995; Irwin et al. 2004). To extract lessons about their origin and evolution, we can look to Earth analogues that display morphologies similar to the fretted terrains both in scale and pattern.

Martian Fretted Terrains and the Terrestrial Skaergaard

Fretted terrains on Mars are analogous to terrestrial counterparts along the east Greenland and Norway coastlines (Figure 32 and 33). Greenland and Norway have coastal archipelagos consisting of thousands of islands, skerries, and small peninsulas that are collectively called the Skjaergaard (Holtedahl, 1970). The Skjaergaard terrains are old, and have been significantly modified through the advance and retreat of valley and outlet glaciers fed by continental ice sheets and polar highland ice caps (Holtedahl, 1970; Johnson et al. 1975). In Greenland cyclical periods of ice-sheet expansion to the continental margin result in warm-based outlet glaciers near the coast, which flow along sinuous valleys with few tributaries. These glaciers cut across a basement geology consisting largely of basaltic rocks overlain in places by younger volcanic and sedimentary rocks (Henriksen et al. 2009). Glacial ice dissects the coastal terrain forming a variety of islands and peninsulas separated by fjords, channels and sounds that extend below sea level to depths of c.a 1000 m (Holtedahl, 1970; Johnson et al. 1975). Topographic profiles across the Skjaergaard reveal the true topography of the valleys separating the islands and peninsulas, as deep and wide (U-shaped), with steep walls, and are identical to topographic profiles that transect Martian fretted terrains along the dichotomy boundary, with islands/mesas separated by deep and wide channels. Larger islands and peninsulas in the Skjaergaard, up to several thousand square kilometres, are relatively flat, have rounded to angular coastlines, and are typically capped by ice, and dissected by glacial and fluvial valleys. These larger islands are comparable to the lager mesas in the Martian fretted terrains. Smaller islands and skerries in the Skjaergaard are closely linked to the larger islands, and also appear within fjords, channels and sounds. These smaller landforms have angular coastlines, and typically form large clusters of islands, which are comparable to knobby terrains and buttes also found in the Martian fretted terrains. The flanks of large islands near the continent in the Skjaergaard mirror the continental shoreline, similar to the flanks of large fretted mesas facing the dichotomy, which closely mirror the opposite terrain. Adjacent islands in the Skjaergaard are portions of larger landforms dissected by glacial valleys, similar to adjacent mesas in the fretted terrains, which appear to be part of larger, dissected units. Finally, glacial valleys and fjords dissecting the coastlines of Greenland penetrate inland towards the ice cap, becoming progressively covered by glaciers, and typically have a small number of tributaries, similar to long, sinuous valleys in areas of Martian fretted terrains, which cut through the dichotomy boundary into the cratered highlands (Figure 33).

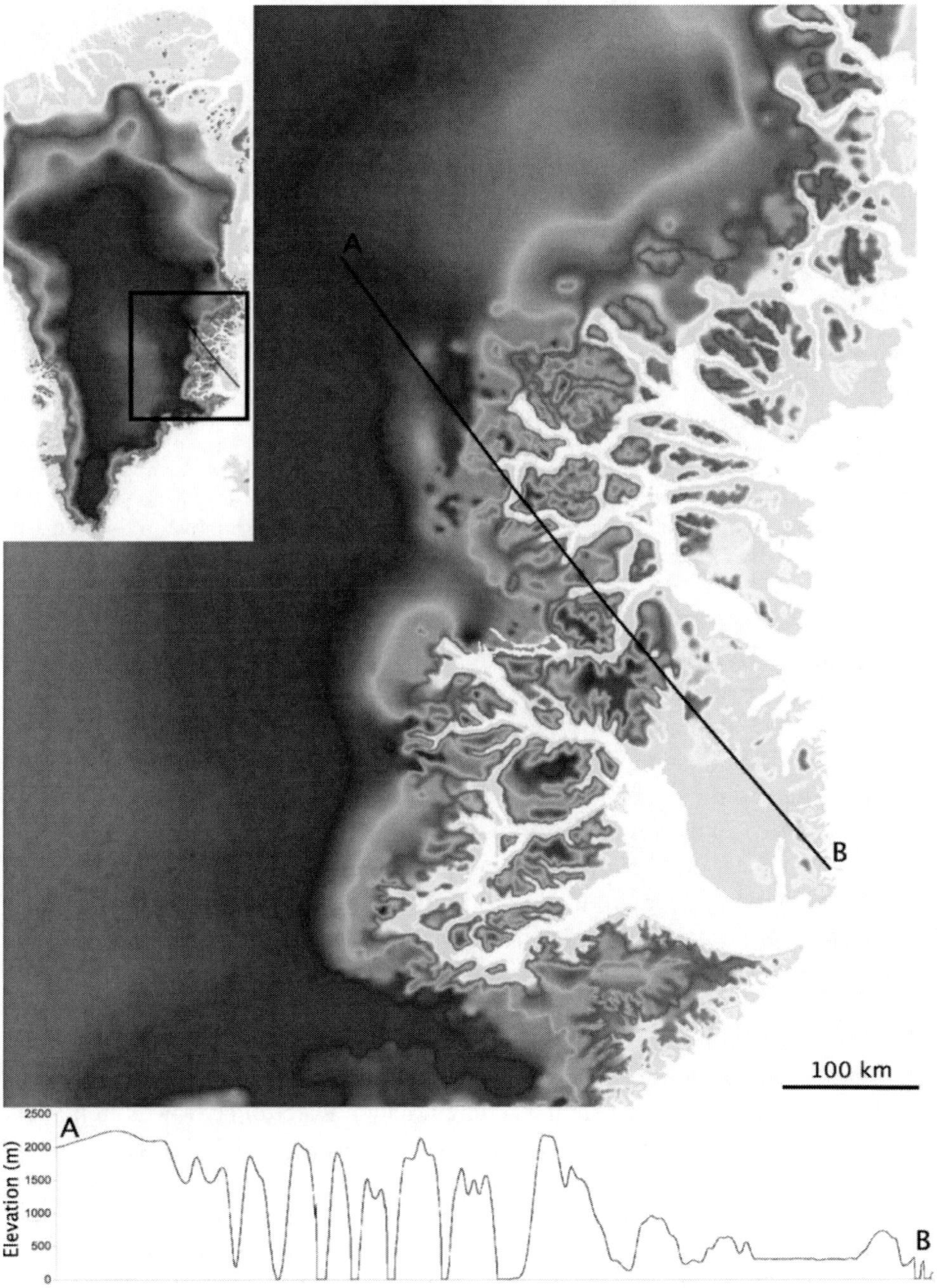

Figure 32. A. DEM of the East coast of Greenland showing the archipelago of islands and skerries. B. Elevation profile across the coastline (black line), showing the morphology of the valleys (note the similarity between this profile and the one shown in Fig. 31).

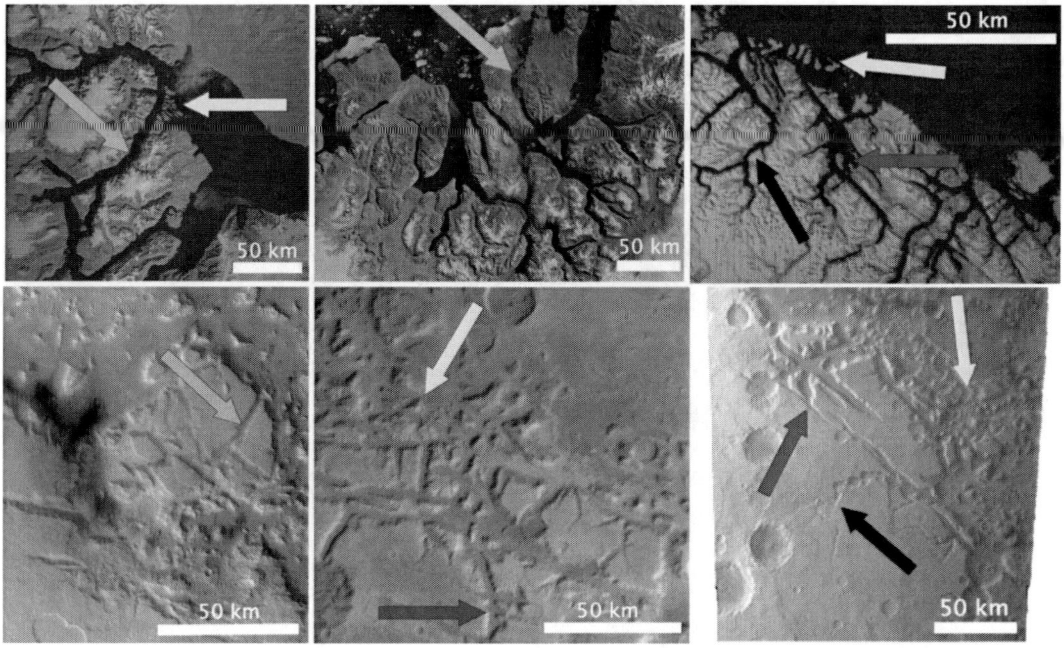

Figure 33. Comparison of Martian fretted terrains and analogue landforms on Earth. Top. Geomorphic features along Greenland's and Norway's coastline. Bottom. Fretted terrains on Mars. Blue arrows point to channels dissecting geomorphic units; Yellow arrows point to islands/knobby terrain; Red arrows point to channel grids; Black arrows point to meandering channels. Images courtesy of ESA and USGS.

A GLACIAL ORIGIN OF MARTIAN FRETTED TERRAINS

The geomorphic similarities presented above indicate that erosion by warm-based glaciers emanating from the cratered highlands could have been responsible for the dissection of the original dichotomy boundary, and the formation of Martian fretted terrains. Outlet glaciers likely flowed along pre-existing fluvial valleys and tectonic structures such as faults, as occurs along the coast of Greenland and in the High Arctic (Bonow et al. 2006). Different lines of evidence support this interpretation. Widespread areas of thumbprint terrain as well as intricate systems of through with medial ridges occur in the northern plains (Kargel and Strom 1992; Kargel et al. 1995). Thumbprint terrain has been interpreted as deposits of ice or rock indicating successive positions of retreating ice (i.e. terminal moraines), or as chains of rootless cones, which are diagnostic features of ice near the surface. On the other hand, curved troughs with central ridges have a similar appearance to terrestrial glacial landscapes where ridges of sediment mark where the sediment-laden water once flowed in tunnels under the ice (Kargel and Strom 1992; Kargel et al. 1995). These features have mapped ages of late early Hesperian to early Amazonian, and therefore overlap the formation of the original fretted terrain. In fact, the distribution of these features follows closely the distribution of fretted terrains along the dichotomy, and point to a glacial origin for the later.

Possible glacio-fluvial landforms are also present in areas of fretted terrain in the Aeolis Mensae region (10ºN-10ºS, 120-150 ºE) (Figure 34). Contrary to fretted terrains at latitudes >30º (i.e. north of Arabia Terra), fretted terrains in Aeolis Mensae have not been substantially modified since their formation in the late Noachian-early Hesperian (Irwin et al. 2004), and geomorphic features in these terrains are indicative of processes taking place at the time of fretting. Preliminary analysis of CTX and HiRISE images has revealed a diversity of possible glacial and glacio-fluvial morphologies in this area. For example, there are delta-type features in some fretted valleys (Figure 34D,H), associated to fluvial features that resemble outwash plains in glacial terrains on Earth. Other lobate features on the floors of fretted valleys resemble wet debris flow or submarine slides (Figure 34A,B). Flow features on the northern flanks of the Gale Crater mound (Figure 34C), share similar morphologic features with terrestrial rock glaciers (Anderson and Bell, 2010), such as elongated tongue-shaped morphology, pressure ridges, and accumulations of large boulders at the flanks.

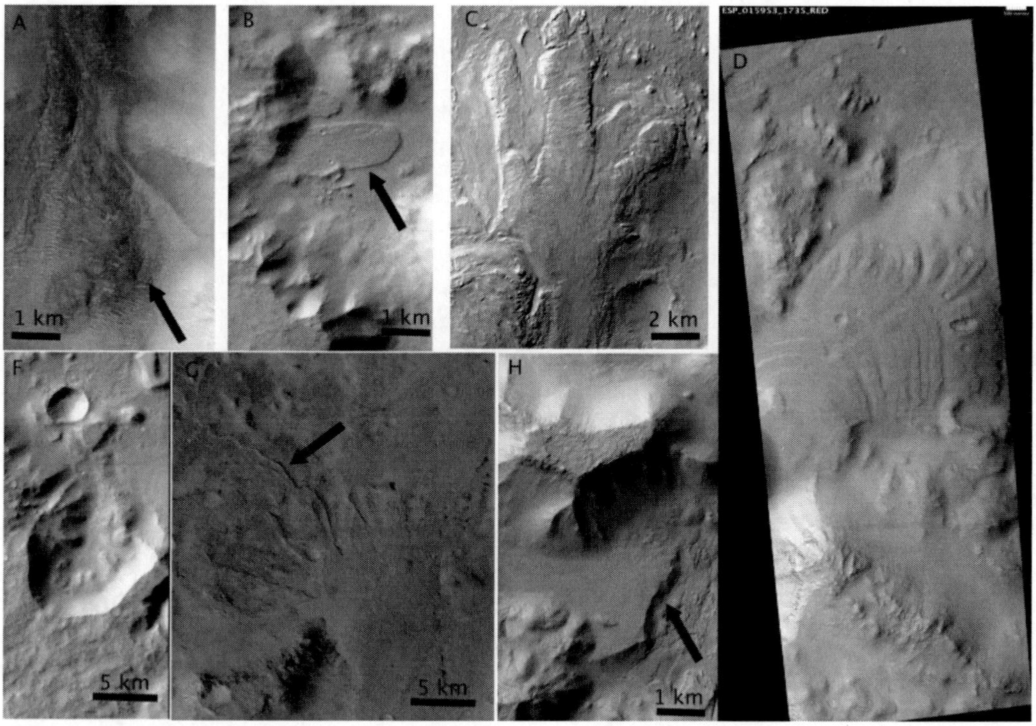

Figure 34. Possible glacial and glacio-fluvial landforms in Aeolis Mensae. A-C. Lobate flow features. D. Layered delta (scale bar is 500 m). F. Cirque valley. G. Inverted channels. H. Layered sediments. Images courtesy of NASA/JPL/ASU.

Estimated erosion rates for the formation of fretted terrains also support a glacial origin. The estimated time of formation of the fretted terrains between the Late Noachian and Early Hesperian (c.a. 10^8 years) implies erosion rates on the order of 10^{-2} mm yr^{-1} to carve a fretted valley 1 km deep (Irwin et al. 2004) and, therefore, the dissection of the plateau and the scale of the valleys imply an efficient erosion and transport mechanism. These erosion rates are 1-2 orders of magnitude higher than those estimated for the degradation of Martian highland craters (Cuffey et al. 2000; Irwin et al. 2004) and cold-based glaciers in the Dry Valleys of Antarctica (Hallet 1996), but comparable to glacier erosion rates in Greenland and Norway (Hallet, 1996).

The model for the origin of the fretted landscape proposed here implies that prior to the formation of fretted terrains, the dichotomy transition was a contiguous surface, possibly a pediment, that connected the highlands with the lowlands following a continuous, more gradual and gentle slope. This is consistent with previous studies that also postulated the emplacement of an aeolian sedimentary wedge along the dichotomy around the Noachian/Hesperian boundary, which later became dissected to form the fretted terrains (Irwin et al. 2004). The matching edges of adjacent mesas and plateaus, and also with the borders of the cratered highlands, along with the observation that some adjacent mesas share portions of the same impact crater, support the idea that the fretted terrains were once a uniform landmass that became dissected by an efficient erosive mechanism, such as glacial activity.

Broader Implications for Early Mars Climate and Hydrological Cycle

A glacial origin of Martian fretted terrains implies that at some point during the Noachian-Hesperian transition, tropical and equatorial ice caps, or ice sheets, occupied portions of the cratered highlands along the dichotomy boundary, and that glacial activity extended well into the northern plains. This period of glaciation would have post-dated the main period of fluvial activity on Mars (Fasset and Head, 2008), in response to increasingly, globally colder conditions on the planet towards the end of the Noachian Period. With the onset of glacial activity, the carving of channels, the mass wasting of the mesas, and the removal of eroded materials towards the lowland plains, modified the original slope of the dichotomy boundary into a series of isolated promontories separated by steep escarpments. Fluctuations in size of the ice sheets and glaciers furthered the degradation of cratered highland materials along the dichotomy, resulting in a transition zone of glacially incised valleys and isolated mesas, similar to terrains in Greenland and Antarctica today. In areas where glacial erosion was more efficient, large mesas were reduced to the smaller and more rounded knobs and buttes. This period of intense glaciation could have been interrupted by warmer spells when localized fluvial activity also reshaped the fretted terrain, as occurs in Polar Regions on Earth. Eventually the ice sheets and glaciers mostly disappeared, likely towards the Late Hesperian-Early Amazonian period, leaving behind mesas with fragmented surfaces and steep scarps, as well as abundant glacial features in the northern plains and in the fretted valleys similar to those found in glacial and periglacial terrains on Earth, including moraines and rootless cones in areas of the thumbprint terrain, systems of troughs with median ridges, lobate flow features in Gale Crater that resemble rock glaciers, eskers, and outwash plains.

The proposed model for the origin of Martian fretted terrains sheds new light on the hydrological evolution of early Mars at a global scale. Large masses of ice in equatorial and mid-latitudes would have required a significant source of moisture. Here, a large body of water in the northern plains appears as the most likely explanation. Moist air from the lowland plains would climb the elevation gradient of c.a 5.5 km across the topographic dichotomy, and precipitate as snow in the cratered highlands, becoming the main source of snow for the ice-sheets and polar caps. A similar process has been proposed to explain the development of valley networks towards the end of the Noachian period (Luo and Stepinski 2009; Irwin et al. 2011). The existence of a northern glacial water mass is supported by our own observation of glacial features in the northern plains, as well as by climate models (Kasting 1991; Colaprete et al. 2003; Wordsworth et al. 2011), geologic observations (Baker et al. 1991; Kargel and Strom 1992; Kargel et al. 1995), and geochemical calculations (Fairén et al. 2009; 2011) for early Mars, and particularly by previous studies suggesting ancient glaciations in the northern plains during the Hesperian period and even into the Amazonian period (Kargel and Strom 1992; Kargel et al. 1995). The proposed glacial origin of fretted terrains and the inference of a large body of water in the northern plains points to globally cold conditions on Mars at the end of the Noachian period, similar to global or continental scale glaciations on Earth, supporting the view that Mars was a glacial planet similar to polar regions on Earth throughout an important part of its geological history.

REFERENCES

Baker D. et al. Flow patterns of lobate debris aprons and lineated valley fill north of Ismeniae Fossae, Mars: Evidence for extensive mid-latitude glaciation in the Late Amazonian. *Icarus* 207: 186-209 (2010).

Baker, V. R. et al. Ancient oceans, ice sheets and the hydrological cycle on Mars. *Nature* 352: 589-594 (1991).

Bibring, J. P. et al., Global mineralogical and aqueous Mars history derived from OMEGA/Mars Express data. *Science* 312: 400-404 (2006).

Bonow, J. M., Lidmar-Bergstrom, K., Japsen, P. Palaeosurfaces in central West Greenland as reference for identification of tectonic movements and estimation of erosion. *Glob. Planet. Change* 50, 161-183 (2006).

Carr, M. H. Mars Global Surveyor observations of Martian fretted terrain. *J. Geophys. Res.* 106, 23,571-23,593 (2001)

Carr, M. H. The Martian drainage system and the origin of valley networks and fretted channels. *J. Geophys. Res.* 100(E4), 7479-7507 (1995).

Colaprete, A., Toon, O. B. Carbon dioxide clouds in an early dense Martian atmosphere. *J. Geophys. Res.* 108: 5025. doi:10.1029/2002JE001967 (2003).

Craddock, R. A., Maxwell, T. A. Geomorphic evolution of the Martian highlands through ancient fluvial processes. J. Geophys. Res. 98: 3453-3468 (1993).

Cuffey, K. M. et al. Entrainment at cold glacier beds. *Geology* 28: 351-354 (2000).

Fairén, A. G., Davila, A. F., Duport, L. G., Amils, R., and McKay, C. P. Stability against freezing of aqueous solutions on early Mars. *Nature* 459: 401-404 (2009).

Fasset, CI, Head, JW (2008) The timing of martian valley network activity: Constrains from buffered crater counting. *Icarus*, 195: 61-89.

Hallet, B., Hunter, L., Bogen, J. Rates of erosion and sedimentation by glaciers: A review of field data and their implications. *Glob. Planet. Change* 12: 213-235 (1996).

Head, J. et al. Tropical to mid-latitude snow and ice accumulation, flow and glaciation on Mars. *Nature* 434: 346-350 (2005)

Head, J. W., Nahm, A. L., Marchant, D. R., Neukum, G. Modification of the dichotomy boundary on Mars by Amazonian midlatitude regional glaciation. *J. Geophys. Res.* 33, L08S03, doi:10.1029/2005GL024360 (2006).

Henriksen, N., Higgins, A. K., Kalsbeek, F., Pulvertaft, T. C. R. Greenland from Archaean to Quaternary. Descriptive text to the 1995 Geological map of Greenland, 1:2 500,000. *2nd edition. Geol. Survey of Denmark and Greenland Bull.* 18, 126 pp. + map (2009).

Holtedahl, O. On the morphology of the west Greenland shelf with general remarks on the 'marginal channel' problem. *Marine Geology* 8, 155-172 (1970).

Irwin III, R. P., Craddock, R. A., Howard, A. D., Flemming, H. L. Topographic influences on the development of Martian valley networks. *J. Geophys. Res.* 116, E02005, doi:10.1029/2010JE003620 (2011).

Irwin, R. P., Watters, T. R., Howard, A. D., Zimbelman, J. R. Sedimentary resurfacing and fretted terrain development along the crustal dichotomy boundary, Aeolis Mensae, Mars. *J. Geophys. Res.* 109. doi:10.1029/2004JE002248 (2004)

Johnson, G. L., Sommerhoff, G., Egloff, J. Structure and morphology of the western Reykjanes Basin and the southern Greenland continental margin. *Marine Geol.* 18, 175-196 (1975)

Kargel J. S., Strom, R. G. Ancient glaciations on Mars. *Geology* 20(1), 3-7 (1992)

Kargel, J. S. et al. Evidence of continental glaciation in the Martian northern plains. *J. Geophys. Res.* 100: 5351-5368 (1995).

Kasting, J. F. CO2 condensation and the climate of early Mars. *Icarus* 94: 1-13 (1991).

Lucchitta, B. K. Ice and debris in the fretted terrain, Mars. *J. Geophys. Res.* 89, B-409-B418 (1984).

Luo, W., Stepinski, T. F. Computer-generated global map of valley networks on Mars. *J. Geophys. Res.*, 114, E11010, doi:10.1029/2009JE003357 (2009)

McGuill, G. E., Dimitriou, A. M. Origin of the Martian global dichotomy by crustal thinning in the Late Noachian or Early Hesperian. *J. Geophys. Res.* 95, 12,595-12,605 (1990).

Sharp, R. P. Mars: Fretted and chaotic terrains. *J. Geophys. Res.* 78, 4073-4083 (1973)

Squyres, S. W. Martian fretted terrain: Flow of erosional debris. *Icarus* 34, 600-613 (1978)

Tanaka, K. L., Skinner Jr., J. A., Hare, T. M. Geologic map of the northern plains of Mars: U.S. *Geological Survey Scientific Investigations Map* SIM-2888, scale 1:15,000,000 (2005)

Wordsworth, R. D. et al. Modelling the past Mars climate and water cycle with a thicker CO2 atmosphere. *Paper presented at the 4th International Workshop on the Mars Atmosphere: modelling and observations*, Paris, France, February 8-11 (2011)

Young D. A. et al. A dynamic early East Antarctic Ice Sheet suggested by ice covered fjord landscapes. *Nature* 474: 72-75 (2011)

In: Mars
Editor: Alberto G. Fairén

ISBN: 978-1-62618-102-1

Chapter 4

TYPES AND FORMATIONAL MECHANISMS OF SOUTH POLAR TROUGHS, MARS

***J. A. P. Rodriguez*[*1] *and T. Platz*[2]**
[1]Planetary Science Institute, Tucson, AZ, , US
[2]U.S. Geological Survey, Flagstaff, AZ, US
and Institute of Geological Sciences,
Freie Universität Berlin, Berlin, Germany

ABSTRACT

The south polar plateau of Mars constitutes the oldest known deposit of surface ice on Mars. The plateau is dissected by spiral troughs, the formation of which remains poorly understood. Our investigation reveals three distinct morphologic types of south polar troughs, each with a distinct formational history involving unique modes of basement and atmospheric controls.

Further stratigraphic and mineralogical of characterizations of these trough types will help us better understand the paleo-environmental evolution of an ancient water-rich landscape.

INTRODUCTION

The Martian polar plateaus were first identified in orbital images obtained by the Mariner and Viking spacecrafts (Murray et al., 1972; Cutts, 1973; Blasius et al., 1982; Howard et al., 1982), and they are thought to consist of CO_2 and H_2O ice intermixed with dust (e.g., Thomas et al., 1992; Mellon, 1996). CO_2 ice forms a residual cap (Thomas, 2000), and may also exist buried within water ice layers (Phillips et al., 2011). The layered strata are thought to have developed in response to Mars' cyclical obliquity, orbital eccentricity, and precession (Thomas et al., 1992).

[*] Planetary Science Institute, Tucson, AZ 85719-2395, USA (alexis@psi.edu),

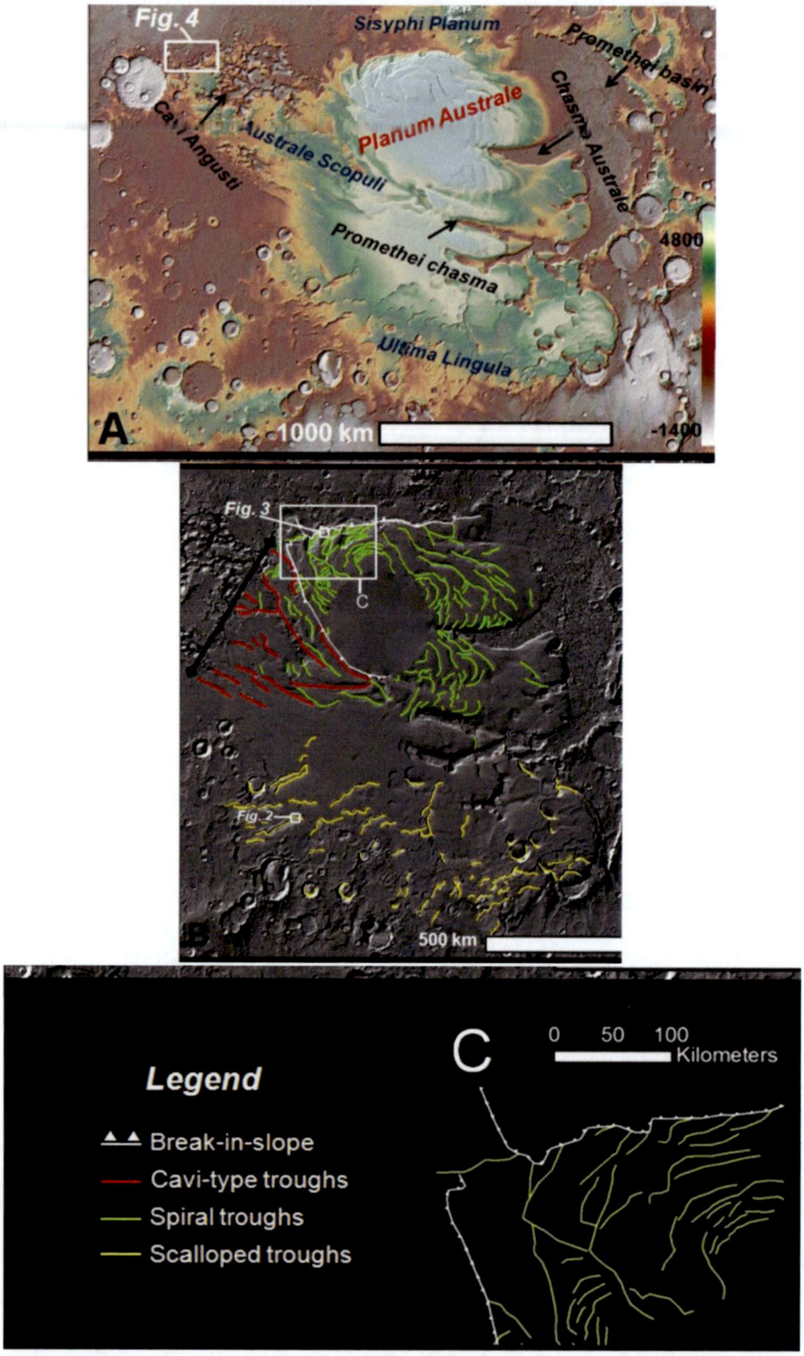

Figure 35. (A, MOLA DEM 512 pixels per degree) Geographic context for Planum Australe. The extent of the proposed morphologic mapping includes the polar plateau and the region of Cavi Angusti outlined in yellow. (B, MOLA-derived shaded relief 512 pixels per degree) Mapping of 4 morphologic types of troughs. Close-up view (C) shows spiral troughs truncated by break in slope. Black line shows the coincidence in the maximum width of resurfacing within Australe Scopuli and Cavi Angusti surprisingly quiet.

Both the south and north polar plateaus show extensive dissection by complex trough systems, however, the south polar troughs display a higher degree of morphologic diversity (Fig. 35).

The Martian polar troughs comprise the most extensive known canyon systems dissected within ice, and consequently, understanding their formational history is pivotal to improving our understanding of the history of surface water cycling and distribution on Mars over the age of these deposits. In addition, the troughs expose thick sequences of layered strata locally throughout the entire thickness of both polar plateaus, and investigations of these deposits have been pivotal to the advancement of Mars polar stratigraphic and climatic studies (e.g., Laskar et al., 2002; Kolb and Tanaka, 2006a, b, in review; Milkovich and Plaut, 2008; Milkovich et al., 2009).

The south polar plateau's surface is significantly older than its north polar counterpart. The surface age of the South Polar Layered Deposits (SPLD) has been estimated to be younger than 100 Ma (Herkenhoff and Plaut, 2000; Koutnik et al., 2002). However, some of the SPLD materials could be just slightly younger than the surface of the Late Hesperian Dorsa Argentea Formation, which they overlie (Tanaka and Scott, 1987; Tanaka and Kolb, 2001).

Thus, the south polar strata may contain stratigraphic and climatic information dating back to perhaps the beginning of the Amazonian Period, about 3.0 to 3.5 Ga (Werner and Tanaka, 2011).

In addition, the SPLD are distributed over topographically complex, diverse, and regionally high terrains, which include rugged cratered surfaces and the relatively flat floor of the shallow Promethei impact basin. On the other hand, the north polar plateau is apparently constructed entirely on relatively flat, smooth plains materials within a nearly hemispheric-scale, several-km deep basin. Thus, the south polar region offers a unique setting that will allow us to assess how basement topography may affect trough development relative to other potential conditions and processes.

Although numerous trough formational mechanisms and controls have been proposed, our knowledge of how these varied through time and with location within the polar plateaus remains incomplete. Polar trough formation has been proposed to have occurred as a consequence of (1) dissection by katabatic winds (Howard, 2000; Koutnik et al., 2005; Kolb and Tanaka, 2006a), locally focused by hydraulic jumps (Smith, 2011), (2) ablation-driven erosion and redeposition during pole-ward trough migration (Howard, 1978; Howard, 2000; Smith and Holt, 2010), (3) *in-situ* erosion (Kolb and Tanaka, 2001; Seu et al., 2007; Rodriguez et al., 2007; Tanaka et al., 2008; Rodriguez and Tanaka, 2011; Rodriguez et al., in prep.), (4) basal melting and jökulhlaup-style discharges (e.g., Howard, 1981; Clifford, 1987; Ghatan and Head, 2004), (5) ductile deformation and glacial flow (e.g., Fisher, 1993; Head and Pratt, 2001; Pratt and Head, 2002; Kargel 2004; Head and Marchant, 2006), (6) glacial erosion and ice streaming (Head, 2000), (7) listric faulting (Grima et al., 2011), (8) cryovolcanism (Tanaka and Kolb, 2001), and (9) basement topographic control (e.g., Kolb and Tanaka, 2006a).

We present the first morphologic classification of troughs in Planum Australe, the south polar plateau of Mars, and describe the likely processes and geologic conditions leading to the unique morphologic character of each trough type.

The Scalloped Troughs

We find three distinct trough morphologic types, namely the scalloped, spiral and cavi-type (Figure 35).These troughs occur exclusively within Ultima Lingula, where polar deposits overlie known densely cratered terrain (Figure 35B). They consist of moats oriented parallel to the equator-facing slopes of scalloped ridges, which interconnect scattered impact crater rims (Figure 35B). Examination of a scalloped trough shows stacks of convex layers along its equator-facing slope (Figure 36), indicating that this particular trough has migrated upwards through time (Smith and Holt, 2010). We propose that trough development and migration would have resulted as ice deposits along the equator-facing slopes were repeatedly removed or thinned by sublimation, which impeded trough infilling during otherwise regional SPLD constructional stages. These troughs show distribution patterns reflecting relatively steep aspects of the subsurface basement topography (e.g., crater margins), which can be explained by the fact that sublimation rates increase with steepening surface inclination.

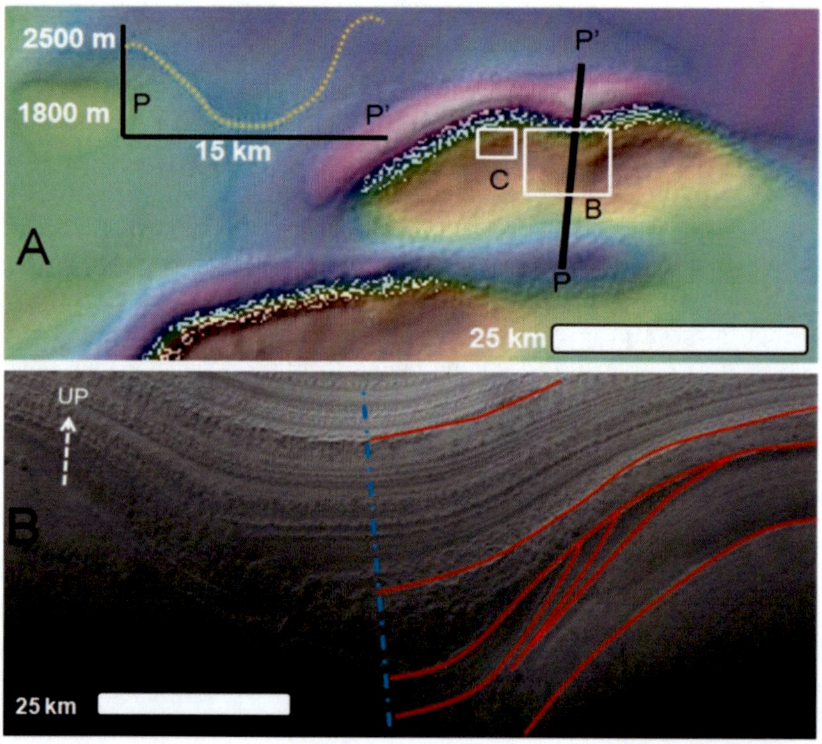

Figure 36. Scalloped trough (A, MOLA DEM, 512 pixels per degree centered at 78°91' S., 248° W., context and location in Figure. 1B) where the equator-facing slope reflects the topography of a sequence of exhumed stacked troughs (B).

The Spiral Troughs

These troughs (Figure 35B) resemble those in the north polar plateau and are arranged concentric to the highest part of Planum Australe. They are particularly well-developed south of Sisyphi Planum, relatively subdued within SLPD materials overlying the Promethei basin floor, discontinuous in the region of Australe Scopuli, and absent in Ultima Lingula. Rodriguez and Tanaka (2011) observed that the trough floors are compartmentalized into coalesced systems of depressions, which display layers that can be traced along their margins.

They suggest that these observations support the existence of a large population of troughs developed *in situ,* predominantly by surface ablation (Rodriguez and Tanaka, 2011). Instead of systems of nested elliptical depressions as observed in Planum Boreum, on Planum Australe we recently observed nested systems of elliptical depressions and trains of linear/curvilinear depressions (respectively blue and yellow arrows in Figure 37), which is consistent with previous investigations suggesting that polar trough development was the result of a combination of katabatic wind erosion and ablation (Smith and Holt, 2010; Rodriguez and Tanaka, 2011). However, we propose that wind erosion played a more important role in south polar trough development than it did in the formation of the north polar counterpart. This hypothesis is supported by our finding of yardangs in alignment with trough orientations (Figure 37, red arrows), some of which cut into the margins of depressions that comprise prominent topographic divides, indicating that the degree of topographic compartmentalization generally decreased over time due to topographic subdual by wind erosion.

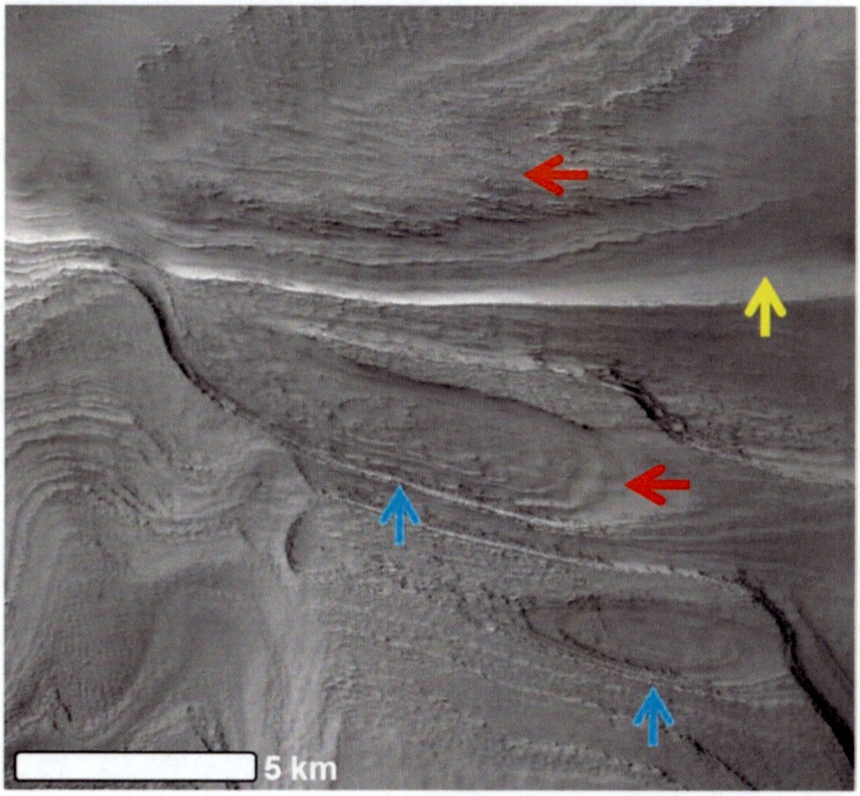

Figure 37. Part of CTX image B08_012598_0958_XN_84S033W, 5 m/pixel) Close-up view reveals extensive wind erosion.

The Cavi-Type Troughs and Planum Australe Slope Breaks

The most prominent slope break in Planum Australe marks an abrupt drop in surface elevation of ~1 km. This slope break extends radially from a location where the rim of the Promethei basin has been hypothesized to be buried underneath the central part of Planum Australe (Rodriguez et al., 2005) to the margin of the polar plateau adjacent to Cavi Angusti, a cluster of depressions within the regional circum polar terrains (Figure 35). The slope break separates a zone of spiral trough development and a lower zone of cavi-type trough occurrence known as Australe Scopuli (Figure 35). The slope break, which hereafter we refer to as the scopuli slope break, ends at a location in close proximity to where the most prominent slope break marking the margin of Planum Australe extends south of Sisyphi Planum (Figure 35B). This slope break, hereafter referred to as the Sisyphi slope break, terminates at the Promethei basin's rim.

The Sisyphi slope break has a maximum relief of ~1500 m and truncates numerous spiral troughs (Figure 35C). The cavi-type troughs comprise nested depression systems that typically have their lengths oriented parallel to the scopuli slope break (Figure 35) and in some places their floors include pockets of exhumed basement bedrock. We observe that the maximum widths of Cavi Angusti and the outer margin of Australe Scopuli coincide (Fig.35B), suggesting at least a partly shared resurfacing history. The formation of the Cavi Angusti depressions has been hypothesized as having resulted from regional volcanism during the Late Hesperian and prior to the emplacement of the SPLD (Ghatan et al., 2003). We counted craters on CTX images and our surface dating points to regional resurfacing occurring ~96 Ma, or 64 Ma when adjusted for an icy target (i.e., Late Amazonian, Figure 38).

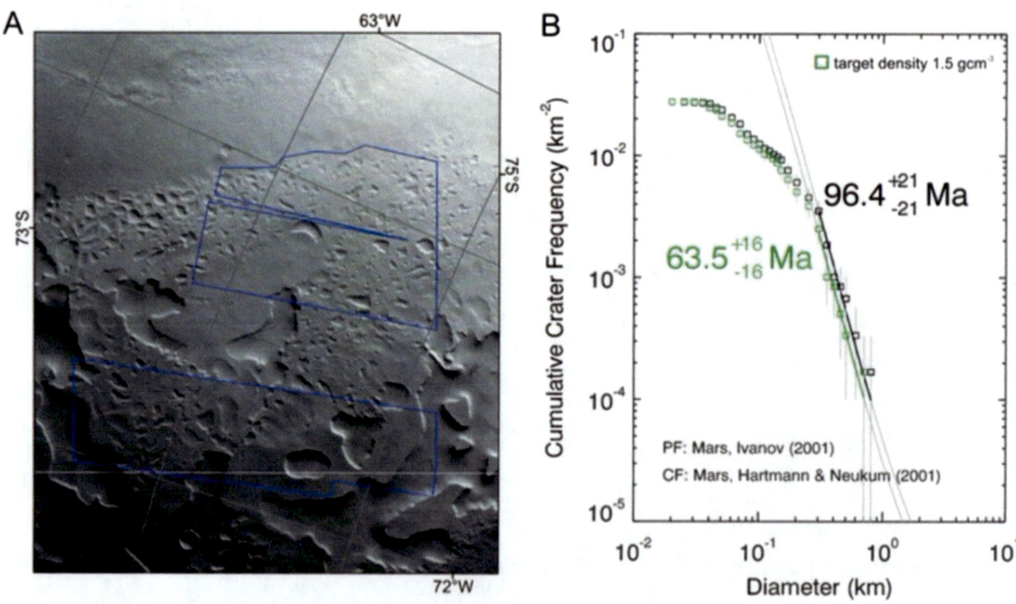

Figure 38. (A) Crater counting area (blue) shown on HRSC nadir image mosaic from a region of collapse within Cavi Angusti and (B) derived model ages. Crater model age adjusted to a target density of 1.5 g cm-3 are shown in green. Location of (A) shown in Figure 1A.

SUMMARY

Our findings open the possibility for some Cavi Angusti development having postdated the formation of Planum Australe. Recently, the development of the cavi-type troughs has been attributed to katabatic wind erosion (Howard, 2000) or the control of ridged basement topography (Kolb and Tanaka, private comm.). However, we propose an alternative hypothesis that links the development of the cavi-type troughs to the formation of the Sisyphi and scopuli breaks-in-slope. As previously suggested we concur with the hypothesis that the Cavi Angusti resulted from volcanic activity, but we also suggest that volcanism (extrusive or intrusive) must have occurred in the south polar region during the Amazonian. In the apparent absence of relatively young volcanic landforms, resurfacing may have occurred in association with high heat flow from intrusive magmatic bodies causing large-scale polar ice removal by the enhanced ablation effect of relatively warmer anabatic winds. Polar ice removal resulted in the exhumation of circum-polar surfaces, SPLD thinning and modification of pre-existing spiral troughs into cavi-type troughs.

REFERENCES

Blasius, K. R., et al., 1982. Topography and stratigraphy of Martian polar layered deposits. *Icarus*. 50, 140-160.

Clifford, S. M., 1987. Polar basal melting on Mars. *Journal of Geophysical Research*. 92. Page number?

Cutts, J. A., 1973. Nature and origin of layered deposits of the Martian polar region. *J. Geophys. Res*. 78, 4231-4249.

Fisher, D. A., 1993. If Martian Ice Caps Flow: Ablation Mechanisms and Appearance. *Icarus*. 105, 501-511.

Ghatan, G. J., et al., 2003. Cavi Angusti, Mars: Characterization and assessment of possible formation mechanisms. *J. Geophys. Res*. 108, 10-1029.

Ghatan, G. J., Head, J. W., 2004. Regional drainage of meltwater beneath a Hesperian-aged south circumpolar ice sheet on Mars. *J. Geophys. Res*. 109.

Grima C. et al. 2011 Large asymmetric polar scarps on Planum Australe, Mars: Characterization and evolution. *Icarus* 212, 96-109.

Head, J. W., 2000. Evidence for Geologically Recent Lateral Migration of Volatile-Rich Polar Layered Deposits. *International Conference on Mars Polar Science and Exploration*, p. 66.

Head, J. W., Pratt, S., 2001. Extensive Hesperian-aged south polar ice sheet on Mars-Evidence for massive melting and retreat, and lateral flow and ponding of meltwater. *Journal of Geophysical Research*. 106, 12275-12300.

Head, J. W., D. R. Marchant, 2006. Regional polar glaciation in the Hesperian Period of the history of Mars: The south circum polar Dorsa Argentea formation as an ancient ice sheet remnant (abstract), *http://www.lpi.usra.edu/meetings/polar2006/pdf/8103.pdf.*

Herkenhoff, K. E., Plaut, J. J., 2000. Surface Ages and Resurfacing Rates of the Polar Layered Deposits on Mars. *Icarus*. 144, 243-253.

Howard, A. D., 1978. Origin of the stepped topography of the Martian poles. *Icarus*. 34, 581-599.

Howard, A. D., 1981. Etched plains and braided ridges of the south polar region of Mars: features produced by basal melting of ground ice? *Reports of Planetary Geology Program*-1981, p. 286-288.

Howard, A. D., 2000. The role of eolian processes in forming surface features of the martian polar layered deposits. *Icarus*. 144, 267-288.

Howard, A. D., et al., 1982. Stratigraphic relationships within Martian polar cap deposits. *Icarus*. 50, 161-215.

Kargel, J. S., 2004. *Mars : a warmer, wetter planet*. Springer; Praxis, London.

Kolb, E. J., Tanaka, K. L., 2001. Geologic History of the Polar Regions of Mars Based on Mars Global Surveyor Data II. Amazonian Period. *Icarus*. 154, 22-39.

Kolb, E. J., Tanaka, K. L., 2006a. Accumulation and erosion of south polar layered deposits in the Promethei Lingula region, Planum Australe, Mars. *Mars*. 2, 1-9.

Kolb, E. J., Tanaka, K. L., 2006b. Geologic Mapping Results of the Australe Scopuli Region Within Planum Australe, Mars. *Fourth International Conference on Mars Polar Science and Exploration,* October 2-6, 2006, Davos, Switzerland. LPI Contribution No. 1323, p. 8085.

Kolb, E.J., Tanaka, K.L., in review, Geologic map of the Planum Australe region of Mars. *U.S. Geol. Surv. Sci. Invest. Map*, scale 1:5,000,000.

Koutnik, M., et al., 2002. South Polar Layered Deposits of Mars: The cratering record. *J. Geophys. Res*. 107, 5100-5112.

Koutnik, M. R., et al., 2005. Eolian controlled modification of the martian south polar layered deposits. Icarus. 174, 490-501.

Laskar, J., Levrard, B., Mustard, J.F., 2002. Orbital forcing of the martian polar layered deposits. *Nature*. 419, 375–377.

Milkovich, S. M., et al., 2009. Stratigraphy of Promethei Lingula, south polar layered deposits, Mars, in radar and imaging data sets, *J. Geophys. Res*., 114, E03002, doi:10.1029/2008JE003162.

Milkovich, S. M., Plaut, J. J., 2008. Martian South Polar Layered Deposit stratigraphy and implications for accumulation history, *J. Geophys. Res*., 113, E06007, doi:10.1029/2007JE002987.

Mellon, M. T., 1996. Limits on the CO2 content of the Martian polar deposits. *Icarus*. 124, 268-279.

Murray, B. C., et al., 1972. Geological framework of the south polar region of Mars. *Icarus*. 17, 328-345.

Pratt, S., Head, J. W., 2002. Lobate flow of the South Polar Layered Terrain Margins: Evidence from MOLA data, *Lunar and Planet. Sci XXXIII,* 1866.pdf.

Phillips, R. J. et al., 2011. *Massive CO2 Ice Deposits Sequestered in the South Polar Layered Deposits of Mars Science*. 332, 838, DOI: 10.1126/science.1203091.

Rodriguez, J. A. P. et al., 2005. Basement topographic control on chasmata initial growth and distribution in the Martian southern polar cap. *Lunar and Planetary Science XXXVI* (2005) 1795.pdf.

Rodriguez, J. A. P., et al., 2007. Recent aeolian erosion and deposition in the north polar plateau of Mars. *Mars* 3, 29–41.

Rodriguez, J. A. P.,Tanaka, K. L., 2011, Evidence for in-situ trough erosion in Planum Boreum, Mars. The fifth international conference on Mars *Polar Science and exploration.*

Seu, R., et al., 2007. Accumulation and Erosion of Mars' South Polar Layered Deposits. *Science*. 317, 1715.

Smith, I. B., Holt J. W., Kocurek G., Howard A. D., Spiga, A., 2011. A theoretical and observational framework for estimating migration rates of spiral troughs in the northern polar layered deposits, *Mars, Fifth Mars Polar Science Conference*, 6050.pdf.

Smith, I. B., Holt, J. W., 2010. Onset and migration of spiral troughs on Mars revealed by orbital radar. *Nature*. 465, 450-453.

Spiga, A., Madeleine, J. B., Forget, F., 2011. Modeling mesoscale atmospheric processes in the Martian polar regions, *Fifth Mars Polar Science Conference,* 6040.pdf.

Tanaka, K. L., 1986, The stratigraphy of Mars. Proc. Lunar Planet. Sci. Conf., 17th, Part 1, *J. Geophys. Res*., 91, suppl., E139-158.

Tanaka, K. L., Kolb, E. J., 2001. Geologic History of the Polar Regions of Mars Based on Mars Global Surveyor Data I. Noachian and Hesperian Periods. *Icarus*. 154, 3-21.

Tanaka, K. L., Scott, D. H., 1987. *Geologic map of the polar regions of Mars. USGS Map I-1802-C*, scale 1:15,000,000.

Tanaka, K. L. et al., 2008. North polar region of Mars: advances in stratigraphy, structure, and erosional modification. *Icarus*. 196, 318–358.

Thomas, P., et al., 1992. *Polar deposits of Mars, in Mars.* edited by H. H. Kiefer, B. M. Jakosky, C. W. Snyder, and M. S. Matthews, pp. 767-795, Univ. of Arizona Press, 1992.

Thomas, P., Malin, M., Edgett, K., Carr, M., Hartmann, W., Ingersoll, A., James, P., Soderblom, L., Veverka, J., Sullivan, R. 2000. North–south geological differences between the residual polar caps on Mars. *Nature* 404, 161–164.

Werner, S. C. Tanaka, K. L., 2011. Redefinition of the crater-density and absolute-age boundaries for the chronostratigraphic system of Mars. *Icarus*. 215, 603-607.

In: Mars
Editor: Alberto G. Fairén

ISBN: 978-1-62618-102-1

Chapter 5

COEVAL SYNTHESIS OF COLD AQUEOUS MINERALOGIES ON MARS

Alberto G. Fairén *

Space Science and Astrobiology Division,
NASA Ames Research Center, Moffett Field, CA, US

ABSTRACT

Several aqueous alteration products have been detected on Mars: iron oxides and oxyhydroxides, hydrated sulfate salts, hydrated silica, phyllosilicates, carbonates and chlorides. It has been hypothesized that there was a sequential synthesis of Martian aqueous minerals in which less soluble phases such as phyllosilicates precipitated early on, when liquid water was likely more abundant, and more soluble phases like salts predominantly precipitated much later, when surface liquid water was substantially diminished. However, the synthesis of phyllosilicates results in a substantial overload of cations that must exist in additional contemporaneous sedimentary salts, which are mostly missing in the Noachian rock record. Here is proposed the alternative of a coeval origin of the different Martian water-derived mineralogies. The bedrock alteration by cold liquid water and the formation of aqueous minerals are modeled for early Mars. Geochemical model calculations show that the aqueous alteration of basaltic rocks resulted in the deposition of phyllosilicates along with either sulfates or carbonates. The results also show that an important proportion of these salts were re-dissolved through a subsequent influx of water into the system and transported to deeper subsurface layers by a diffusion process, leaving the less soluble phyllosilicates exposed on the ancient surface. In conclusion, Martian aqueous mineral deposits do not necessarily reveal a sequential mineral formation within geochemically homogeneous time periods. Instead, the model presented here is consistent with the contemporary formation of all major water-derived mineralogies in different aqueous environments during the wet Noachian/Hesperian times, revealing an active and heterogeneous early Mars.

* alberto.g.fairen@nasa.gov

1. INTRODUCTION

Understanding the geochemical conditions on early Mars requires an explanation for the presence of sulfates and phyllosilicates, the two major families of aqueous-derived mineralogies that have been detected on the surface, which must be also consistent with the absence of widespread sedimentary carbonates (Bandfield et al., 2003; Bibring et al., 2006; Ehlmann et al., 2008). Whereas the identified hydrated sulfate salts include kieserite and other polyhydrated Mg-sulfates, Fe-sulfates, and gypsum (Squyres et al., 2004; Gendrin et al., 2005; Bibring et al., 2006), the Al- and Fe-rich phyllosilicates include kaolinite and Fe/Mg-OH smectites such as nontronite, saponite, chlorite and montmorillonite (Bibring et al., 2006; Poulet et al., 2005; Mustard et al., 2008).

On Mars, both groups of minerals do not generally occur together, which has been interpreted as a marker for detached mineral formation due to differing planetary environmental conditions changing globally, either in time or in space, allowing the formulation of three different and mutually exclusive theories. In the first hypothesis, the one that finds more community acceptance today, the distribution of aqueous minerals is proposed to indicate a clear temporal separation between sequential different geochemical conditions, including a neutral to alkaline Early Noachian time period, when phyllosilicates formed, followed by an acidic Noachian to Hesperian era when sulfates precipitated (Poulet et al., 2005; Bibring et al., 2006).

While also indicating chronological disengagement, the second hypothesis suggests a reverse time sequence: massive layered sulfate-rich sediments would have been deposited during the Late Noachian, followed by a synthesis of phyllosilicates starting in the Hesperian with carbonates forming patinas in a later stage, after the collapse of the atmosphere (Bullock and Moore, 2007). The third hypothesis is the only one that suggests a contemporary deposition of phyllosilicates and sulfates, but in this case the disconnection is spatial, in a scenario in which clays generally would be the result of subsurface circum-neutral aqueous activity, while sulfates would represent the acidic near-surface environment (Bandfield et al., 2008; Ehlmann et al., 2011).

However, a separate deposition of the different aqueous minerals, whether in time or in space, is geochemically inconsistent: the chemical weathering of the basaltic crust and the subsequent formation of phyllosilicates yields an excess of cations in amounts equal or greater to the number of moles of clays (Milliken et al., 2009). These cations must reside in additional coeval alteration products to meet mass balance constraints, most likely in the form of sedimentary salts.

Here I propose that the major water-alteration products on the Martian surface were deposited simultaneously in space and in time, creating diverse geochemical conditions over the entire surface of a cold Mars during the wet Noachian/Hesperian times. A comprehensive geochemical model is offered thereof, explaining (*i*) the mineral diversity on Mars, accounting for the coeval synthesis of phyllosilicates and sulfates in different environments, and the absence of widespread carbonate deposits; and (*ii*) the general depletion of salts associated with clays in the Noachian sedimentary rock record, as a result of the post-depositional downward migration of cations.

2. METHODS

The Geochemist's Workbench software (Bethke, 2002) was used to develop the aqueous thermodynamic calculations which recreate the expected mineralogy on Mars. The concentrations of dissolved species in the Martian aqueous solutions modeled are detailed in Table 1. Equally, the presence of CO_2, sulfur and iron as the main acidifying agents in early Martian waters (Fairén et al., 2004) was considered, as they drive acidification by

$$CO_{2(g)} + H_2O \rightarrow HCO_3^- + H^+ \quad (1)$$

$$SO_2 + \frac{1}{2}O_2 \rightarrow S^0 + SO_3 \quad (2)$$

$$SO_3 + H_2O \rightarrow H_2SO_4$$

$$Fe^{2+} + 2H_2O \rightarrow Fe(OH)_2 + 2H^+ \quad (3)$$

$$Fe^{3+} + 3H_2O \rightarrow Fe(OH)_3 + 3H^+$$

Table 1. Soil composition used to recreate the expected mineralogy on early Mars. Estimated after Fairén et al. (2004), except: (a) obtained assuming a solution at 273 K in equilibrium with jarosite for $a(_{Fe}{}^{3+}) = 10^{-4.3}$ and $a(_{Fe}{}^{3+}) = 10^{-3.1}$, respectively, as governed by the equation $\log(a_{SO4}{}^{2-}) = 3\cdot\log(a_H{}^+) - 0.5\cdot\log(a_K{}^+) - 1.5\cdot\log(a_{Fe}{}^{3+}) - 2.91$; and (b) assumed here to promote phyllosilicate formation. HCO^-_3 is not a fixed parameter here, as it has been swapped for CO_2(g), the Y-axis variable in Figure 39a

Constituent	Molality (x10^{-3} mol kg^{-1})
SO_4^{2}-	0.18 (6.5 – 13.5)[(a)]
Mg^{2+}	1.0
$SiO_{2\ (aq)}$	0.18 - 1.0
Na^{2+}	0.8
Fe^{2+}/Fe^{3+}	0.05 - 0.8
Cl^-	0.65
Ca^{2+}	0.5
Al^{3+}	0.1 [(b)]
K^+	0.07

In the case of reduced and oxidizing iron (equations 3), these reactions keep the chemical stability in the aqueous systems by releasing H^+ into the solutions and thus buffering a long-term acidity, neutralizing the pH-raising capacity of the basaltic minerals in the host rock. The ferrous iron was photochemically oxidized to ferric, through photochemical pathways or radiolytical-sourced oxidants (Krasnopolsky, 1993). Alternative models consider sulfur (Halevy et al., 2007) or the oxidation of Fe^{2+} sulfates at low temperature (Tosca et al., 2008) as the main acidifying agents. But sulfur volatiles have brief residence times in an optically thin atmosphere (Johnson et al., 2008); and the occurrence of ferric mineralogies suggests that iron played some role in controlling the pH under strong acidic conditions. Temperatures in the calculations were assumed to be 273 K, typical of a cold early Mars, in the presence of a CO_2-dominated atmosphere with unfixed pressure (CO_2 pressure is the y-axis variable in

Figure 39A). Supplementary cations were included, such as Mg^{2+}, Ca^{2+}, K^+ or Na^+, and silica, a by-product of the acid dissolution of ferromagnesian silicates, typical of Martian igneous rocks in such acidic conditions. For the analysis of the synthesis of phyllosilicates, aluminum was also added, in a concentration $[Al^{3+}] = 10^{-4}$ M.

Table 2. Mineral phase composition used to examine the downward migration of salts

Mineral	Moles/l H_2O
Anorthite	12
Diopside	6
Enstatite	2
Fayalite	1.378
Forsterite	1.376
Ilmenite	0.3
Spinel	0.01
Troilite	0.01
Hydroxylapatite	0.001
Magnesite	0.001

The Phreeqc software (Parkhurst and Appelo, 1999) was used to model the migration of cations downward into the subsurface, implementing an algorithm of ionic transport through vertical diffusion in *n* cells (representing *n* solutions). The model considers multidiffussion, and the specific diffusion coefficients utilized for each ion are those included in the Phreeqc database. The composition of martian basalts has been extracted from orbital observations (Hamilton and Christense, 1997; Bandfield et al., 2000; McSween et al., 2009). As mean values, the mineralogy used consisted of 65% volume plagioclase (andesite), 27% pyroxenes (orto- and clinopyroxenes), 6% olivine (forsterite and fayalite) and 2% of other components. Sediment porosity ($cm^3_{water}/cm^3_{sediment}$) of 0.3 represents a mixture of minerals and water consistent in 0.5 cm^3 water per 1 cm^3 basalt. Initial conditions were fixed as pH=7, Eh=4, and T=273 K, leading to the mineral composition shown in Table 2. The equilibrium of water with these mineral phases conduces to the emergence of a number of ions in solution and to the synthesis of phyllosilicates reported to exist on Mars (Mustard et al., 2008), including kaolinite, nontronite, chamosite, saponite and smectites. The diffusion of water was modeled in a set of 20 identical cells of 1 cm each, representing different layers of solutions in the subsurface. At t=0, the conditions are the same in all cells, representing the water/basalt mixture before the beginning of any process of auto-organization. The model starts with a process of evaporation + freezing in the upper part of the system (the upper cell), representing the surface solution, conducing to a rise in the ionic concentration as the water is lost. This will result in a net vertical concentrations gradient promoting the diffusion of salts to the lower cells, as the engine of diffusion is the difference between concentrations. The overall process is governed by the Fick's laws of diffusion (Fick, 1855; Brogioli and Vailati, 2001). The evolution of the solution through time in the upper layers will be dependent upon the velocity of evaporation + freezing, the diffusion coefficient of each ion, and the temporal relationships between the processes of evaporation + freezing and diffusion. The precipitation of salts is inhibited by imposing a velocity of evaporation + freezing lower than the velocity of vertical diffusion.

The influence of the adsorption process on the surface of the clays was determined modeling the negatively charged surface of clays defining a reactive surface able to adsorb ions in its diffuse double layer (DDL). The DDL occurs at the interface between the clay surface and the soil solution, and it is made up of the permanent negative charge of the clay and the exchangeable cations in the soil solution that balance the negative charge (Mojid and Cho, 2006). The net value of the adsorption depends on each cation, on the quantity of surface in weight, and on the number of reactive centers.

3. Results

3.1. Coeval Formation of Aqueous Minerals

Thermodynamic equilibrium calculations were used to determine the boundaries for carbonates (siderite), phyllosilicates (minnesotaite, low Fe-Mg smectites, nontronite), ferrous and ferric sulfates (including jarosite) and iron oxyhydroxides (goethite) precipitation on early Mars, at different atmospheric CO_2 pressures and both under reducing and oxidizing conditions. Sedimentation models in depositional environments establish that siderite ($FeCO_3$) is always the first major carbonate to precipitate from solutions enriched in iron, because it is the most insoluble of the major common carbonates. Minnesotaite ($Fe_3Si_4O_{10}(OH)_2$) is the simplest of the possible ferrous phyllosilicates, as well as a stable phase in terrestrial sediments. The pH for waters on the surface of Mars varied in function of the quantity of available water and as a consequence of different ionic input, basically related to iron, sulfate and silica concentrations, as well as the final oxidation state of the iron itself, which largely determined the fluid chemistry. The stability boundaries for carbonates, phyllosilicates and sulfates on ancient Mars are represented in Figure 39A.

Phyllosilicates formed in mildly acidic to alkaline aqueous solutions, with a $pH>4$ for nontronite ($Ca_{.5}(Si_7Al_{.8}Fe_{.2})(Fe_{3.5}Al_{.4}Mg_{.1})O_{20}(OH)_4$) and a $pH>6$ for other smectites with low content in Fe and Mg ((Ca, Na, H)(Al, Mg, Fe, Zn)$_2$(Si, Al)$_4O_{10}(OH)_2$ - xH_2O). Sulfate deposition dominates in solutions moderately to highly acidic, with a $pH<6$ conducive to the synthesis of kieserite ($MgSO_4$). In the overlapping phyllosilicates/sulfates pH range, between 4 and 6, the formation of nontronite would be favored in areas where SiO_2 activity in surface waters was high as a result of intense weathering of the early basaltic crust, as aqueous solutions evolve in response to continued silicate weathering processes and mineral precipitation (Figure 39B). In this case, a competition for Mg between nontronite and kieserite is expected, therefore allowing coeval synthesis of phyllosilicates and sulfates in the same stratigraphic sequence. Carbonates formed at $pH>6$, overlapping with the synthesis of low-Fe-Mg smectites, and therefore the model anticipates the co-precipitation of smectites and siderite ($FeCO_3$) or any alteration product that could have resulted from the later substitution of Fe in siderite, such as Mg- or Mn-carbonate ($MgCO_3$, $MnCO_3$). The case for magnesite is particularly appealing, as kieserite and nontronite seem to be among the most abundant sulfates and phyllosilicates on Mars, respectively, suggesting that Mg may be readily available from the weathering of olivine (forsterite, Mg_2SiO_4). In these conditions, the model anticipates a competition for Mg between magnesite and low-Fe-Mg smectites, such as montmorillonite or saponite, permitting the contemporary synthesis of phyllosilicates and carbonates.

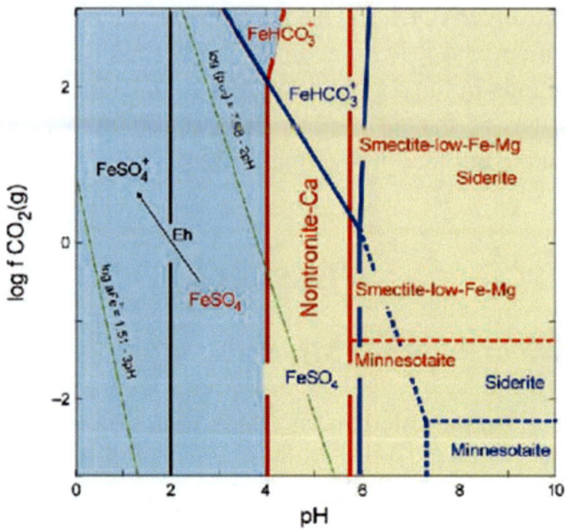

(a)

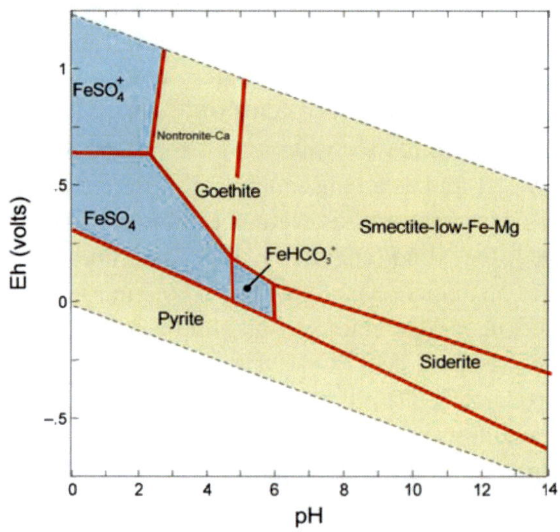

(b)

Figure 39. Relationship between mineralogies and pCO_2-Eh-pH conditions for Mars. A: Results obtained as a function of CO_2 pressure and pH, and calculated at 273 K. Red lines represent the limits of the mineral phases under oxidizing conditions (Eh = 0.15 V), and blue lines under reducing conditions (Eh~ - 0.15 V). Solid lines shape the limits of the stability fields of the major mineral phases as nontronites and smectites, as well as major ions; dotted lines separate the fields for the secondary mineralogies to be stable, as siderite and minnesotaite, which occur at low Al activity. Mineral phases synthesized under oxidizing conditions are labeled in red, while mineralogies formed under reducing conditions are identified in blue. Note that siderite and minnesotaite would occur at very low Al activity, locally associated to the other phases. Black line marks the averaged boundary (for lower and higher iron concentrations) between iron oxyhydroxides and those conditions driving to the synthesis of acidic sulfates through $SO_4^-Fe^{3+}/Fe^{2+}$, which are associated. B: Results obtained as a function of pH and Eh. This graphic emerges from the combination of a diagram removing SiO_2 and including SiO_2, at pCO_2 = 1 atm and T = 273 K. Blue indicates phase solutions, light brown indicates minerals. The stability fields of jarosite (acidic oxidizing solutions with $FeSO_4^+$), siderite and various phyllosilicate minerals are represented. Credit: D. Fernández-Remolar.

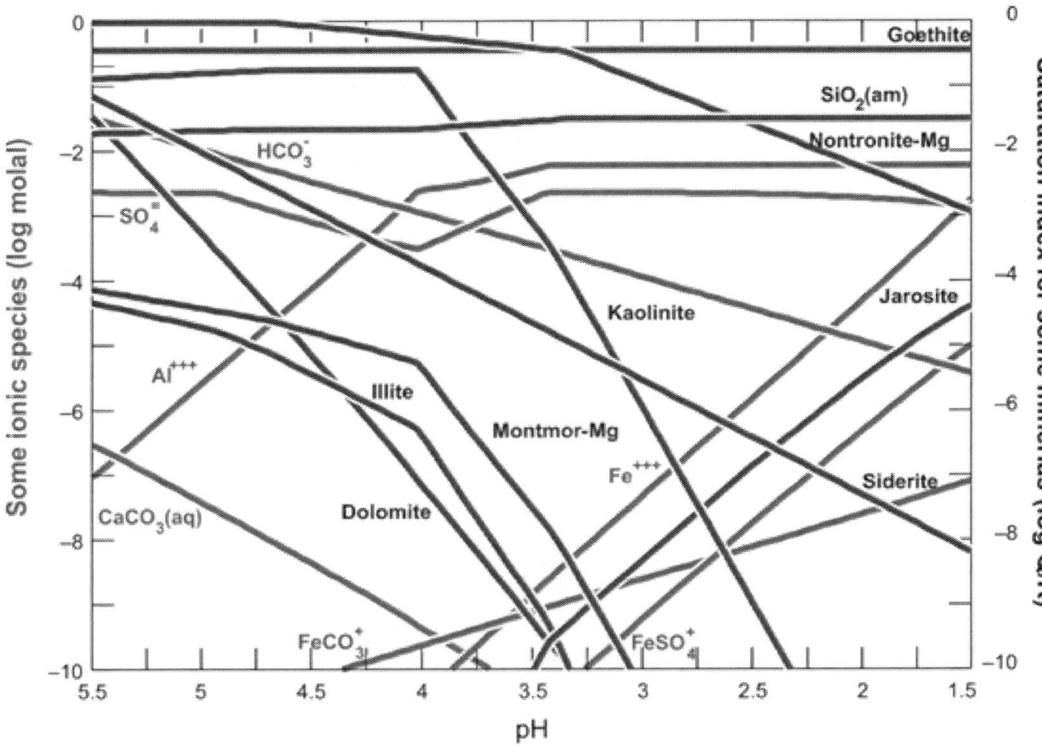

Figure 40. Evolution of the ionic concentrations in aqueous solutions (red lines) and the mineralogical phases precipitated (blue lines), ranging from mildly acidic and low oxidizing to strong acidic and high oxidizing conditions (Eh = 0.15 - 0.75 V, and pH=5.5 - 1.5). The pH variations can be used to describe different environmental conditions in separate locations during the same early Noachian to Hesperian time period, or to simulate the global environmental changes through time that occurred in Mars as the planet dried out from the Early Noachian to the Hesperian. The simulation has been constructed from a solution composition as follows: $[CO_2] = 4$, $[Al^{3+}] = 10^{-3}$, $[Fe^{2+}] = 10^{-3}$, $[Ca^{2+}] = 10^{-3}$, $[K^+] = 10^{-3}$, $[Mg^{2+}] = 10^{-3}$, $[SO_4^{2-}] =$ 10-, $[SiO_2] = 10^{-4}$, and T = 273 K. $[Na^+] = 10^{-3}$ and $[Cl^-] = 10^{-3}$ were fixed for maintaining the solution charge balance. Three main observations can be obtained based on this model: (*i*) subsaturation of phyllosilicates prevent precipitation of clays under highly acidic conditions, although slightly acidic conditions allow phyllosilicate formation; (*ii*) subsaturation of carbonates like siderite and dolomite is a general condition, but local precipitation is allowed at the highest pH levels; and (*iii*) the concentration of Fe^{3+} sulfate ionic complexes, that are the backbone of some acidic sulfates like jarosite, increases with decreasing pH. Interestingly, iron oxides can eventually precipitate under any pH and Eh condition considered in this model. Credit: D. Fernández-Remolar.

As expected, the model does not predict coeval synthesis of carbonates and sulfates, and therefore the missing deposits needed to ion-complement the phyllosilicate sequences in ancient Mars are expected to mainly consist of one or the other salt, depending on the geochemical characteristics of the local hydrological system: sulfates overlapping nontronites and carbonates interbeded with low-Fe-Mg smectites. Goethite (FeO(OH)) and other oxyhydroxides precipitate at pH below 2, a range at which jarosite ((Na, K, Rb, NH_4, H_3O, Ag, Tl)(Fe, Cr, V)$^{3+}{}_3(OH)_6(SO_4)_2$) and goethite are the expected iron-bearing phases.

As the geochemical environment is a dynamical, not a static system, the evolution of the model over time is represented in Figure 40, linked to the changes in the atmosphere: during the Noachian, the atmosphere was relatively thicker and composed primarily by CO_2 (Carr, 1999; Fairén et al., 2004), and was later diminished in the Late Noachian to Hesperian by several mechanisms such as sputtering, weathering and impact erosion (Carr, 1999). The progressive drying of Mars associated with the evolution of the atmosphere explains the subsequent major changes from Si-rich deposits to S enriched salty materials.

Volcanic S-bearing gases and Cl_2, inputting into such a declining CO_2 atmosphere exposed to solar radiation, would have turned the system globally into more acidic conditions in the Late Hesperian, after the hydration of the S chemical species (Bullock and Moore, 2007).

The change of geochemical parameters affecting the surface mineralogy through time is basically regarded here as an increase in acidity, redox potential and salinity, related to aridity. In these conditions, mineralogical data suggest phyllosilicate subsaturation for a pH below 4 and a continuous and strong carbonate subsaturation, which contrasts to a high increase in the ferric and sulfate ionic complexes linked to a rising saturation index for jarosite.

As such, during the earlier, wetter period of Mars' history, the excess of cations expected after the synthesis of phyllosilicates (Milliken et al., 2009) formed complementary salts, both widespread sulfates during the Noachian-Hesperian, and limited carbonates only during the Early Noachian epoch. Chlorides would be synthesized after sulfate minerals are precipitated. These results support the idea that the codeposition of different aqueous minerals could have been a global phenomenon on early Mars and, therefore, explaining the location or fate of the original sulfates and carbonates in the Noachian sedimentary record must be a first-order objective in the ongoing exploration of Mars.

3.2. Ultimate Location of Noachian Salts

Here is analyzed the differential mobility of chlorides, sulfates and carbonates with respect to phyllosilicates, once they are formed, therefore helping to explain the chemical imbalance reported in Noachian sediments (Milliken et al., 2009).

The starting point of the model is a basalt/water equilibrium solution, from which phyllosilicates and complementary salts in the form of sulfates, carbonates or chlorides have precipitated as solid phases, and water has been ultimately lost in a process of evaporation + freezing (Fairén et al., 2009), as modeled above. Subsequently, a limited addition of water to the system causes the re-dissolution of salts, because the solubility of these salts is much higher than that of clays and other low-temperature silicates (Tardy and Fritz, 1981; Nagy, 1995; Köhler et al., 2005).

The ionic excess remains in solution, where it can be remobilized downwards into the subsurface by means of a concentrations gradient generated by the differential saturation state between surface and subsurface layers. The results are presented in Figure 41. The model predicts that soluble salts formed in the primeval wetter period (Chevrier et al., 2007; Fairén et al., 2009) would have been re-dissolved and transported to deeper layers shortly after formation or throughout the history of the planet due to downward migration of water and differential salt mobility.

This was the general trend all around the planet, though some locations did not have subsequent water input, and consequently phyllosilicates and salts can still be observed to occur together in outcrops at these locations (see Discussion section).

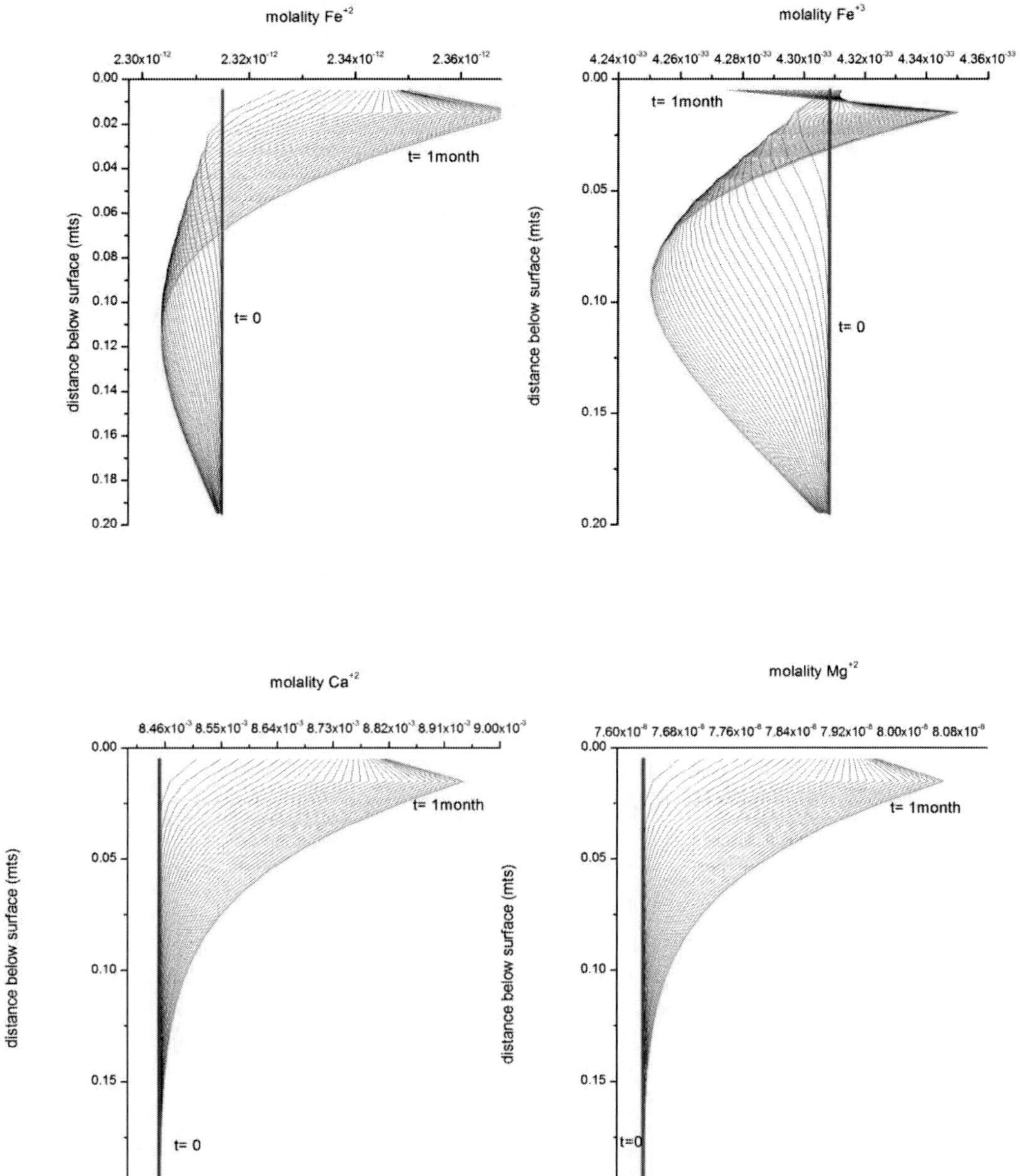

Figure 41. Evolution of concentration in depth for various dissolved cations during 1 month. At the starting point (t=0), all concentrations are the same. In the initial steps, the quantity of cations tends to decrease in the surface by diffusion to lower levels, but at the same time a significant quantity of water is lost by evaporation + freezing, therefore transiently increasing the ionic concentration in the surface. This increase in molality does not mean a parallel increase in the number of moles.

Included in the calculations is the complementary reservoir of cations that must exist attached to the phyllosilicates themselves, derived from the selective adsorption of cations to the negatively charged surface of clays, and the results are detailed in Figure 42. This is a very effective system to immobilize ions, and does not imply the incorporation of the cations to the structural building of the mineral. As such, these adsorpted cations would be undetectable with the usual orbital spectroscopic techniques, as they are based on the recognition and identification of specific functional groups using diagnostic molecular vibrational features: the crystal structure varies from mineral to mineral according to its functional groups and the type of ligands formed, therefore producing different absorptions and making specific mineral identification possible from spectroscopy (Clark, 1999).

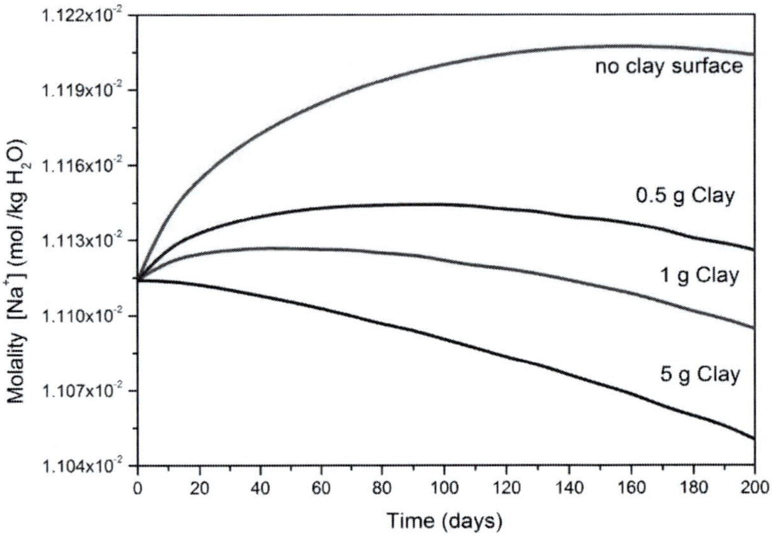

(a)

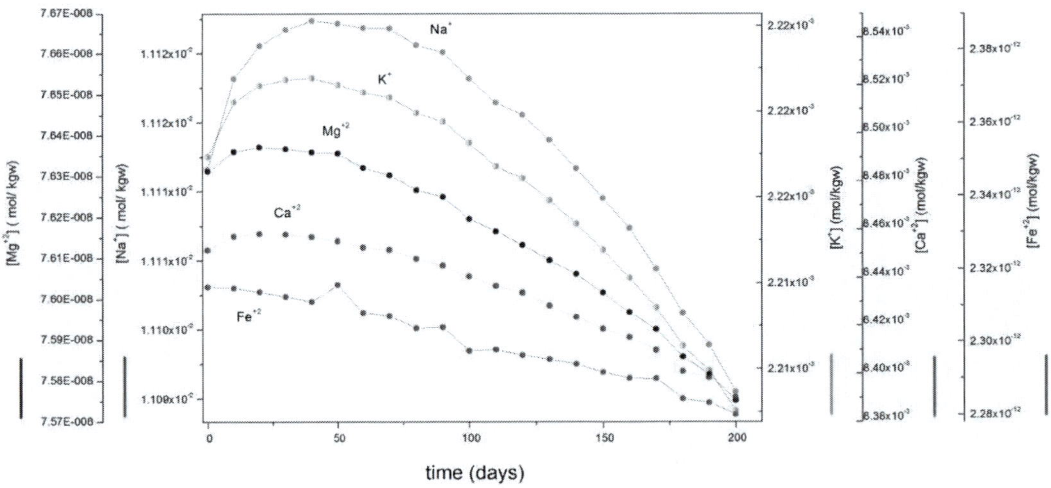

(b)

Figure 42. A: Evolution of Na^{+} molality on the surface during 200 days in a system subjected to evaporation + freezing. Four different scenarios for the reactive clay surface are shown, from no ionic absorption (red line) to progressively more surface of negatively-charged clays. The lowering of ionic concentration is faster as the quantity of clays increases. B: Evolution of the molality of different cations on the surface during 200 days in a system subjected to evaporation + freezing. In this case, the reactive clay surface has a constant value (1 gr clay. 0.1 mol of surface-active sites). The concentration of monovalent cations decreases faster than that of divalent, as monovalent cations are more easily adsorpted in the surface of clays.

The results obtained suggest that the cations that formed the salts needed to balance the stoichiometry of the Noachian sediments characterized by the presence of clays (Milliken et al., 2009) are mostly buried some cm below the phyllosilicate layer, and also a reduced proportion is attached to clays themselves. In both cases they still remain undetected, because (*i*) the visible-infrared spectrometers orbiting Mars have a limit of penetrability which makes them sensitive only to the minerals directly exposed on the surface or down to just some microns depth; and (*ii*) the spectroscopic techniques currently used are based on the identification of functional groups, and not of isolated cations attached to exposed minerals.

4. DISCUSSION

4.1. Coeval Synthesis of Aqueous Minerals on the Earth

A degree of heterogeneity is expected where the chemistry of an aquatic reservoir is governed by both an atmospherically well-mixed weak acid and the saturation of a mineral containing a derivative of that acid. Earth's hydrosphere is a good example: though the majority of Earth's aquatic reservoir has a composition and pH dominated by calcite saturation, a large variety of environments exist, such as hydrothermal vent systems, acidic fluvial-lacustrine environments or hyper-alkaline lakes. The occurrence of substantially different but coeval water geochemistries is frequent in locations commonly cited as Martian analogues, as for example the perennial springs in Axel Heiberg island (Andersen et al., 2002) or the headwaters of Rio Tinto (Amils et al., 2007). Clays, sulfates, and chlorides have been observed to precipitate out together in terrestrial acid-saline lakes and ground waters in southern Western Australia (Benison et al., 2007). In the high Andes of Chile, phyllosilicates precipitate directly from modern acid saline lakes and groundwaters where sulfates are also abundant (Risacher et al., 2002). And episodic excursions in acidity, as those reported in lakes in southeast Canada where pHs have been measured to drop from 7.0 to 3.2 in less than 2 months (Tranter et al., 1994), could have also occurred on Martian water masses.

4.2. Coeval Synthesis of Aqueous Minerals on Mars

Heterogeneity in the early Martian weathering environment and the resulting mineralogy is also expected due to spatiotemporal variations in the availability of water, climate, temperature, weather patterns, moisture conditions, contact with the atmosphere, climate fragmentation due to topographic effects, and primary mineral assemblage. While the model presented here suggests that the majority of surface salts were redissolved and buried in the immediate subsurface, relics of the original mineral assemblages can still be identified in some places on Mars, implying that water has been absent from these locations after the synthesis of the aqueous minerals. Such relic mineralogies could have formed contemporarily in diverse aqueous environments, appearing today sometimes within meters or even in spatial juxtaposition of each other, suggesting mixtures: this has been shown by the MER Spirit in the Columbia Hills at Gusev crater and by MEX-OMEGA and MRO-CRISM in several places on Mars, including regional coexposures of phyllosilicates with either carbonates

(Ehlmann et al., 2008), sulfates (Poulet et al., 2005; Wiseman et al., 2008; Murchie et al., 2009) or chlorides (Murchie et al., 2009) in different locations. This diverse mineralogy denotes variations in geological and geochemical environments (different pH-dependent environments) in a small spatial and/or temporal range, pointing to an overlap in the synthesis of phyllosilicates and carbonates, and phyllosilicates and sulfates, on a wet early Mars (Noachian and Hesperian). The cases for Nili Fossae and Valles Marineris are shown in Fig. 43, where deposits of phyllosilicates and carbonates (Ehlmann et al., 2008), and phyllosilicates and sulfates (Murchie et al., 2009), respectively, have been suggested to occur together. In addition, meteorites from Mars support the existence of diverse subsurface fluid compositions and pHs percolating through the materials of the different source environments of nakhlites and shergottites (Rao et al., 2008).

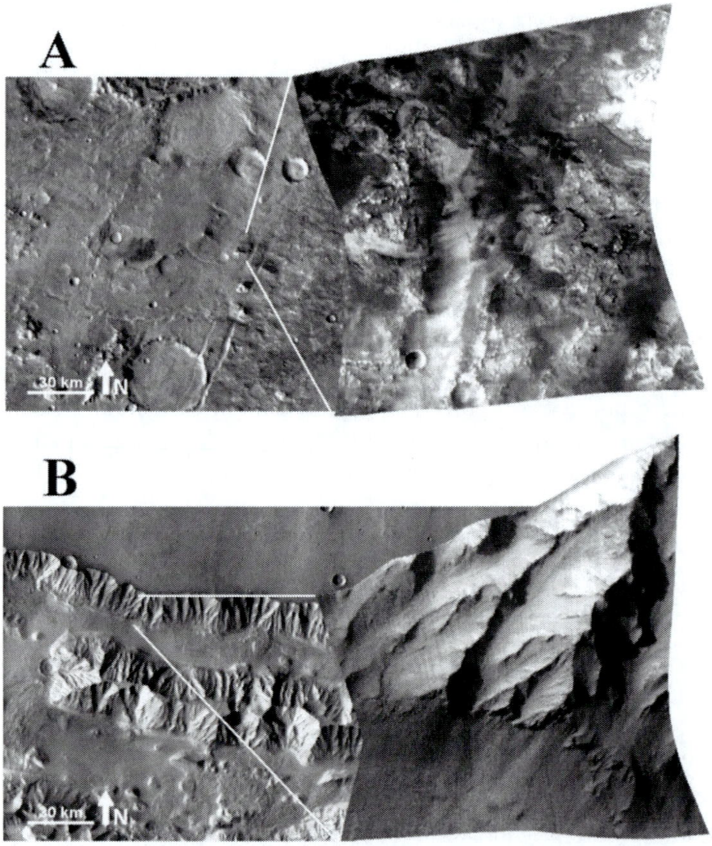

Figure 43. Mixtures of aqueous minerals on Mars. Summary products after Pelkey et al. (2007). A: CRISM observation FRT000097E2 and its context from a THEMIS IR day photomosaic at 256 deg/pixel of a portion of the Nili Fossae region. The CRISM observation, centered at 76.42 E, 19.20 N and ~15 km across, is shown in false colors: green are Fe/Mg-phyllosilicates (D2300), blue are carbonates (BDCARB), and red are low-calcium pyroxenes (OLINDEX). B: CRISM observation FRT0000B8E2 and its context from a THEMIS IR day photomosaic at 256 deg/pixel of a portion of Valles Marineris. The CRISM observation, centered at 304.27 E, 13.90 S and ~15 km across, is shown in false colors: green are Fe/Mg-phyllosilicates (D2300), blue are hydrated sulfates (BD1900), and red are low-calcium pyroxenes (LCPINDEX). Note the presence of sulfates in upper stratigraphic levels than phyllosilicates. Credit: G. Marzo.

4.3. Redox Conditions on Early Mars

Model results predict the precipitation of nontronite, a Fe^{3+}-bearing dioctahedral smectite clay, as expected in an oxidizing environment. Nontronite has been detected by CRISM in several locations on Mars (Mustard et al., 2008), and its presence points to the existence of oxidants in the environment, thereby casting doubts on the hypothesis of exclusively reducing conditions at the time of phyllosilicate synthesis (Bibring et al., 2006). Alternatively, nontronite may not have been a primary precipitate, but an oxidation product of a ferrous iron precursor. But, if this was the case, nontronite is not a primeval Martian mineral, as it would be a secondary alteration product formed in oxidizing conditions. Consequently, the presence of nontronite strongly suggests oxidizing conditions over large regions of early Mars. As an effect of the prevalence of oxidizing and acidic conditions, large-scale carbonate precipitation was prevented all around the planet (Fairén et al., 2004), although the results obtained here predict more regional and/or local identifications of carbonates adding to those already reported (Elhmann et al., 2008).

4.4 Atmospheric Pressure on Early Mars

The results presented here remain consistent and do not experience substantial excursions at any reasonable atmospheric pCO_2 for early Mars. Only below 0.01 bar, the synthesis of smectites and carbonates is totally inhibited, a scenario expected at the end of the wet primeval age of Mars (Noachian and Hesperian). Noteworthy, nontronite and sulfates would still precipitate at very low atmospheric pCO_2, implying that some of the nontronite deposits reported on Mars could actually have been deposited at the very end of the wet period, likely interbedded with kieserite horizons. Consequently, these results open a window to the synthesis of phyllosilicates during the Hesperian in regions where the input of water would compensate the ionic supersaturation derived from the general planetary dryness, and therefore pH could rise over 4. One such location may be Mawrth Vallis, where Fe/Mg smectites, Al-phyllosilicates, hydrated silica and possibly sulfates have been detected by CRISM (19); although initially dated as Noachian (Bibring et al., 2006), detailed geological analyses suggest that the clays in Mawrth Vallis are drape deposits formed in the Hesperian, postdating the excavation of the channel (Bullock and Moore, 2007).

4.5. Downward Migration of Salts

Differential mobility transport between surface salts is evident in arid, hyper-arid and cold desert soils on Earth. In arid and hyper-arid deserts, dew, fog and sporadic precipitation events enable water and solutes to percolate into the soil (Phillips, 1994; Amundson et al., 2008). Relatively thick and widespread deposits of gypsum and halite are consequently often found several decimeters under the surface, whose composition is otherwise dominated by windblown salt-rich and basaltic rock materials (Amundson et al., 2008; Ewing et al., 2006). In cold regions, differential freezing point depressions of saturated solutions of salts can also cause salt mobility (Clark and Van Hart, 1981). The direction of salt movement in cold desert soils can be both upward and downward (Ugolini and Anderson, 1973). However, the presence of SO_4 and NO_3 of atmospheric origin at depths of up to 2 m in soils of the Dry Valleys, Antarctica (Boa et al., 2000; Michalski et al., 2005), clearly shows that atmospherically derived salts migrate in a net downward direction in Earth environments that approach the temperature and humidity conditions of Mars (Amundson et al., 2008).

4.6. The Geochemical Evolution of Mars

As Mars has dried out, a transition from environments with high water activity to others with less water activity is obviously expected, corresponding to a transition from clay-dominated aqueous systems to sulfate-controlled environments (Bibring et al., 2006). Sulfates would globally emerge as a result of a major CO_2 atmosphere imbalance produced by the core cooling. At the same time, episodic volcanism would have provided the main source for the planetary high acidification, by the introduction of substantial amounts of SO_2 into the declining atmosphere, and therefore overcoming the mild acidic conditions maintained by a primitive CO_2-rich and thick atmosphere. The hydrologic system probably started with mildly acidic regimes as the prevailing environments in the Early Noachian, and ended with highly acidic, oxidizing and more salty environments dominating the end of the Hesperian. But this picture of a changing Mars would include different environmental conditions between regions, as well as variable conditions through time. At the end, a sequential transition occurred, and a collage of different watery regimes created the puzzling mineralogy we can see now on the surface. In this scenario, the typical environments on Mars would have been varying acidic, all coexisting together over the surface in the same time periods and increasing their acidity and salinity through time. Therefore, the results presented here open a window to heterochrony in the geochronological relationships between materials, implying a real coexistence of the two types of water systems, namely mildly and high acidic, during hundreds of millions of years in the Noachian and the Hesperian.

CONCLUSION

The model presented here considers all Si-C-Fe-S mineralogies, indicating the stability boundaries for the synthesis of sulfates and phyllosilicates, explaining the absence of widespread carbonates as a result of mildly to highly acidic water regimes, and accounting for the extensive presence of iron over the Martian surface, describing a heterogeneous planet with changing weathering regimes. The prevailing environments on Mars would have been cold and varying acidic, all coexisting together over the surface of Mars during the wet Noachian to Hesperian geological times. Model results predict the precipitation of regional deposits of low-Fe-Mg smectites and sparse carbonates defining the Early Noachian sediments; and large volumes of sulfates and nontronite overlapping through the entire Noachian and Hesperian rock record. As water was resupplied to the dried surface after the precipitation of the aqueous minerals, the most soluble salts were remobilized down into the subsurface, leaving the less soluble phyllosilicates exposed at the surface.

REFERENCES

Amils R, et al. (2007) Extreme environments as Mars terrestrial analogs: the Rio Tinto case. *Planet. Space Sci.* 55: 370-381.

Amundson R, et al. (2008) On the in situ aqueous alteration of soils on Mars. *Geochim. Chosmochim. Acta* 72: 3845-3864.

Andersen DT, Pollard WH, McKay CP, Heldmann J (2002) Cold springs in permafrost on Earth and Mars. *J. Geophys. Res.* 107: 10.1029/2000JE001436.

Bandfield JL, Hamilton VE, Christensen PR (2000) A global view of martian surface compositions from MGS-TES. *Science* 287: 1626-1630.

Bandfield JL, Glotch TD, Christensen PR (2003) Spectroscopic identification of carbonate minerals in the Martian dust. *Science* 301: 1084–1087.

Bandfield JL (2008) A complex history of martian surface mineralogy and aqueous processes. *Lunar Planet. Sci. Conf.* XXXIX: 1741.

Benison KC, et al. (2007) Sedimentology of acid saline lakes in southern Western Australia: newly described processes and products of an extreme environment. *J. Sediment. Res.* 77: 366–388.

Bethke CM (2002) *The Geochemist's Workbench, Release 4.0: A User's Guide to Rxn, Act2, Tact, React, and Gtplot*, ed. University of Illinois, 224 pp.

Bibring JP, et al. (2006) Global mineralogical and aqueous Mars history derived from OMEGA/Mars Express data. *Science* 312: 400-404.

Bishop JL, et al. (2008) Phyllosilicate diversity and past aqueous activity revealed at Mawrth Vallis, Mars. *Science* 321, 830-833.

Boa H, Campbell DA, Bockheim JG, Thiemens MH (2000) Origins of sulphate in Antarctic dry-valley soils as deduced from anomalous ^{17}O compositions. *Nature* 407: 499–502.

Brogioli D, Vailati A (2001) Diffusive mass transfer by nonequilibrium fluctuations: Fick's law revisited. *Phys. Rev.* E 63: 1-4.

Bruland KW (1993) Trace elements in seawater. In *Chemical oceanography*, 8, eds. JP Riley, R Chester (Academic Press, London) pp. 157-220.

Bullock MA, More, JM (2007) Atmospheric conditions on early Mars and the missing layered carbonates. *Geophys. Res. Lett.* 34, L19201, doi: 10.1029/2007GL030688.

Carr MH (1999) Retention of an atmosphere on Early Mars. *J. Geophys. Res.* 104: 21897-21909.

Catling DC (1999) A chemical model for evaporites on early Mars: possible sedimentary tracers of the early climate and implications for exploration. *J. Geophys. Res.* 104: 16,453-16,469.

Chevrier V, Poulet F, Bibring JP (2007) Early geochemical environment of Mars as determined from thermodynamics of phyllosilicates. *Nature* 448: 60-63.

Clark BC, Van Hart DC (1981) The salts of Mars. *Icarus* 45: 370–378.

Clark RN (1999) Spectroscopy of rocks and minerals, and principles of spectroscopy. In *Manual of remote sensing, Volume 3, Remote sensing for the Earth Sciences*, ed. A.N. Rencz (John Wiley and Sons, New York) pp. 3- 58.

Ehlmann BL, et al. (2008) Orbital identification of carbonate-bearing rocks on Mars. *Science* 322: 1828-1832.

Ehlmann, BL, JF Mustard, SL Murchie, J-P Bibring, A Meunier, AA Fraeman, Y Langevin (2011) Subsurface water and clay mineral formation during the early history of Mars. Nature, 479, 53-60, doi: 10.1038/nature10582.

Ewing SA, et al. (2006) A threshold in soil formation at Earth's arid-hyperarid transition. *Geochim. Cosmochim. Acta* 70: 5293–5322.

Fairén, AG, Fernández-Remolar D, Dohm JM, Baker VR, Amils R (2004) Inhibition of carbonate synthesis in acidic oceans on early Mars. *Nature* 431: 423-426.

Fairén AG, Davila AF, Duport LG, Amils R, Mckay CP (2009) Stability against freezing of aqueous solutions on early Mars. *Nature* 459: 401-404.

Fick A (1855) On liquid diffusion. *Philos. Mag. J. Sci.* 10: 31–39.

Gendrin A, et al. (2005) Sulfates in Martian layered terrains: the OMEGA/Mars Express view. *Science* 307: 1587-1591.

Halevy I, Zuber MT, Schrag DP (2007) A sulfur dioxide climate feedback on early Mars. *Science* 318: 1903-1907.

Hamilton VE, Christensen PR (1997) Interpreting the origins and evolutions of martian basalts from pyroxene composition: II. Vibrational spectroscopy of clinopyroxenes and terrestrial basaltic rocks. *Lunar Planet. Sci. Conf. XXVIII*. 1560.

Holland HD (1973) The oceans: a possible source of iron in iron-formations. *Econ. Geol.* 68: 1169-1172.

Johnson SS, Mischna MA, Grove TL, Zuber MT (2008) Sulfur-induced greenhouse warming on early Mars. *J. Geophys. Res.* 113: E08005, doi:10.1029/2007JE002962.

Köhler SJ, Bosbach D, Oelkers EH (2005) Do clay mineral dissolution rates reach steady state? *Geochim. Cosmochim. Acta* 69: 1997-2006.

Krasnopolsky VA (1993) Photochemistry of the martian atmosphere (mean conditions). *Icarus* 101: 313-332.

McSween HY, et al. (2009) Elemental composition of the martian crust. *Science* 324: 736-739.

Michalski G, Bockheim JG, Kendall C, Thiemens M (2005) Isotopic composition of Antarctic Dry Valley nitrate: implications for NO_y sources and cycling in Antarctica. *Geophys. Res. Lett.* 32: L13817, doi:10.1029/2004GL022121.

Milliken RE, Fischer WW, Hurowitz JA (2009) Missing salts on early Mars. *Geophys. Res. Lett.* 36: L11202, doi:10.1029/2009GL038558.

Mojid MA, Cho H (2006) Estimating the fully developed diffuse double layer thickness from the bulk electrical conductivity in clay. *App. Clay Sci.* 33: 278-286.

Murchie SL, et al. (2009) A synthesis of Martian aqueous mineralogy after one Mars year of observations from the Mars Reconnaissance Orbiter. *J. Geophys. Res.*, 114, E00D06, doi:10.1029/2009JE003342.

Mustard JF, et al. (2008) Hydrated silicate minerals on Mars observed by the Mars Reconnaissance Orbiter CRISM instrument. *Nature* 454: 305-309.

Nagy KL (1995) Dissolution and precipitation kinetics of sheet silicates. *Rev. Min. Geochem.* 31: 173-233.

Parkhurst DL, Appelo CAJ (1999) *User's guide to PHREEQC (Version 2)--a computer program for speciation, batch-reaction, one-dimensional transport, and inverse geochemical calculations*, ed. U.S. Geological Survey, Water-Resources Investigations Report 99-4259, 312 p.

Pelkey SM, et al. (2007) CRISM multispectral summary products: Parameterizing mineral diversity on Mars from reflectance. *J. Geophys. Res.* 112: E08S14, doi: 10.1029/2006JE002831.

Phillips FM (1994) Environmental tracers for water movement in desert soils of the American southwest. *Soil Sci. Soc. Am. J.* 58: 15–24.

Poulet F, et al. (2005) Phyllosilicates on Mars and implications for early martian climate. *Nature* 438: 623-627.

Rao MN, Nyquist LE, Wentworth SJ, Sutton SR, Garrison DH (2008) The nature of Martian fluids based on mobile element studies in salt-assemblages from Martian meteorites. *J. Geophys. Res.* 113: E06002, doi:10.1029/2007JE002958.

Risacher F, et al. (2002) Hydrogeochemistry of two adjacent acid saline lakes in the Andes of northern Chile. *Chem. Geol.* 187: 39-57.

Squyres SW, et al. (2004) In situ evidence for an ancient aqueous environment at Meridiani Planum, Mars. *Science* 306: 1709-1714.

Tardy Y, Fritz, B (1981) An ideal solid solution model for calculating solubility of clay minerals. *Clay Min.* 16: 361-373.

Tosca NJ, McLennan SM, Dyar MD, Sklute EC, Michel FM (2008) Fe oxidation processes at Meridiani Planum and implications for secondary Fe mineralogy on Mars. *J. Geophys. Res.* 113: E05005, doi:10.1029/2007JE003019.

Tranter M, Davies TD, Wigington PJ, Eshleman KN (1994) Episodic acidification of freshwater systems in Canada. *Water, Air, Soil Poll.* 72: 19-39.

Ugolini FC, Anderson DM (1973) Ionic migration and weathering in frozen Antarctic soils. *Soil Sci.* 115: 461–471.

Wiseman SM, et al. (2008) Phyllosilicate and sulfate-hematite deposits within Miyamoto crater in southern Sinus Meridiani, Mars. *Geophys. Res. Lett.* 35, doi:10.1029/2008GL035363.

In: Mars
Editor: Alberto G. Fairén
ISBN: 978-1-62618-102-1

Chapter 6

NOACHIAN ENVIRONMENTS ON MARS AND THE POSSIBLE EMERGENCE OF LIFE

Julio E. Valdivia-Silva*, Hugo Beraldi, Maria Colin-García, José L. García-Martínez, Alejandro Heredia-Barbero, Héctor J. Durand-Manterola, Alicia Negrón-Mendoza, Fernando Ortega-Gutierrez and Sergio Ramos
Seminario Interdisciplinario del Origen de la Vida (SIOV),
Universidad Nacional Autónoma de México, México D.F., México

ABSTRACT

Life seems to make planet Earth a unique place in the universe, but favorable conditions for its development certainly exist and must have existed elsewhere and almost at any time across the Universe. In the Solar System, Mars is the nearest planet that seems promising for finding a second example where life may have surged and evolved. In contrast to Earth, present Mars is probably too cold and dry, and with a thin atmosphere to permit life forms to duel on its present surface. However, early in its history, Mars was climatically much like the young Earth, when life emerged. In spite of the much smaller size compared to Earth and its greater distance to the Sun, Mars could have harbored life if certain fortuitous conditions coincided there at the right time and place. Favorable habitable conditions must include a denser and warmer atmosphere that buffers global climate, permanent bodies of liquid water in which chemical evolution can occur, a global magnetic field to protect the planet from damaging radiation, chemical availability, volcanism that recirculates gases and water and creates new heterogeneous environments, and impacts that melting ice and can form microenvironments with hydrothermal activity. The scars leftover by abundant running and stagnant water on the Noachian lands of Mars support the full attainment of those conditions and hence, most probably also the emergence of Martian life.

* Correspondent author: Julio E. Valdivia-Silva MD., PhD., NASA Ames Research Center, Space Sciences & Astrobiology Division, N245, M/S 245-3, Room 213A, Moffett Field, CA, 94035, US, E-mail: julio.e.valdiviasilva@nasa.gov

In this work we examine four central issues of the Noachian history (~4.5 to 3.7 Ga) of Mars, germane to the eventual appearance of life: 1) The nature and evolution of early global processes such as planetary accretion, building of a dense atmosphere, the presence of a global magnetic field and a magma ocean, and the effects of the Late Heavy Bombardment, that brought abundant organic compounds and volatiles to the early Mars; 2) the thermal evolution of the crust, including plate tectonics and volatile-rich volcanism; 3) the Martian liquid paleohydrology (lakes, rivers and putative oceans); and 4) the role of minerals in prebiotic chemical evolution potentially leading to the appearance of living organisms on Mars.

1. Introduction

Why should we focus on the Noachian of Mars? The answer is because the geologic record of this time period could keep essential clues to the origin of life in that planet. Noachian conditions were apparently similar to those of the primitive Earth when life arose, and because Mars seemed to have harbored, in spite of several limiting factors (e.g. size of the planet, the ephemeral magnetic field, composition and density of the atmosphere, and amount and permanence of liquid water), many of the essential conditions that are expected to have intervened in the emergence of life on Earth.

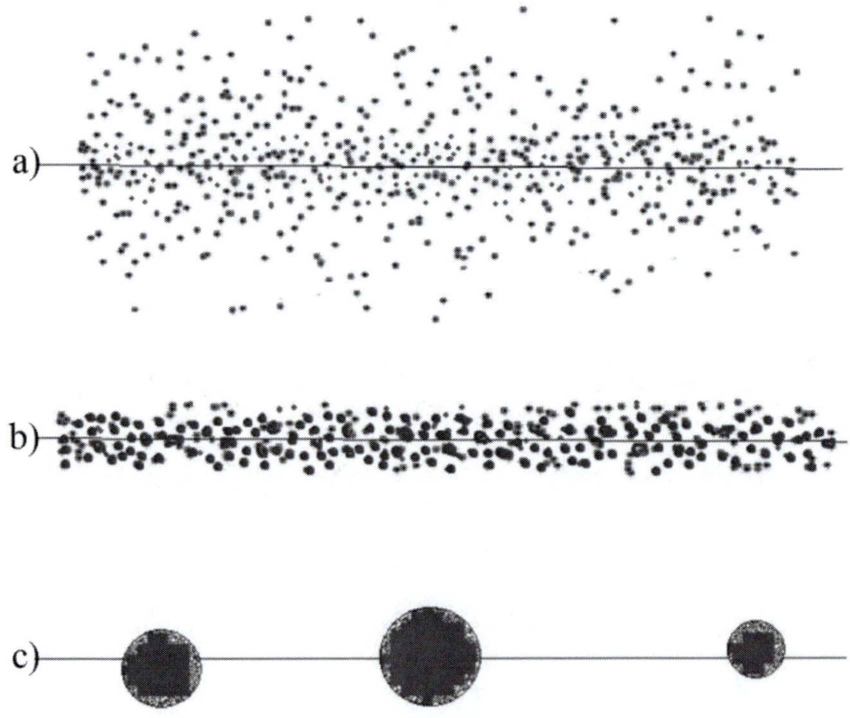

Figure 44. Stages of the Nebular hypothesis for the accretion at Mars orbit. a) accumulation of dust particles into the middle orbital plane; b) formation of planetesimals via gravitational instability fragmentation; c) runaway accretion of the largest planetesimals to form embryos. A large embryo would result in Mars formation.

The planetary history of Earth and Mars diverged drastically after the first few hundred million years of their formation (Schulze-Makuch et al., 2007), and it is evident that subsequent environmental conditions indeed favoured Earth for the appearance and evolution of life. While Earth has had abundant liquid water on its surface almost since its formation, Mars only had it early in its history, with oscillations to warmer and wetter stages that finally turned into the dry and cold present conditions. Thus, if life emerged on Mars independently, it must have been early and during these warm and wetter periods. Life as we know it requires a flow of energy through a system of complex chemical interactions occurring within an aqueous environment. Energy, complex chemistry, and water were surely available on both, Earth and Mars in their early histories. Therefore, the origin of life may have occurred separately on both planets. By reviewing the known early history of life on Earth, and contrasting the environmental and climatological conditions around it with those hypothesized for Mars, the possibility exists for life to have arisen there if liquid water was continuously abundant for millions of years, before a dramatic cooling in the ~mid Hesperian.

We focus here on the Noachian - Early Hesperian transition time (Figure 44), because these are epochs with solid evidence of liquid water, and because the origin of life under this scenario might be achievable within a few hundred million years (Lahav, 1999) or much less (Lazcano and Miller, 1994). These early environments of Mars seem to have been similar to those on Earth during the Hadean-Archean transition, when life likely appeared (Valdivia-Silva et al., 2011). Ingredients and requirements for life (e.g. liquid water, above freezing temperatures, dense atmosphere) are also seen in later Martian periods, such as the Hesperian or even the Amazonian (Barlow, 2008). Nonetheless, it is in the Noachian when there was a magnetosphere, active degassing and volcanism, a denser atmosphere, and less acidic water conditions, perhaps being these more favorable for life than the more acidic conditions under an evaporitic regime, and with a loss of atmospheric pressure that dominated Mars in the Hesperian-Amazonian times. Thus, the optimal conditions for the potential emergence of life on Mars most probably occurred in the Late Noachian-Early Hesperian transition, and again, this time is approximately equivalent to the Hadean-Archean transition (4.0-3.7 Ga), a time when life probably already existed in our planet. Yet, because planetary and climatic evolutions are so different among terrestrial planets, perhaps we should abandon the idea that these processes on Mars and Earth shared all temporary events including the "windows" for the origin of life. The answer to the question *"did life emerge in Noachian-early Hesperian Earth-like scenarios on Mars?"* must rely on integrative knowledge on the evolution of Mars from the standpoint of astrophysical, global, and local phenomena that might have determined or exerted a strong influence in the possible appearance of life (see Table 1).

Section 1 reviews the astrophysical phenomena, and discusses the context of Mars during the Late Heavy Bombardment (LHB), the composition and surface pressure of the Martian atmosphere, and the possibility and importance for life of an early magnetosphere. *Section 2* deals with the volcano-tectonic phenomena, including the early presence of a magma ocean and plate tectonics on Mars, explosive volcanism, and degassing of greenhouse gas species of biotic importance. *Section 3* focuses on the paleohydrological phenomena, and discusses the particular aqueous environments on early Mars as potential niches for life. *Section 4* reviews processes of the prebiotic chemical evolution for different environments inferred for the Noachian-Hesperian of Mars, focusing on the interactions between minerals and organic molecules as precursors of more complex compounds of astrobiological importance, and their chances of survival. It is suggested that if organic chemical evolution was accomplished on

Mars and continued for sufficient time, life might have emerged on the planet. It is not discussed, but not discarded, whether Mars and Earth interchanged living organisms through expelled impact ejecta (local panspermia) (Davies, 1996; Nicholson et al., 2000).

Table 1. Global processes and its main physical and chemical effects on planetary evolution eventually leading to the origin of life on Earth and possibly on Mars

Geospheres of terrestrial planets	*Global scale Processes and their effects on the Geospheres*					
	Accretion	Giant impacts	Planet Radioactivity	Gravity	Interior cooling	Solar radiation
Core-Mantle	Core-mantle differentiation and Magma ocean	Planetary growth and Magma ocean	Melting and Convection in the Mantle	Convection and Plate tectonics	Magnetic field endurance	None
Atmosphere	Primitive and reduced if any	Erosion or Growth	Photo-dissociation	Retention	Growth by exsolution from magmas	Limited erosion and climate regulation
Hydrosphere	Retained in the Mantle	Erosion or Growth	Radiolytical processes may occur	Sedimentary processes	Growth by exsolution from magmas	Water cycle Limited erosion
Lithosphere	Primitive crust	Large basins and Crustal thinning	Crustal melting and Plate tectonics	Plate tectonics and Petrologic differentiation	Crustal growth and Plate tectonics	Light and heat absorption Photolysis of chemical bonds
Biosphere	Absent	Total extinction if present at all	Mutations	Oceanic moon tides on Earth	Increased habitability	Photosynthesis

Intersections of columns and rows refer to the main influences or consequences of one over the other.

2. Early Planetary Conditions and Implications for Life

Accretion and Formation of the Core

The accretion of a planet is intimately linked to its chemical composition, and the later functioning of its magnetosphere, its atmosphere, and its hydrology, all important factors for habitability. The models of the Solar System formation propose that, as with the rest of the planets, Mars was formed during an accretionary process. With a mass less than one-ninth that of Earth, Mars accretion, appears to have been shorter and less catastrophic than Earth's. The nebular hypothesis states that at the beginning, the history of Mars was the same as for the rest of the terrestrial planets, up to the embrionary stage when its accretion stopped. Such embrionary size is the main observation that led to the development of the rapid accretion hypothesis, of probably less than 10 Ma, compared to more than 50 million years for the case of Earth and Venus. This short accretion time certainly prevented Mars from suffering major catastrophes like the Moon-forming event at Earth. Likewise, it prevented Mars from accreting more mass that could help the planet retain its atmosphere and magnetic dynamo for a longer period of time, which is important for keeping molecules of liquid water and consequently potential life.

The Nebular Hypothesis

The nebular hypothesis maintains that the Solar System formed from the gravitational collapse of a dense fragment of a giant molecular cloud, surrounded by a protoplanetary disk of gas and fine dust particles. Three broadly overlapping stages are recognized (Figure 44) (Canup and Agnor, 1998). In the first stage, dust and ice from the circumstellar disc co-adhere and form bodies with gravitational interactions (Benz, 2000), where particles several millimeters in size (or larger), slowly settle near the middle plane of the disk, forming a very thin (less than 100 km) and dense layer (Figure 44). These mm-sized particles evolve into 1- to 10-km-diameter planetesimals in 10^5 years. This layer is gravitationally unstable and may fragment into numerous clumps, which in turn collapse into planetesimals. The second stage involves runaway gravitational growth of the largest kilometer-sized objects (Lin and Papaloizou, 1985) to form from 50–100 Moons- to Mars-sized planetary embryos in 10^5 to 10^6 years (Figure 44). As the runaway stage consumes most of the smaller material in the inner Solar System, the subsequent accretion and growth (the third stage) of the terrestrial planets becomes dominated by collisions between these embryos and, ultimately, with other planets, as proposed by Safronov (1954). Since accretion is a stochastic process, one can not readily predict particular growth histories for the terrestrial planets. However dynamic models for the accretion of Mars and the other planets, based on Monte Carlo simulations, generate some solutions that result in terrestrial planets with the expected sizes and distribution, which include Mars-size solutions. Particularly, given the embrionary size of Mars, clearly suggests that this planet did not participate in the last accretion stage of embryo collisions and accumulation to reach a major sized like Earth or Venus.

The mass and density of Mars are also important consequences of the accretion, and may affect habitable conditions in the following way: the lesser the planetary mass, the lesser the amount of volatiles (including water), but the smaller density may imply more volatiles in the composition of Mars. In any case, Mars seems to have been accreted from the same materials found on Earth. Although accreted materials may evolve into different new materials as they interact and exchange mass and energy, their proportions may not be equal on Earth and Mars given their drastic differences in volume and hydrous conditions. The stored energy after Mars accretion, either by impacts, the decay of radioactive elements, or gravitational potential energy producing heat at the moment of the inner planet differentiation (Barlow, 2008; Wetherill, 1990), would seem to last for thousands of millions of years, with the capacity to create subsurface thermal conditions where an active 'ecosphere' could remain active thanks to the stability of liquid water under some of these primitive conditions. If life was capable of evolving on Earth with similar primary materials, there is no reason to think that Mars composition may be a limiting factor for life.

Accretion Time for Mars

The accretion time for Mars is taken as the time of core formation. On this regard, the simplest model is to assume that Mars remained undifferentiated until the time of core formation, at which point the planet was fully accreted and differentiated into a core and a mantle. Taking into account the uncertainty in the Hf/W ratio of the Martian mantle, previous studies had concluded that Mars' core could have formed anytime between 0 and > 10 Ma after Solar System birth (Dauphas and Pourmand, 2011). Dauphas and Pourmand (2011) calculate a model age of core formation of ~4 Ma after the Solar System was formed. This short accretion time is difficult to reconcile with an Earth-like mode of accretion involving collisions between planetary embryos over several tens of millions years. For comparison, Earth's accretion is best dated by the Moon-forming impact, which is estimated to have occurred >50 Ma after Solar System birth (Touboul et al., 2007).

During the process of accretion Mars melted due to three sources of heat: the kinetic energy of impacting bodies, heat generated from decay of radioactive elements and the transformation of gravitational energy into heat at the moment of differentiation. The energy necessary to melt a Mars-size body is in the order of $2x10^{30}$ J (Barlow, 2008). On Mars the accretion process generated ~ $4x10^{30}$ J (Barlow, 2008; Wetherill, 1990) and among the short life radionucleides ^{26}Al contributed with $2x10^{30}$ J (Elkins-Tanton et al., 2005). Hence, there was enough energy to melt the whole planet during its early formation. During the first few million years of Mars accretion, the heavy siderophile materials sank creating an iron core and the lighter lithophiles materials rose and formed the crust.

The analysis of the ^{182}W and ^{182}Hf isotopes in the Martian meteorites can give us important clues of the moment in which this differentiation occurred. The ^{182}W is the product of the radioactive decay of ^{182}Hf and being a siderophile element it would have followed the iron towards the core during its formation. So, if ^{182}W is found in meteorites that come from the crust or mantle of Mars it is because it was formed through radioactive decay of ^{182}Hf after the formation of the core. As the half life of ^{182}Hf is 9 Ma then almost all ^{182}W (94%) should have been produced in 36 Ma. So, the core was formed within this time. Other studies suggest that the accretion and the differentiation of Mars were completed between 10 and 15 Ma after Solar System formation (Barlow, 2008; Kleine et al., 2002; Lee and Halliday, 1997).

The Fossil Magnetic Field

On Earth, sea floor spreads apart slowly at the mid-oceanic ridges as new magma flows up from Earth's hot interior. Meanwhile, the direction of Earth's magnetic field reverses occasionally, resulting in alternating magnetized stripes in the newly-formed crust, a finding that validated the once controversial theory of plate tectonics (Tarbuk and Lutgens, 2004). NASA's Mars Global Surveyor discovery of patterns of banded magnetic fields on a portion of the Martian surface suggests past movement of the Martian crust, back when the planet was more dynamic. The adjacent stripes opposite in polarity give a preserved similarity to patterns seen in the crust of Earth's sea floors. The bands on Mars are wider than those on Earth, perhaps because the Martian crust was generated at a greater rate, causing a given magnetic field to be imprinted over a wider area before it reverse direction. Alternatively, perhaps the Martian magnetic field reversed direction less frequently, which would have given more time for any field direction to imprint itself in a steady moving crust, resulting in wider bands. In order to call this pattern a crustal spreading center, like that observed in the mid-ocean on Earth, we need to find a point of symmetry, which has not been found yet, where the pattern on one site matches the pattern on the other. This pattern on Mars could revolutionize current thoughts of the planet's evolution. If the bands on Mars are imprints of plate tectonics activity, this is extinct today. Apparently, on Mars, the magnetism was distributed regularly over the highlands of the planet with alternating maxima and minima and with banded structure.

The map of Martian magnetic regions may help solve another mystery: the origin of the striking difference between the smooth, sparsely cratered northern lowlands of Mars and the heavily cratered southern highlands or, Mars Dichotomy. This map reveals that the lowlands are largely free of magnetism, indicating the northern crust formed after the dynamo died. The map also identifies an area in the southern highlands as the oldest surviving unmodified crust on Mars. This area is where the magnetic stripes are most prominent. The bands are oriented approximately east to west and are about 160 kilometers wide and 965 kilometers long; though the longest band stretches more than 1930 kilometers.

At present, Mars does not have an active magnetic field. For a planet to generate its own magnetic field it must have a conductive fluid with a convective movement in its interior. The terrestrial planets have a conductive iron core. If parts of this core are melted, as happens in the case of Earth, the planet can generate a magnetic field.

That Noachian terrains on the southern hemisphere exhibit a fossil crustal magnetic field exists, indicates that in that period the planet had a planetary dynamo, and probably was dipolar (Acuña et al., 1999). While magnetized rocks surrounds the Hellas basin, this fossil magnetic field is not present in its interior, suggesting that the disappearance of the planetary dynamo occurred before the formation of this basin. The time in which Hellas was formed is estimated by Frey (2006) at ~ 4.05-4.13 Ga, this is before the end of the Noachian, located by Frey (2006) at 3.70 Ga. Therefore, the Martian dynamo only worked during two thirds of the first Martian geological period.

Variation of the Magnetic Field Throughout Time

Maxwell's equation of electromagnetism and Ohms Law, tell us how the magnetic field varies with respect to time. The following induction equation may be derived (Parks, 1991)

$$\frac{\partial \mathbf{B}}{\partial t} = \nabla \times (\mathbf{v} \times \mathbf{B}) + \left(\frac{1}{\mu_0 \sigma}\right)\nabla^2 \mathbf{B}$$

Where B is the magnetic induction, v the velocity of convection of the conductive fluid and $1/\mu_0\sigma$ the magnetic diffusivity.

The first term on the right describes the generation of the magnetic field by the movement of the fluid, and the second term describes the diffusion of this magnetic field.

The fact that the magnetic dipole disappeared was probably due to the fact that the Martian core solidified. So, the velocity of convection became null (v=0). Hence, the first term of the induction equation becomes zero, leaving only the diffusion term. Due to this diffusion the magnetic field decayed: The decay time for a magnetic field of a terrestrial planet is of the order of 10^4 years (Barlow, 2008). This tells us that, on a geological timescale, the disappearance of the magnetic field is almost instantaneous as soon as the convection in the core ceases.

It is well known that the presence of a planetary dynamo protects the atmosphere from the direct interaction with the solar wind. When the solar wind affects directly the atmosphere, it erodes it; though, the planetary magnetic field acts as a shield which protects the atmosphere from being lost to space. Thus, when convection in the core of Mars ceased in the Noachian, the erosion of the atmosphere must have started immediately. Nowadays the Martian atmosphere loses ~ 1 kg/s (Lundin et al., 1990).

Table 2. Location, size and age of the 20 largest impact basins on Mars; data are from Frey (2008)

Basin	Center	Diameter estimate	Age (Ga)	Impactor size
Amazonis	27° N, 172° W	2873	4.14	348.5
Daedalia	26°S, 132° W	2639	4.18	338.7
Hellas	42°S, 294°W	2070	4.09 *	312.4
Scopolus	7°N, 278°W	2250	4.13	321.2
In Amazonis	29.3N, 167.5 W	1156	4.15	257.2
Sirenum	67°S, 155°W	1069	4.18	250.6
Amenthes	1S°, 249° W	1070	4.2	250.7
Zephyria	12° S, 196° W	1193	4.2	260.0
N Tharsis	17.6N, 116.4 W	1347	4.14	270.7
Isidis	13°N, 273°W	1352	3.92 *	271.0
SE Elysium	3.7N, 189.7	1403	4.135	274.4
Utopia	45°N, 245°W	3380	4.105	367.8
North Polar	76°N, 185°W	1600	4.12	286.7
Solis	23.8S, 84.7 W	1663	4.14	290.4
SW Daedalia	29°S, 146°W	1278	4.16	266.0
Argyre	49°S, 43°W	1315	4.03 *	268.5
Chryse	25°N, 42°W	1725	4.14	294.0
Acidalia	58°N, 19°W	3087	4.13	357.0
Ares	4°N, 16°W	3300	4.19 *	365.0
Hematite	3°N, 2°W	1065	4.14	250.3

*Ages are from Frey (2006); the rest, from Frey (2008).

In Frey´s (2006) model, the Martian lowlands formed at 4.13 Ga or perhaps slightly earlier (Table 2). This is close to the time when the global magnetic field died, according to the anomalies seen in large and ancient basins. Indeed, some of the largest impact basins, such as Hellas, Isidis and Argyre, have no obvious magnetic anomalies (Langlais et al., 2004). Less obvious, but previously recognized large basins like Utopia, Chryse and Acidalia (termed the ''lowland-making'' basins), have weak anomalies. Only the two most subtle large basins, Daedalia and Ares, have prominent anomalies lying within their main rings (Frey, 2004). Solely on the basis of their ''visibility,'' it appears that only the oldest basins retain significant magnetization features in their interiors. Thus, basins younger than ~4 Ga formed after the main magnetic field shut off (Langlais et al., 2004). The lowland-making basins probably formed when the magnetic field was shutting down, or shortly after the dynamo turned off, depending on how long it would take for the field to decay to where it could no longer magnetize the cooling crust. The apparent early loss of the Martian magnetic field, probably before the end of the LHB, would have affected the conditions for the appearance of life in The Noachian.

The Noachian Martian Atmosphere

The atmosphere on Mars, as well as that on other terrestrial planets, has suffered an evolution from the possible primary system. The so called secondary atmosphere, which has evolved through the Martian geological history, had its origin most probably in the degassing of the planetary body through volcanism plus the contribution of impacting bodies like meteoroids, asteroids and comets.

During its evolution an atmosphere has different sources and sinks of volatiles. The main sources of these volatiles are, as mentioned above, volcanism and impactors. On the other hand, sinks of volatiles are of two types, losses to space that imply the volatiles are lost definitely from the planet and the absorption of volatiles by the planetary crust or mantle. The latter, remove the volatiles from the atmosphere but are stored in the planet's solid body so that the material is not lost definitely from the planet and can return later to the atmosphere by other mechanisms.

A better understanding of the loss mechanisms of volatiles into the space and the atmospheric erosion by impactors throughout the Noachian history could help in further analyses of the conditions that allowed or inhibited habitability.

Loss Mechanisms

In the Noachian, the Martian atmosphere had different ways to escape the planet into space: Jeans escape, drag by solar wind, blowoff (hydrodynamic escape) and erosion by impact.

The atmospheric loss of volatiles to the crust or mantle that Mars may have had in the Noachian period are, for example, chemical reactions with the solid surface of the planet and the volatiles buried by subduction, if Mars ever had it.

Jeans escape. The fastest molecules in the atmosphere of a planet may have velocities greater than the planetary escape velocity. If these molecules were to be found above or in the exobase they would be lost to space. This mechanism is known as thermal escape or Jeans evaporation. Currently the exobase on Mars is at a height of 220 km (Bauer and Lammer, 2004) but it is not known at what height it was during the Noachian. Due to the greater solar ultraviolet radiation during this period, the warming of the layers closest to the exobase probably caused an important thermal escape.

The light components, such as hydrogen and helium, are most prone to be lost; this is why they do not exist in abundance on terrestrial planets. At the level of the exobase the molecules mean free path is equal to the height scale, which means that a molecule that moves up at escape velocity or a greater velocity will never collide with another molecule and will escape from the planet. This is the process by which hydrogen escapes from Earth today.

In a secondary atmosphere where CO_2 predominates, like that hypothesized for early Mars, water is lost with preference over carbon dioxide by the following reactions

$$CO_2 + h\nu \rightarrow CO + O$$

Oxygen is lost in space dragged by the solar wind as O^+ or dragged hydrodynamically by hydrogen or it reacts with the rocks on the surface. Other reaction, which contributes to the loss of water, is

$$H_2O + h\nu \rightarrow HO + H$$

where hydrogen is lost due to the Jeans escape. And finally, in the reaction

$CO + HO \rightarrow CO_2 + H$

Again hydrogen is lost to space due to the Jeans escape.

Although both, carbon dioxide and water, were photo dissociated the CO_2 was recovered and the final result is that the water was lost from the atmosphere, then Jeans escape on Mars produce a loss of water mainly.

Drag by the solar wind. Actually the solar wind drag is really a series of mechanisms such as: dragging through the convective electric field, sputtering, charge exchange, etc., all of them caused by the absence of a planetary magnetic field that prevents the interaction between the solar wind and the ionosphere and exosphere of the planet, as well as the large amount of energy that represents a plasma with supersonic flow. During the Early Noachian the Sun emitted a faster and denser wind than it currently does (Lammer et al., 2008); powerful enough to drag great quantities of volatile elements. Yet Mars was probably protected by its magnetic field that seemingly had at the time. Although, some authors suggested that during the Early Noachian, the magnetic dipole of Mars weakened or disappeared (Frey, 2006), and at this moment the solar wind affected the Martian ionosphere; the most erosive action of the solar wind had already passed, since the velocity and density of the solar wind of its first days decrease rapidly. This is why the drag of the solar wind must have been a minor mechanism during the Early Noachian and acted mainly at the end of this period, and during the Hesperian and Amazonian (Durand-Manterola, 2001).

Blowoff or hydrodynamic escape. This mechanism consists of the drag of a heavy gas (one that cannot escape through of the Jean mechanism) driven off the planet by the viscous drag of other lighter gases (normally hydrogen) which can escape. In the Noachian more hydrogen was available due to the photodissociation of water by the more intense ultraviolet radiation from the young Sun, and by the bursts of high energy of the impactors that were more frequent in this period.

The escape flow of the heavy gas is

$$\Phi_2 = \frac{N_2}{N_1}\left(\frac{m_c - m_2}{m_c - m_1}\right)\Phi_1$$

Where Φ is the flow, N is the density, m the molecular mass, the subscripts 1 and 2 refer to the light and heavy gases respectively, and m_c is

$$m_c = m_1 + \frac{NkT\Phi_1}{bg_p}$$

Where N is the total number density, k is the Boltzmann constant, T the temperature of the gases, b is the binary collision parameter and g_p is the acceleration due to gravity for this planet.

The condition for blowoff is

$$m_c = m_1 + \frac{NkT\Phi_1}{bg_p} > m_2$$

$$\Rightarrow$$

$$NT\Phi_1 > (m_2 - m_1)\frac{bg_p}{k}$$

During the Noachian period, with a denser atmosphere ($N > N_{actual}$), with a greater temperature ($T>T_{actual}$) and a greater hydrogen flow ($\Phi_1 > \Phi_{1actual}$) than at present, it is probable that hydrodynamic escape was an important factor in the loss of atmosphere (de Pater and Lissauer, 2001).

Comets and Asteroids Impacts

Cometary and asteroid impacts are important for planetary evolution, and eventually for habitable conditions, because they bring water and volatiles to the planet, because they form basins (often with reminiscent active hydrology), and also because they inflict destruction, which is detrimental for life. Impacts played an important role during the formation of Mars early atmosphere, they also played a significant role in further evolution by causing its erosion, mainly by the catastrophic effects of large impacts like Hellas, catastrophe that was repeated at least twice more when Argyre and Isidis formed 45 and 60 million years after Hellas. With the disappearance of its protective magnetic field at the same epoch, the steady atmospheric loss through scattering by impacts and solar wind has been constant up to the present epoch.

The balance between delivery and loss of atmosphere depends on the size, velocity, and composition of the impactor as well as on the size of the impacted planet. San Pham et al. (2009) investigated the effect of impacts on the atmospheric evolution of Mars (Figure 45) by means of a Tangent Plane Approach (Melosh and Vickery, 1989) in which impactors above a critical mass (m_{crit}) are capable of eroding the Martian atmosphere above a plane tangent to the impact site (m_{tan}); impactors with mass less than the critical mass ($m_{crit} = 10\ m_{tan}$) are assumed not to erode the atmosphere. The parameters in the tangent plane model are varied according to the numerical impact simulations and impactor flux estimates given in the literature. For completeness, these authors have taken into account the effect of atmospheric pressure as a function of time. Comparison of the Tangent Plane Approach with other atmospheric evolution models, which contain a larger number of parameters (Svetsov, 2007), has shown proficiency in modeling the atmospheric mass evolution, given appropriately adapted parameters. Thus the impacts on Mars were an important factor of atmospheric erosion, assuming the impactor flux during the LHB was 2 orders of magnitude larger than the background, exponentially declining flux. This modeling yields an atmospheric pressure loss of 1 bar in the last four billion years. To explain the loss of a speculatively more massive atmospheric since the Late Noachian, other atmospheric erosion mechanisms such as solar wind stripping (Lundin et al., 2007) need to be considered. In Melosh and Vickery's model (1989), the volatile contribution of cometary impactors to the Martian atmosphere is neglected.

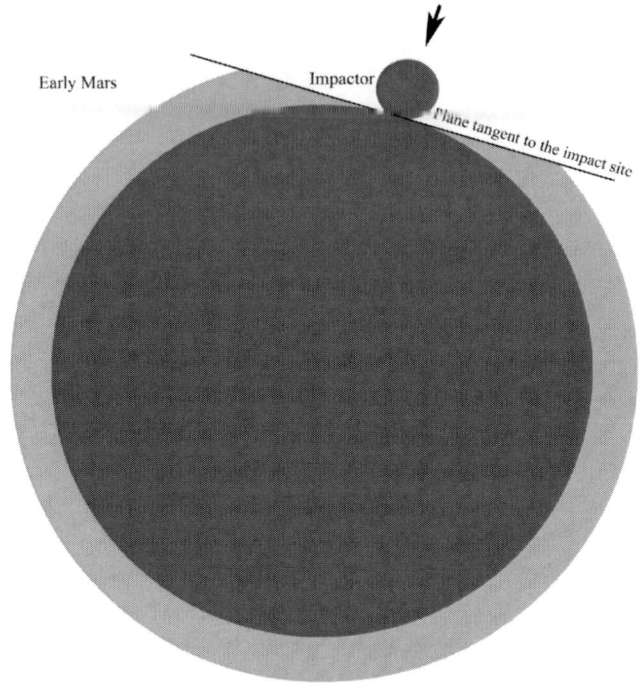

Figure 45. Erosion of Martian atmosphere by impacts of comets and asteroids. Impactors with critical mass are capable to erode the atmosphere (see text for explanation).

Alternatively, impact erosion, or the gas expelled to space from an atmosphere, can be modeled as follows: the amount of atmospheric gas expelled by an impactor is

$$M = \frac{\pi R^2 P_0 E}{g_p}$$

where R is the radius of the impactor, P_0 is the atmospheric pressure on the surface of the planet, g_p is the acceleration due to gravity of this planet, and P_0/g_p is the atmospheric mass per unit area, and E is

$$E = \frac{v_i^2}{v_e^2(1 + E_v)}$$

where v_i is the velocity of the impactor with respect to the planet, v_e is the escape velocity of the planet and E_v is the evaporative loading parameter, which is inversely proportional to the latent heat of evaporation of the impactor.

When E > 1 and R > H (scale height) erosion by impact occurs, and if E < 1 or R < H then the gas does not acquire sufficient velocity to escape the planet (de Pater and Lissauer, 2007). Importantly, during the Noachian there were enough impactors to fulfill the condition R>H, and the effects of those that produced large basins (Table 2).

Atmospheric Mass and Greenhouse Effect

Sedimentary and geomorphological features that indicate the presence of liquid water in the Noachian terrains on Mars require atmospheric temperatures above the freezing point. This is paradoxical because at that time the Sun had less luminosity than at present and, hence, less heating capacity (Gough, 1981; Kasting and Grinspoon, 1999). However, taken together, pressure and temperature maintaining liquid water could have been caused by large amounts of greenhouse gases in the Noachian atmosphere. Important sources of these volatiles are represented by extensive Noachian plains of volcanic origin (Tanaka, 1986; Tanaka et al., 1988).

The atmosphere of Mars nowadays is formed of 95.32 % of CO_2 (Faure and Mensing, 2007), which is a greenhouse gas, and the most abundant volatile compound on Earth after water, expelled by volcanoes (Tarbuk and Lutgens, 2004). Thus, CO_2 was probably one such gas responsible for maintaining favorable temperatures during the Noachian. Other greenhouse gases like O_3 (ozone), CH_4 (methane), SO_2 (sulfur dioxide) and N_2O (nitrous oxide) are destroyed by ultraviolet radiation from the Sun and removed from the atmosphere in a few millennia. Hence, they are less prone to maintaining a greenhouse effect in the long term (in the order of Ga), unless there is a constant input of these volatiles into the atmosphere.

Based on measured volcanic areas of different periods on Mars, Durand-Manterola (2003) estimated that the amount of volatiles produced during the Early Noachian was between 9.42×10^{19} and 9.54×10^{19} kg, and the total amount of volatiles degassed in the Noachian would vary from 2.78×10^{20} and 3.06×10^{20} kg. Estimates for the Early Noachian, however, may be too low due to the fact that the extent of volcanic surfaces calculated by Tanaka et al. (1988), are uncertain since many of these surfaces have been buried under more recent formations. Nevertheless, most of the volatiles generated by Martian volcanism were expelled during the Noachian.

Pollack (1981) estimated that the ratio of H_2O to CO_2 on Earth is ~10:1. Assuming that this ratio is the same on Mars then, from the total amount of degassed volatiles on Mars throughout its history (3.73×10^{20} kg), the total amount of degassed CO_2 would be ~3.39×10^{19} kg, that is ~ 6.4 terrestrial atmospheric masses (TAM; 1 TAM = 5.28×10^{18} kg or 8.6 bars of pressure). The amount of degassed CO_2 during the Noachian would be between 2.53×10^{19} and 2.78×10^{19} kg (between 4.79 and 5.27 TAM or between 6.42 and 7.06 bars). This amount is sufficient to produce an increase of temperature, by greenhouse effect, above 0° C and, thus, important for the maintenance of habitable conditions on Mars.

Impact Basins

Large Noachian impact basins are reminiscent of an epic catastrophic era with uncommon large impacts during the LHB, between 3.9 Ga and 4.26 Ga (Table 2). Large diameter, visible and buried impact basins, seen as ''quasi-circular depressions'', provide a self-consistent chronology for major events on early Mars in terms of crater retention ages (crater counts). The large-diameter crater retention ages provide constraints for the relative age of the largest impact basins, when the lowlands formed, and when the global magnetic field died. This coincides with the formation of the so called Mars dichotomy, perhaps fueled by one or more large impacts. Even when there is no clear link between large impacts and the cessation of the Martian magnetic field, the fact is that the demise of the field also took place by the same time. It is possible that for a relatively small planet like Mars, the large impacts affected the inner core-mantle structure, with concomitant consequences for the functioning of the dynamo. Along with the early disappearance of the magnetic field, impacts must have eroded much of the Martian atmosphere. Thus basing-making impactors ranging in size from 140 km to 400 km (close to Vesta's size) during the LHB set the grounds for the later development of the Martian frigid desert of our days. Large basins are what left from that epoch.

Table 3. Volcanism in the history of Mars

~Age (Ga) at the base	>4.51	4.51	4.1 or 3.8	3.9 or 3.6	3.7 or 3.5	3.2 or 2.0	
Magmatic System	Pre Noachian	Early Noachian	Middle Noachian	Late Noachian	Hesperian	Amazonian	Petrologic composition
Magma Ocean	▬						Komatiitic and basaltic
Intercrater lavas		▬	▬				Basaltic
Claritas Rise			▬				Basaltic
Syrtis Major					▬		Basaltic and non-basaltic
Syria Planum				▬	▬		Basaltic
Elysium Mons						▬	Basaltic and non-basaltic?
Apollinaris Patera					▬		Basaltic
Alba Patera					▬	▬	Basaltic
Smooth Plains					▬	▬	Basaltic
Tyrrhena Patera				▬	▬		Basaltic and non-basaltic?
Hadriaca Patera				▬	▬		Basaltic and non-basaltic?
Ulysses Patera				▬	▬		Basaltic
Olympus Mons						▬	Basaltic
Ascraeus Mons						▬	Basaltic
Pavonis Mons						▬	Basaltic
Arsia Mons						▬	Basaltic
Tharsis Tholus				▬			Basaltic
Hecates Tholus				▬	▬		Basaltic and non-basaltic?
Ceraunius Tholus					▬		Basaltic

General age intervals (bars) of main phases of volcanism in Mars selected and combined from several sources e.g. (Dohm et al., 2009; Hartmann, 1999; Hartmann, 2005; Lillis et al., 2006; Plescia and Saunders, 1979; Robinson et al., 1993; Schulze-Makuch et al., 2007; Tanaka, 1986), among many others. Note that the limits for the Martian epochs are subject to large uncertainties (>1Ga), particularly for the upper and lower limits of the Hesperian (see Hartman, 2005).

Crater retention ages of the 20 largest (D > 1000 km) impact basins on Mars (Table 2) suggest most formed in a relatively short time, perhaps in less than 200 million years. Conversion to the Hartmann-Neukum model chronology (Hartmann and Neukum, 2001) suggests absolute ages between 4.10 and 4.25 Ga for all but the three youngest basins Hellas, Argyre and Isidis, with most falling within an even narrower interval of 4.12-4.14 Ga (Frey, 2008). The sharp peak in formation ages for the largest impact basins on Mars suggests the possibility of a cataclysmic Late Heavy Bombardment (LHB) on Mars. The short time is consistent with the Nice model (Gomes et al., 2005) and a "Terminal Lunar Cataclysm" (Ryder et al., 2000). However, the inferred ''absolute'' age of the Martian bombardment peak is, based on the conversion to Hartmann-Neukum ages, within 4.1–4.25 Ga, whereas the LHB peak on the Moon is roughly dated at 3.8–4.0 Ga (Tera et al., 1974).

Of the 560 basins with diameter > 200 km, the great majority (>85%) have little or no characteristic circular, topographic relief. In the Martian highlands, buried basins outnumber visible basins by a factor of six. In the lowlands, buried basins are 20 times more abundant than visible basins in this size range. A significant number of very large basins (D > 1000 km) exist, roughly equally divided between the two hemispheres. The most obvious of these are Hellas, Argyre and Isidis (all in the highlands). Others, not so obvious and now recognized as giant impacts, include Chryse, Acidalia and Utopia, in the northern lowlands (Frey, 2006). These large basins were probably filled with water and sediments by the time of Hellas impact.

Importantly, on Earth has been demonstrated that asteroid and comet impact events substantially altered the localized availability and flow of liquid water. Impact basins can localize water due 1) circulation of water into hydrothermal systems, 2) ponding of water in the hydrologic depression of the crater due to excavation below the water table, 3) entrapment of water in the fractures of shocked rocks, or 4) melting of existing frozen water sources by the het of impact. Interestingly, from the onset of high temperature impact to the sustained existence of crater lakes and water-containing impact-fractures over several millions of years, craters can provide locally potential environments for the emergence of life.

However, on the other side of the coin evidence of large impacts might show evidence of annihilation events of emergent life, which is recently a hotly debated topic (Abramov and Mojzsis, 2009).

Axis Obliquity

Today, Mars's obliquity, the angle that the axis of rotation of the planet forms with the vertical to the plane of its orbit, is of 25.19º, very similar to that of the Earth (23.45º). This similarity is once again fortuitous. In the case of Earth, due to the presence of the Moon, the obliquity suffers no great changes throughout time (~ 3°) (de Pater and Lissauer, 2007; Faure and Mensing, 2007). Mars on the other hand, due to the fact it does not have a massive moon to stabilize its axis; it can have an obliquity which goes from 0º to 80º (Barlow, 2008). Over geologic time these changes have had an important impact on the Martian climate which would affect any biota that might have evolve at any given time. For example, at moments when the Martian axis has an obliquity of ~ 80°, each hemisphere would have daylight with the length of half a Martian year and a night of the same length. If during this half Martian year of daylight the temperatures were sufficiently high for the presence of liquid water and the eventual development of life forms, these same life forms would have to face up to the other half Martian year at night, where temperature would fall well below freezing point. However, the possibility does exist that the atmosphere with a maximum temperature at the subsolar point and a minimum at the antisolar point would form important convection cells, like those observed in the thermosphere of Venus, (Milone and Wilson, 2008) between the daylight and the night sides. These thermal cells would transport heat to the dark hemisphere moderating its temperature and helping to maintain liquid water for longer periods of time around the planet.

3. Tectonothermal History of Early Mars

Volcanism (the extrusive component of magmatism) is the main agent of surface change on planetary surfaces, and it comprises all natural processes that involve the presence of molten rock at or near the surface; this definition is used here to frame the nature of volcanism in early Mars as a fundamental process for its thermal evolution and the possible origin and conservation of life in that planet. However, because Mars in the Noachian-Hesperian transition periodically may have enjoyed rather gentle environmental conditions such as abundant and enduring liquid water, as well as a dense, hence warm atmosphere, all of which in many ways may have resembled those of the present Earth (Jakosky and Phillips, 2001; McKay and Stoker, 1989), the timing and inherent problems of measuring the absolute ages of events in Mars are briefly analyzed below.

Mars Chronology

The measurement of relative time for the geologic history Mars relies on stratigraphy, a discipline that simply tells us that rock-forming events are orderly preserved in lithological units, with the lower member at the bottom and the younger at the top of any rock column. Absolute time, on the other hand, is measured on the basis of decay with time of certain radioactive elements present in many rocks. However, because rocks from Mars are only available as meteorites, whose stratigraphic position in the rock record is unknown, the building of an absolute time scale for Mars has relied on indirect methods essentially related to crater-counting, and its comparison with the time scale built on that basis for the Moon, where abundant rocks, and glasses have been isotopically dated and whose stratigraphic position is well known. Figure 46 shows time scales for the Earth and Mars, the divisions of which are based on significant geologic events that marked globally the surfaces of the planetary bodies. However, in the case of Mars, the scales for the limiting eons may vary greatly and up to 2 billion years (Doran et al., 2004), due to poorly known factors such as impact rates through time, secondary cratering, and periodic resurfacing of the planets by erosion, volcanism and sedimentation. Nevertheless, for the general purposes of this work, the figure shows a recently revised range for the main stratigraphic limits proposed for the divisions of the Mars geologic history (Hartmann, 2005), in comparison with the time scale for the Earth, for which the post-Hadean absolute timing has been much better determined (Gradstein et al., 2004)

The geology of Mars has been divided in three chronostratigraphic systems: Noachian, Hesperian, and Amazonian, estimated to occupy about 50, 25 and 25 % respectively of the Martian surface, in marked contrast with the Earth, whose approximately equivalent epochs Hadean, Archean and Proterozoic-Phanerozoic extend over quite different proportions of our planetary surface; in particular, the Hadean system is virtually absent from the Earth's register, whereas the Noachian is the dominant rock system of Mars. Note that the time boundaries of their divisions in the two scales do not match because the history of the Earth is based on orogenic episodes caused by the continuous action of plate tectonics, as well as on the history of life in changing the face of the Earth, whilst the geologic history of Mars, where these factors had a minor role if any, is mainly based on its cratering register and the development of fluvial network systems presumably caused by large amounts of liquid water running from time to time across the surface of that planet.

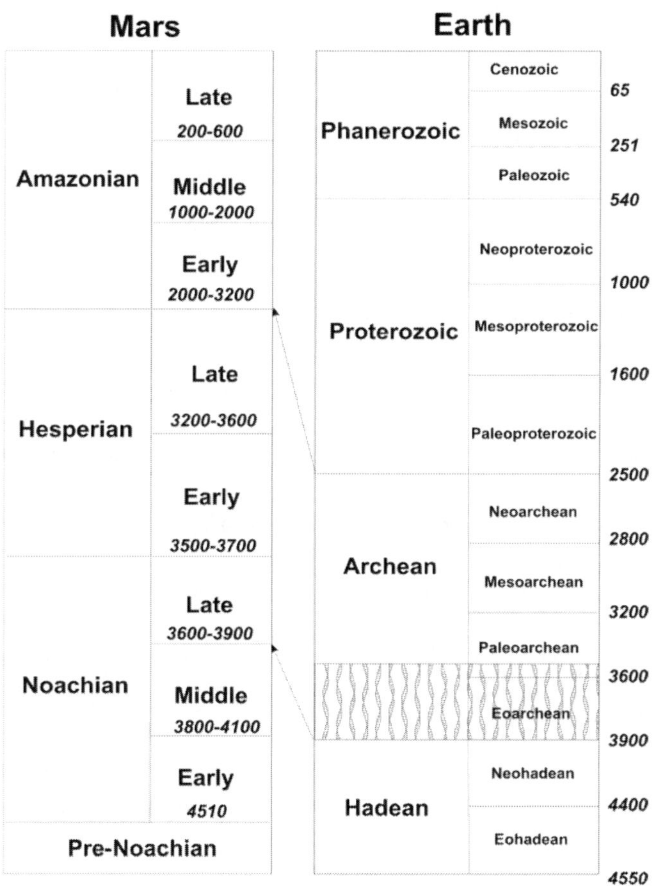

Figure 46. Compared geologic time scales for Mars and Earth, in which the Eon and Era time divisions are based mainly on major events that marked their respective geologic evolution. The time limits for Mars were taken from Hartmann (2005), and from Gradstein et al. (2004) for the Earth. The last two eras (Neohadean and Eohadean) of Earth's history are proposed in this work, with a limiting age at 4.4 Ga, as suggested by the oldest crystallized mineral phase (zircon) so far dated in our planet (Wilde et al., 2001). The ornamented band corresponds to the epoch when life probably appeared on Earth.

Magma Ocean

Terrestrial planets including the Moon and probably Mercury appear to have transited through an epoch when a global ocean of magma dominated their thermal evolution (Wood et al., 2006). Mars is not the exception in spite of its embryonic dimensions, and the importance of this process for the eventual emergence of life in the Earth, Venus or Mars rests on the role that it had for degassing the planet, to form an atmosphere and other hospitable environments where liquid water was able to exist for considerable periods of time. On Earth, a magma ocean probably existed a few tens of million years after the beginning of accretion as a consequence of precipitation of its metallic core, and it continued to exist or was regenerated after the formation of the Moon (Wood and Halliday, 2005). On Mars, a similar early magmatic evolution is deemed possible based on theoretical grounds (Elkins-Tanton, 2008). Apparently Mars accreted within 29 My of the origin of the Solar System, based on Nd isotopic studies of Martian meteorites (Harper et al., 1995) and, although the red planet underwent no further accretionary impacts as those which formed the Earth-Moon system, the separation of the core of Mars would have melted part of the silicate mantle and formed a magma ocean. The presence, volume, and duration of an early magma ocean on Mars, as on other terrestrial planets, depend on many factors, but the accretion rate was crucial for the process to be affective for forming a magma ocean hundreds or thousands of kilometers thick. The time taken for cooling of the igneous ocean, where life simply could not emerge, depends on the mass of the atmosphere present above the surface and may vary from a few thousand to a few million years.

The relevance of magma oceans for the origin of life rests specifically on their capacity to dissolve and then degas volatiles essential for organic chemistry processes such as water, carbon dioxide, nitrogen and sulfur dioxide. Modeling by Elkins-Tanton (2008) arrived to the conclusion that solidification and cooling of a 2,000 km thick hydrous magma ocean on the Earth or Mars takes place within 5 Ma, and only 0.1 Ma if anhydrous. This process is capable of producing thick atmospheres (up to 100 bars) by degassing water and CO_2 from original concentrations of those volatiles in the magma as low as 0.05 %. Thus, organic chemistry leading to the possible origin of life on Mars could have taken place very early in the history of the planet, when it had a thick atmosphere and a warm surface that permitted the existence of abundant liquid water for relatively long periods of time. Table 3 illustrates the main magmatic events and structures that have marked the evolution of Mars throughout its geologic history spanning from its accretion about 4,550 Ma (pre-Noachian) ago to the present time (Amazonian). We note that this chronological chart of the main magmatic provinces of Mars is only indicative, and the bars depict epochs when the volcanic province was fully active, but does not include much younger or earlier but and volumetrically insignificant events, some of which may extend to very recent times of the Martian history.

Characteristics of Martian Volcanism

Post-magma ocean volcanism played a fundamental role in the emergence and development of life on Earth and probably in other planets. Natural phenomena associated with this process include degassing of volatiles essential for prebiotic chemistry (H_2, N_2, H_2O, H_2S, NH_3, CH_4, CO, CO_2, and HCN), hydrothermal activity, hydration of silicates, fracturing, and heating of the upper crust and lightning in the atmosphere. All of these factors are beneficial for the emergence of life on planetary bodies provided that other conditions occur, as discussed elsewhere in this chapter. In fact, the common belief that the last universal common ancestor (Dohm et al., 2009) on Earth is placed towards the bottom of the life tree, and that it probably belonged to the thermophile-type organisms (Stetter, 2006), stresses the relevance of volcanism in the early history of life in our Solar System. However, if the primary atmosphere was lost from Mars by giant impacts and enhanced solar wind in the absence of a magnetic field, it seems that post-magma ocean degassing from volcanoes throughout its history even if it was not sufficient to preserve true water oceans for millions of years in the Noachian, the existence of lakes where life could have emerged and evolved given other favorable conditions, has been widely documented (Cabrol and Grin, 2010; Fassett and Head, 2008).

On Earth, volcanism acquires different types that may be associated with two major eruption mechanisms: effusive (lava fields and shield volcanoes), and explosive (pyroclastic). The first type is associated with low viscosity, hot magmas most commonly of basaltic composition forming the oceanic crust and other igneous provinces of the Earth. Pyroclastic rocks occur when the main mode of volcanism works through a central edifice with explosions in their summits or flanks, or a through fissures open in the crust, accompanied by the violent extrusion of magmas rich in volatiles, essentially water, carbon dioxide, and sulfur dioxide. In fact this later type of volcanism requires abundant fluids in the interior of a planet. However, central volcanism in Mars (subdued basaltic shields and paterae) is substantially different to subduction-related volcanism on Earth (conspicuous chains of stratovolcanoes of andesitic composition). Nevertheless, the high water content of CNC meteorites (Zimbelman et al., 1991) and geomorphologic studies of volcanic zones imply extensive explosive eruptions in Mars, process that is also favored by its diminished gravity force. Central edifices in Mars include the gigantic shields of the Tharsis Bulge (Arsia, Pavonis and Aescreus), and the Olympus Mons, together with huge, dish-like structures such as Alba Patera, whose summital craters (calderas) and surrounding debris aprons suggest a major component of explosive volcanism in the history of Mars (Hynek et al., 2003), and hence a constant and prolific source of volatiles to the atmosphere.

However, the composition of volcanic gases in Mars probably differed (more SO_2-rich) from that associated to volcanoes of the Earth (Halevy et al., 2007; Wanke et al., 1994). This is supported by the sulfur-rich composition of the Martian regolith and dust, and the apparent absence or scarcity of carbonates, which has been explained by the acidity of the environment associated with high SO_2 and H_2S degassing of Mars (Gaillard and Scaillet, 2009). Tharsis volcanism, for example, contained an average of 10-100 times more sulfur than magmas of the Earth (Gaillard and Scaillet, 2009). These, sulfur-rich system would produce an equivalent of 20-60 meter sulfate layer across the entire surface of Mars, and will create chemistry with abundant sulfurous species favorable for the emergence of life (Halevy et al., 2007).

Plate Tectonics on Mars?

Plate tectonics on Earth has played a crucial role in the appearance of life and its further evolution through its inherent global mechanisms of mobility and recycling of materials, including the volatiles essential for life, particularly water. Andesites on Earth bear undisputed evidence of hydrous magmas that evolve from the upper mantle and the lower crust in subduction zones, which are global structures always associated with plate tectonics. Thus, the reported presence of andesites in the highlands and lowlands of Mars (Bandfield, 2002), and the Mars Pathfinder "andesites" (Nikolaeva and Abdrakhimov, 1999) should in principle indicate similar Earth-like tectonic regimes for Mars. However, the "andesite" spectra found in the northern lowlands of Mars have been considered an artifact of aqueous alteration of basaltic surfaces (Wyatt and McSween, 2002), and evidence from meteorites (shergotites) and geochemistry strongly suggest that the Martian crust is essentially basaltic (Mustard et al., 2005). This dominant crust has been identified in many places such as the picritic basalts (20-30 % olivine) found in Gusev crater (Gellert et al., 2006; McSween et al., 2004; McSween et al., 2006), and several other regions of Mars (El Maarry et al., 2009). The assumption of plate tectonics activity on Mars is based essentially on, a) theoretical analysis of the cooling history of that planet from a magma ocean stage to a thick solid lid (lithosphere) that initiated the process of subduction once the magma ocean solidified completely (Nimmo and Stevenson, 2000; Sleep, 2000), b) the spectacular display of remnant magnetized crustal bands in the southern highlands (Connerney et al., 1999), and c) the apparent geologic structures in the surface of Mars, which are typical products of plate tectonics on Earth (Anguita et al., 2001). Naturally, if this tectonic regime indeed was real for the Noachian Mars, the inevitable consequence would be that life may have flourished abundantly in early Mars. Baker et al. (2007) concluded that plate tectonics acted only during the LHB, and not into the Middle Noachian as proposed by Sleep (1994); nonetheless, this later view was implicitly accepted by Baker et al. (2007), who concluded that plate subduction caused cooling of the core, which in turn shot off the magnetic field of Mars, followed by atmospheric loss and a dramatic climate change from a warm-wet to cold-dry conditions after mid-Hesperian times.

Baker et al. (2007) further speculated that an equivalent ocean 130 m deep was transported to the mantle by subduction beneath Tharsis, together with other volatile-rich material such as carbonates and sulfates, and thus plate tectonics should have played a fundamental role in Mars chemical evolution up to the Middle Noachian, forming all kind of compressive structures such as fold-thrust belts and sedimentary basins similar to those in the Earth and during a long period of time extending 500 Ma (4500-4000 Ma) since accretion of the planet. In fact, Sleep (1994) proposed that the Martian dichotomy defined by the thick crust of the southern highlands and the thin crust of the northern plains was caused by subduction of the thick crust and sea floor spreading in the northern region during the Late Noachian and Early Hesperian, with up to four coexisting plates in the "oceanic" crust of the lowlands. The degassing of volatiles transported to the mantle-core boundary by plate tectonics in Mars would have occurred from huge plumes of partially molten hydrous mantle that approached the surface, episodically returning volatiles to the atmosphere. These volatiles, in the absence of further plate tectonics action in Mars, remained trapped in permafrost of the Martian soils.

Indeed, plate tectonics in early Mars, to be validated, would need to address these critical questions: 1. How abundant, if any, are andesites in Mars? 2. Are there rocks and structures in Mars that are typical of Earth subduction zones such as volcanic arcs, suture mélanges, mylonites or ophiolites, including their characteristic pillowed basalts? In fact, recent models on the subject do not accept true plate tectonics for Mars at any time in its history (Breuer and

Spohn, 2003; Harder, 2000; Neukum and Hiller, 1981), while many others take a more conservative view regarding the timing of the process (Lenardic et al., 2004). Thus, the eventual finding of abundant andesites in Mars would constitute a crucial element for the geochemical and tectonic evolutionary models of that planet. If only basalts were present, volcanism would have degassed most water, together with sulfur and carbon dioxides, directly to the atmosphere, but if abundant andesites were also present implying plate tectonics and subduction in early Mars, recycling of water and other volatiles from the mantle and crust to the former atmosphere would provide more favorable conditions for the inception of life.

Volcanism, Volatiles, Minerals, and the Carbonate Paradox on Mars

Volcanism is always accompanied by hydrothermal activity, and this process alters the primary mineralogy of the solidified magmas.

Table 4 shows the kind of secondary minerals that may be derived under variable physicochemical conditions at the surface of present or ancient Mars from the alteration of volcanic rocks, considered mainly of basaltic composition.

Table 4. Primary igneous phases and their secondary minerals caused by hydrothermalism under varying physicochemical conditions in the surface of Mars

Primary basaltic	*Main secondary minerals that may develop from the primary basaltic phases*					
phases	*Clays*	*Zeolites*	*Serpentines*	*Carbonates*	*Silica*	*Hematite*
Olivine (Fo_{90})	═══		═══			
Orthopyroxene	═══		═══		═══	═══
Clinopyroxene				═══		
Plagioclase (An)						
Fe-Ti oxides	═══	═══		═══	═══	═══
Glasses	═══	═══		═══	═══	═══

The actual identification of all the secondary minerals listed in Table 4 (Bibring et al., 2006; Poulet et al., 2005) in the surface of Mars implies the prolonged action of hydrothermal fluids, essentially water, in the upper crustal levels and hence the high probability for chemical evolution favorable for the generation of living systems.

Tharsis Superplume

By far, the Tharsis magmato-tectonic structure (see (Johnson and Phillips, (2005) for a recent summary) is the largest expression of central volcanism in Mars (Dohm et al., 2001; Wilson et al., 2001) that delivered magmas and volatiles to the Martian surface since the Noachian; it extends over 25 % of the surface of the planet in its western hemisphere and it is constituted by up to $3x10^8$ km^3 of volcanic products (Phillips et al., 2001) extruded during a lifetime of at least one billion years.

Independently of its debated origin (first order mantle convection, multiple plumes, or impact-related superplume), a geologic system of this magnitude, undoubtedly provided unique scenarios for the inception of life by degassing abundant H_2O, SO_2, H_2S and CO_2 to the vanishing atmosphere, and creating probably thousands of hot springs within a tectonically fractured and wet crust, thus accelerating the pace of chemical evolution compared to other regions of Mars where this heat and chemistry were not available.

If superplumes developed early in the Mars history fueled by formation of the core, then the massive access of hot magmas to the surface would cause prolonged excursions of a warm climate in the otherwise cold planet that prevailed since the Late Heavy Bombardment.

The Tharsis bulge includes three major central volcanoes that from north to south are Ascreaus Mons, Pavonis Mons and Arsia Mons, but Olympus Mons and Alba Patera in the periphery of the bulge may also form part of the structure. These volcanic constructs rise over 12 km from the reference Martian level and the region was the provider of aqueous fluids and potential energy necessary to carve the huge canyons associated with Valles Marineris later in the Amazonian.

The Tharsis petrological composition is essentially basaltic, with mean silica (SiO_2) content of 42.0 wt. % (Baratoux et al., 2011), namely, several units below the silica mean composition of basaltic rocks on Earth (47-52 wt. %). The reason for this is not clear, but one possibility is that Martian basalts are much richer in sulfides and iron oxides than those of the Earth.

The Late Noachian inception and further evolution of the Tharsis structure represented the most important product of cooling and degassing of the planet, and the process may have continued to recent times, as evidenced by ages of basaltic Martian meteorites as young as ca. 500-175 Ma (Borg et al., 2003; Nyquist et al., 2001). Tharsis, together with the Sun, may have brought life to Mars at one time of its history.

Summarizing, the most relevant role of volcanism for the origin of life in the terrestrial planets is as a powerful provider of volatiles (mainly water, CO_2, and SO_2) essential for pre-biotic chemical evolution and eventually life.

Craddock and Greely (2009) calculated that about 0.8 bar of these species were degassed from all Martian volcanoes to the atmosphere from the Late Noachian to the Amazonian, but according to Carr (1986) Mars interior outgassed up to 20 bars of CO_2, which is equivalent to hundreds of meters of globally spread layers of carbonates.

And yet, because Mars had abundant liquid water in the Late Noachian, and this condition requires minimum atmospheric pressures of 2-3 bars, with CO_2 as the most abundant volatile species in the primitive atmosphere, a major problem arises because C or CO_2 have not been found in sufficient amounts on the surface or in the atmosphere of Mars to account for its abundance in the past.

Several possibilities may exist to resolve this paradox: a) Mars never had a thick atmosphere, b) its composition was not rich in CO_2, c) CO_2 was lost from the planet (impacts and solar wind), and d) CO_2 was sequestered in the subsurface, where it would be still present in the shallow crust as carbonates (Morris et al., 2010) methane clathrates in the permafrost, or as carbonate scapolites in the deep crust. On the other hand, the abundant sulfur dioxide and H_2S, also degassed from Martian volcanoes but now absent from the atmosphere, may be easily accommodated by the global layers and hydrothermal deposits rich in several hydrated sulfates of calcium, iron and magnesium identified in the surface of Mars (Bibring et al., 2006).

4. LIQUID WATER ON MARS

Early Presence of Water on Mars

As many other planetary bodies in the Solar System, Mars accreted water from icy bodies and chondritic materials in its primitive geospheric and atmospheric realms (Leshin, 2000; Leshin et al., 1996; Owen et al., 1992). Degassing and condensation of this water would be distributed in different proportions and physical states (solid, liquid, and gaseous) in the early atmosphere, the surface of the planet, and the planetary interiors. The present Mars still holds part of that early water inventory (Feldman et al., 2004) although much has been lost into space (Jakosky and Phillips, 2001).

It is likely that Mars, another water world of the inner Solar System, had greater atmospheric temperatures and pressures, and small and big water bodies could have existed. Nonetheless, sedimentary and geomorphological evidence to support true oceans (with currents, tides, recycling, etc.) is lacking. Overall, the presence of an early hydrosphere in the Noachian of Mars is revealed essentially by the relative abundance of water-formed geomorphological features (water-carved valleys) on the oldest terrains, and minerals and sedimentary deposits that imply liquid-water chemistry for their formation [e.g. carbonates, sulfates, and phyllosilicates; (Di Achille and Hynek, 2010; Ehlmann et al., 2011a)].

In the Noachian terrains, dendritic valley networks are widely distributed, but are rare in younger terrains (Hesperian and Amazonian) (Luo and Stepinski, 2009). Thus, climatic conditions in the Noachian were radically different from the Hesperian and Amazonian. These networks of dendritic valleys may have been formed by rain or snow melting. However, some valleys are abruptly terminated, suggesting collapse via underground melting or sapping (Hartmann, 2003). In both cases, liquid water flowed on the surface of Mars during the Noachian period. Interestingly, the rate of discharge (~ 300-3000 $m^3 s^{-1}$) is in the same order for common terrestrial systems (Barlow, 2008; Irwin et al., 2005).

The onset of the widespread fluvial erosion that formed valley networks on Mars must have occurred after the Late Heavy Bombardment (LHB), once water could have condensed at the surface. The oldest evidence of liquid water on Mars (based on crater counting – diameter and frequency age estimations) comes from widespread valley networks that cross-cut highland Noachian terrains, and may be as old as ~3.9 Ga (Fassett and Head, 2008), although much older terrains may have been eroded by water as well [e.g. ~4.2 Ga Amenthes and Zephyria craters; (Frey, 2006; 2008)]. Nevertheless, obliteration by impacts and partial burial makes this interpretation less concise.

The greatest activity of liquid water at the Martian surface seems to have lasted from the Noachian to the Hesperian eons (> 3 Ga); however, whether this liquid condition was sustained or episodic is unknown. Because terrestrial life is expected to have originated in a liquid matrix, two essential conditions should be met for life to have originated elsewhere: that water remains liquid, and that time allows for chemical and biotic evolution to occur while this state prevails.

Estimates of water abundance propose equivalents of global depths of ~40 m (Greeley, 1987) to >100 m of water (Carr, 2006), and up to ~1000 m (Durand-Manterola, 2003), consistent with volume estimates of ~10^7 km^3 (Carr and Head, 2003; Durand-Manterola, 2003). The clue for the inception of life on Mars lies on whether this water reservoir could have remained liquid for at least 10 Ma years, which may be a subjective time period for a theoretical biogenic minimum (Lazcano and Miller, 1994; Oberbeck and Fogleman, 1989), or a contrasting figure for the 300-400 Ma between the end of the LHB and the oldest Earth's fossils (Lahav, 1999).

Role of Liquid Water for the Emergence of Life

For chemical evolution to occur fast enough so the time is right for origination of life, the ideal state of water is liquid, because this state is the medium for stabilization, transportation, delivery, and promoter of chemical reactions in life's biochemistry. In a cold (< 0 °C) planet, molecules could hardly interact with each other and assemble complex structures and protocells within a matrix of permanent ice. In the atmosphere, molecules and suspended solids can travel and interact very rapidly, but solids and condensed liquids would tend to settle down onto the surface and would leave the atmosphere without these potential life-constituent reactants. Instead, in a liquid phase on a rocky planet, minerals and organic molecules can coexist and interact with each other easily and permanently at various scales, and under different physical (depth, currents, turbulence, flow velocity, temperature) and chemical conditions (pH, salinity, dissolved gases, redox conditions). Thus, the preferable scenario for life's development would be either at the gaseous-liquid-solid, the ice-liquid-solid, or the liquid-solid interface. This includes niches in shallow waters and near-shore environments (either supported by a dense and greenhouse atmosphere or by surface hydrothermal convection), salty brine waters underneath permanent ice covers, and deep hydrothermal systems. The latter scenario can also sustain liquid water circulation regardless of cold surface temperatures, thin atmospheres, and lack of magnetosphere. It is possible that even today these systems are active on Mars where geothermal gradient is locally enough as to maintain temperatures above 0° C around water sources (permafrost).

Paleohydrology of Mars

Currently, the presence of abundant frozen water on Mars is beyond question and recent spacecraft are unveiling the manifold appearances of Martian water largely hidden from visual inspection (Bibring et al., 2006; Boynton et al., 2002). Although Martian water is not liquid today, and has probably not been so throughout much of its history, bodies of liquid water likely existed at other times in different reservoirs, with important biological implications. Periodic liquid water on Mars could have provided opportunities for biogenesis, as well as evolutionary progress at the surface during short-lived climatic perturbations. Although our knowledge about water on Mars is certainly incomplete, we may gain further insight through future Mars missions, and observational, experimental and/or theoretical studies.

It is consensus that liquid water, standing and flowing, existed in extensive regions of Mars during the Noachian and Hesperian. Although the exact functioning of the early Mars hydrology and the initial water inventory are under study, certain clastic and chemical deposits that necessarily require liquid water for their formation can be recognized all across the Martian surface. This cumulative evidence includes lakes that formed after impacts or by ponding, channels, gullies and valleys carved by water (Carr, 2006), clay formation and alteration in aqueous solutions (Ehlmann et al., 2011a; Ehlmann et al., 2011b; Mustard et al., 2008), and weathering and chemical sediment formation [e.g. carbonates, sulfates, silicates; (Grotzinger et al., 2011; Morris et al., 2010; Squyres et al., 2008). In fact, episodic processes driven by ice and transient liquid water seem to have endured all throughout the Amazonian (Dickson et al., 2008) and may be still going on today during 'hot' Martian summers.

Despite the presence of liquid water on early Mars, some geomorphological features and sedimentary processes involving water may occur geologically fast (<10,000 years) or episodically (Harrison and Chapman, 2010), so the time length at which water on Mars stayed liquid may have been geologically short. In this case, if life requires liquid water continuously for millions of years to emerge, Mars would have been sterile under this scenario of pulsed and short-lived, rather than continuous liquid water. Permanent sources of liquid water would depend on the atmospheric conditions (greenhouse gases and pressure) and the presence of active heat sources (impacts, magmatism, volcanism, hydrothermalism, radioactive decay) that could melt and re-circulate water reservoirs. Some authors estimate that water could have remained liquid at the surface for 260-700 Ma (McKay and Davis, 1991). However, the timing for the formation of geomorphological features and mineral deposits on Mars can only be estimated today with ample range of error, even for thick sedimentary sequences that can be deposited geologically fast (<10 Ma). Note that even for stratigraphic sequences for which estimates consider tens of millions of years of formation, hiatuses have not been assessed. Thus, a thick sedimentary sequence could have been formed in pulses rather that in a single, continuous event. The same may apply for hydrothermal systems, which tend to exist for >1 Ma on Earth (Cathles et al., 1997). Nevertheless, the different geology of Mars and its overall lack of Earth-like tectonics may have resulted in plumes or hot spots (such as Olympus Mons) that could have maintained hydrothermal systems active for hundreds of millions of years since the Noachian. These magmatic/volcanic systems could have provided adequate conditions for life to have arisen.

Water Chemistry and Niche Heterogeneity

The first hydrologic systems on Mars must have been originated by degassing of volatiles from the mantle into the atmosphere as in other rocky planets. Volatiles (e.g. CO_2, H_2O, SO_2) degassed this way would interact with each other and form new compounds as water condensed and reacted with the dominantly-basaltic Martian surface. The latest spectroscopic observations reveal that, overall, the aqueous alteration and weathering of the original basaltic crust produced large amounts of phyllosilicates (Bibring et al., 2006; Ehlmann et al., 2011a; Milliken et al., 2008; Mustard et al., 2008). Their formation implied the release of cations that would later, toward the Hesperian, form vast amounts of salts (e.g. sulfates, chlorides, halides), enhanced by volcanic degassing of SO_2 (Barlow, 2008), the main constituent of Hesperian sulphate deposits, and evaporitic processes. Particularly, clay minerals distributed on Noachian terrains suggest that liquid water tended to be enriched in soluble ions (e.g. K^+, Na^+, Ca^{2+} or Mg^{2+}), leaving less soluble ions (e.g. Fe^{3+}, Al^{3+}, and H_4SiO_4) to precipitate as phyllosilicates and oxides (Chevrier et al., 2007). This early chemistry seemed to have occurred in slightly acidic to alkaline waters (pH 4 to 12) and variable temperatures (Chevrier et al., 2007). A thick, CO_2-rich atmosphere (along with H_2O and SO_2) is conceived for the early Mars. Under these conditions, carbonates likely precipitated widely within Noachian water bodies (Morris et al., 2010), but may be hindered from spectroscopic observations due to burial under a dust cover (Bishop et al., 2011). These carbonates would be an important reservoir of atmospheric CO_2.

Another early alteration product on Mars surface was the oxidation of iron. From the Middle to Late Noachian, water reacted with Fe-rich species to form oxides [e.g. $Fe^{2+} + H_2O \leftrightarrow Fe^{3+} + \frac{1}{2}H_2 + OH^-$, followed by $Fe^{3+} + H_2O \leftrightarrow FeOH^{2+} + H^+$; (Chevrier et al., 2007; Dreibus and Wänke, 1987; Lammer et al., 2003)], which are pervasive in the Martian regolith. Iron likely reacted also with Cl^-, Br^-, OH^-, as well as weak anionic ligands, such as NO^{3-} and ClO_4^- that hydrolyze in aqueous solution forming $FeOH^{2+}$ and hydroxyl-type complexes; from this state, further oxidation to $Fe(OH)_3$ makes Fe an insoluble colloidal suspension (Rabinowitch and Stockmayer, 1942). Oxidation reactions of Fe suggest that the resultant gaseous H_2 was gradually lost into space, mainly by solar wind and impacts. This implies a net loss of water that would increase concentration of solutes and overall salinity in the early Martian waters (Knauth and Burt, 2002). It is unlikely that all the water bodies in the Noachian were homogeneous in composition. Different terrains would provide different primary constituents, as they were already heterogeneous in essence, and climatic changes would influence the distribution and recurrence of confined and unconfined water bodies.

Assuming that a thick, greenhouse-rich atmosphere endured all throughout the Noachian (see above), precipitation, lake formation and fluvial discharges must have been active processes. This implies movement and transport of material, as well as weathering and alteration of pre-existing materials. Organic and inorganic materials, interacting locally in a confined area, could also be transported long distances by running water and promote heterogeneity of environments and chemistry. This in turn would increase the possibility of more and more heterogeneous microenvironments, boosting the chances for chemical evolution to encompass a greater variety of possible reactants that could potentially result favorable for life origination and maintenance. In case that life emerged on Mars, it should have been necessarily adapted to pre-existing conditions of that particular environment, and probably evolved, adapting itself to the changing climates and local conditions. By the end of the Noachian and the beginning of the Hesperian (~3.7 – 3.5 Ga), wet conditions apparently persevered, as indicated by extensive chemical sedimentary deposits (Cabrol and Grin, 2010). However, global climatic conditions, likely triggered by the loss of the magnetosphere and the atmosphere, seem to have become dry and cool, which also influenced Mars hydrology, turning water more acidic and changing its state from liquid to an even more icy state most of the time, with increasing ice-dominated processes, sapping, and mass flows, responsible for the formation of features that do not require a permanent liquid-water component (Pieri, 1980).

Implications of Liquid Water for Life

It is clear that Mars went from a denser atmosphere to a thinner one, from more magmatism and volcanism to lesser activity, from more ice and liquid water at the surface to an evaporitic drier system. All this happening on a planet that, although with signs of tectonism, apparently did not develop plate tectonics as on the Earth and, therefore, did not have such concomitant planetary consequences, including crustal recycling and rehydration of the mantle. The loss of magnetosphere to preserve its atmosphere, maintain humid environments free of cosmic radiation reaching the surface, the low planetary gravity, and the lack of tectonics, set apart Mars from Earth. Although this dims the chances for heterogeneity and diversification of niches where life could have developed on Mars (as compared to the more surface area, amount of water, tectonics, and lunar and magnetic field protection of Earth), life may have occurred in warmer, subsurface environments where water could have existed liquid over geologic times [e.g. hydrothermal systems, post-impact basins; (Cockell, 2006)]. Nevertheless, if life needed diversity of environments, movement and recycling of the continental crust, vertical and horizontal local and global redistribution of water, continental-scale topographic changes, wind currents, a humid atmosphere, and an abundant liquid hydrosphere, Mars would not be a candidate for habitability. On the other hand, if life on Mars ever came to existence, and if such life was biochemically similar to the one we know from Earth, then it surely started as prokaryotic, diversified rapidly, and developed adaptability to the gradual changes occurring in the respective ecological niches. The fossil record on Earth shows that massive extinctions have a decreasing effect of diversity only on macroscopic life, but surely not at the same degree on microbes. Thus, even with global cataclysmic changes of extreme drying and cooling on Mars, it is possible for life to have adapted to ecological niches that may even exist today (e.g. in the sub-surface). On Mars, however, less heterogeneous niches may have implied less biotic diversity and, thus, perhaps less chances of long-term survival under current planetary conditions. Nevertheless, if these putative microbial communities were eventually transported by underground water movement or at the surface by flows and winds, then mixing, competition, and perhaps conjugation may have taken place, which in turn could have promoted the acquisition of novel genetic tools and, thus, the evolution of more plastic metabolisms.

Life on Earth has shown to be extremely rapid in evolving and adapting to changing environments, and also hard to eradicate. The almost continuous fossil record of prokaryotes and eukaryotes since their appearance ~3.5 Ga ago, and the magnificent examples of life diversity we see today in every possible niche on Earth, testify for the endurance and adaptability life has as a whole. However, life has left an imprint since the earliest geological record, which speaks of abiotic and biotic physical and chemical exerts functioning at the same time. If life ever existed on Mars, we should sooner or later be able to find this type of imprint or biosignatures in that planet.

5. Surface Chemical and Mineral Evolution

The search for life in outside the Earth is based on the assumption that life is the consequence of some physical and chemical specific conditions, if these conditions are fulfilled, life emerged. Another assumption is that life leaves undoubted remains and imprints of its presence (Southam et al., 2007).

One of the main scientific interests regarding Mars is the possible existence of life past or present. There are three hypothetical possibilities in this context: first, life emerged on the planet but it went extinct. Second, life emerged and continued its evolution, and third, life never emerged on the planet. The understanding of the conditions that could have facilitated or impeded the emergence of life on Mars is of capital concern to accept or discard these hypotheses. In the first two cases, if life emerged on the planet, the identification of potential physico-chemical environments in which life could have appeared is of main importance, and also the understanding of the role of molecular, organic-inorganic interactions in the whole process. Specifically, minerals were likely related to the emergence of life on Earth, and if the same course of events occurred on Mars, minerals may have played a key role for life's emergence there too. The putative existence of habitable environments on Mars and their extent is a main scientific question that must be addressed.

Since the only place in the Solar System that undoubtedly harbors life is Earth, comparisons are made having our planet as reference. If life emerged on Mars but disappeared, it is also important to detect the imprints that it might have left. On Earth, those imprints are recorded on rocks, and as a parallel process, if life ever existed on Mars, the identification of the geologic deposits could potentially preserve this evidence making sampling and widening of Mars analysis a current essential activity. On Earth, this preservation requires specific processes like sedimentation, mainly clays that entombed biological materials (Farmer and Des Marais, 1999). The assembly of organic molecules, their chemical reactions, the catalysis, and the shielding of some molecules, are processes known to be performed by minerals, but once life emerged, biological activity could have also be able to modify the mineral surface. Having this idea in mind, in this section we try to present a review of the minerals that could have been present on Mars during the Noachian and to understand their possible contribution to chemical evolution.

The Mineral Record on Mars

Minerals and their occurrence can tell us about the chemistry, pressure, and temperatures of past environments on Mars and therefore some inferences about the possibility to develop complex molecules and life.

According to the geological record, the history of Mars can be divided in three epochs. First is the Noachian (~4.5-3.7 Ga), characterized by large impact craters and possibly extensive flooding by liquid water (Carr, 1987; Hynek and Phillips, 2001; McKay and Davis, 1991). This period was followed by the Hesperian (~3.7-3.2 Ga), dominated by extensive lava plains, and, finally the Amazonian (~3.2-0.0 Ga), with minimal craters and some lava flows. The remote mineralogical analysis, conducted along years of scientific missions on the planet, indicates that sequential eras can be recognized on its surface. If sedimentary mineralogy is used to understand the history of the planet (Grotzinger et al., 2011) the history of Mars is separated into the Phyllosian, Theiikian and Siderikian Eras (Figure 47). The Phyllosian represents a period with neutral pH and abundant water; during this period large deposits of clay minerals were formed that occurred during the Noachian. In the Hesperian the planet should have been still wet and the pH acidic; those conditions favored the deposit of large amounts of sulfate that corresponds to the Theiikian. Finally, the Siderikian characterized by the formation of ferric oxides; during this period a very dry environment dominated (Bibring et al., 2006).

Minerals evolve, and mineral evolution is a continuum, and at least, on Earth, Hazen et al. (2008) suggested that physical, chemical and/or biological processes diversified mineralogy in three ways: first by increasing the range of composition inherited from the solar nebula; second because the geological processes change the physical and chemical conditions that altered minerals; and finally, by the action of organisms, that allowed the formation of new mineral phases. On Mars, the first two events were accomplished and surely contributed to the mineral evolution (Hazen et al., 2008).

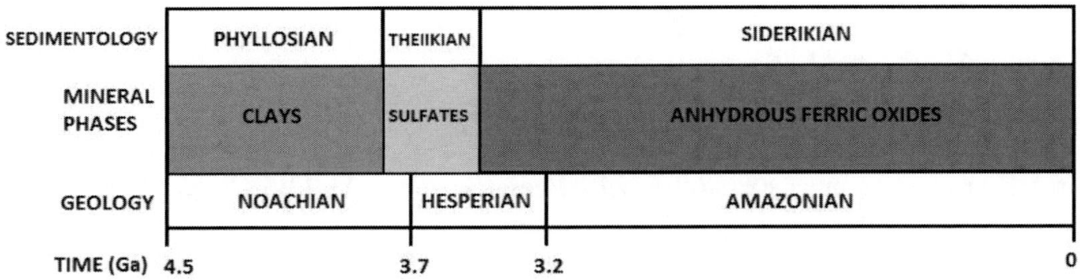

Figure 47. The History of Mars, from the geological (Noachian, Hesperian and Amazonian) and sedimentary (Phyllosian, Theiikian and Siderikian) points of view. Modified from Bibring et al. 2006.

The history of minerals on terrestrial planets starts with the accretionary phase. Our Solar System began approximately 4.57 Ga ago, when a dense core of a molecular cloud, composed mainly of gases and some dust, collapsed and formed the planetesimals that evolved to form the planets (Irvine, 1998). The dust in these clouds is in the form of grains, called presolar, and can be identified because of their anomalous isotopic composition (Nittler, 2003). Their analyses have been possible through the examination of some chondritic meteorites, and it revealed that presolar phases include: diamond, silicon carbide, graphite, silicon nitride, corundum, spinel, hibonite, titanium oxide (Nguyen et al., 2007; Nittler, 2003), and silicates (Messenger et al., 2003). The most characteristic objects of chondrites are chondrules –small, once molten droplets formed in space, in the form of spherical objects, close to 1 mm of diameter (Rubin, 2000), calcium-aluminum-rich inclusions (CAI's), amoeboid olivine aggregates (AOA) and some other refractory objects (Hazen et al., 2008). The analyses show that approximately 60 primary mineral phases are recognized on chondrites, and at least 20 occur in micrometer or nanometer scale grains (Hazen et al., 2008). It is important to mention that most of the 83 stable elements in nature were present at this time reflecting an almost ubiquitous and ancient chemical complexity.

When planetesimals accreted they suffered both aqueous and thermal alterations that lead to a new generation of minerals. In total, at least 250 different mineral species are present in all types of meteorites, the bodies that could have accreted to form the terrestrial planets. In general terms, the minerals brought by meteorites are dominated by magnesium silicates, iron sulfide and Fe-Ni metal. Chlorite, serpentine, talc, and other phyllosilicates are the product of low temperature alteration (less than 100 °C) of olivine and pryroxene in CI-, CM- and CR-type chondrites (Brearley and Jones, 1998), and also hydroxides, sulfates and carbonates, oxides, sulfides, halides and oxyhydroxides (Brearley, 2006). Thermal alteration below 950 °C yielded new species like albite, feldspathoids, micas, amphiboles and pyroxenes, among others. After accretion, the surface of the four terrestrial planets was altered by the crystallization of igneous rocks, originated by internal processes and the continuous bombardment of asteroidal and cometary bodies.

On Earth, the evolution of minerals strongly depended on crust and mantle recycling (plate tectonics), whereas in the case of Mars, evolution was probably different as discussed above. Nevertheless, during the initial history of Mars, the inventory of minerals was perhaps similar to that of the early Earth.

Martian Surface and Its Dominant Minerals

The Martian surface shows abundant traces of volcanic activity, a possible active hydrologic past with liquid water, and a complex mineralogy (Shearer et al., 2008; Skok et al., 2009). The mineral data offered by recent studies (spacecrafts and telescopes) on the Martian surface and other extraterrestrial bodies (Bland et al., 2011), might bring light to ancient environmental conditions on the planets and the nature of extraterrestrial chemical processes. Recent observations on Mars revealed phyllosilicate minerals (*i.e.*, clays) (Fairén et al., 2010; Poulet et al., 2005), opaline silica (Carter et al., 2010), carbonates (Michalski and Niles, 2010; Morris et al., 2010) and others, suggesting a more intricate, wet, and chemically-complex physico-chemical environment. As we point out in our manuscript here, the Noachian epoch is more complex than it was formerly envisioned (Johnson, 2008). The presence of hydrated minerals provides a record of water-related processes, and the presence of phyllosilicates by the OMEGA instrument has been reported over large areas. For some authors (Poulet et al., 2005), this is a clear evidence of the past presence of water because phyllosilicates represent highly altered products involving the action of this liquid (Poulet et al., 2005). For other authors, like Ehlmann et al. (2011), Martian clays were formed by hydrothermalism from groundwater circulation, while the surface was cold, with arid harsh conditions and only transient surface water for over 4 Ga. This implies that an aqueous potentially habitable environment may have been present in a subsurface environment (Ehlmann et al., 2011b).

Mars has a surface area of about 145 million km^2, similar to the continental surface on Earth; thus, the missions only have observed a very small part of the Martian surface. These observations of the Martian solids or the mineral content are the main source of information to explain the potential organic-inorganic prebiotic evolution, and thus are of extreme importance. The missions exploring the surface of Mars give the opportunity to have direct experimental data (Farmer and Des Marais, 1999). The following discussion seeks to bring up mineral groups important for the development of complex molecules and further self-assembled systems as primitive organic complex structures (Brack, 1998). Mineral growth can give information on the grain scale and their fabrics should record a range of processes occurring in the Earth or other extraterrestrial bodies (Bland et al., 2011). Spectroscopy, electronic microscopy and atomic force microscopy studies would give information at the nanoscale for mineralogical features and insights into the potential of catalysis, self-assembly and housing of life possibilities that also might be studied on Mars (Staufer, 2011). The strategy of identifying minerals on Mars brings the possibility to recognize the building blocks for complex chemical evolution.

Plagioclase

This silicate feldspar constitutes the greater part of all basaltic and intermediate rocks on Earth and also in the surface of Mars (Bandfield, 2002). Because this phase is the main bearer of the cations Ca^{+2} and Na^{+} in those rocks, it should be the most important provider of these metals to aqueous solutions during low temperature alteration, from which most carbonates, sulfates, clays, zeolites, and many salts present in the surface of Mars may have precipitated (Gooding and Keil, 1978), indeed enriching the chemistry of most Martian environments.

Olivine, Pyroxene and Serpentinization

High spatial resolution infrared data has been used to characterize Mars surface, highlighting olivine $(Mg, Fe)_2SiO_4$ and pyroxene $(Ca,Mg,Fe)_2Si_2O_6$ as important representative minerals in Martian surface (Hoefen et al., 2003). In other words, olivine and pyroxene are widespread on Mars. Since these minerals cannot remain intact in the presence of water (Olsen and Rimstidt, 2007) Mars must have been dry for a long time. Olivine is unstable in aqueous environments and it is the most soluble silicate mineral in certain organic acids (Barman et al., 1992). Because of these properties, olivine should have played a key role for chemical evolution. Olivine (Figure 48) has been used in simulation experiments to analyze the formation of complex organic matter; for example, in the simulation of hydrothermal vents, Berndt et al.(1996) demonstrated the reduction of CO_2 to short hydrocarbons by olivine serpentinization at 300ºC and 500 bar (Berndt et al., 1996). McCollom and Seewald (2001) studied the reaction of CO_2 in the presence of olivine at 300ºC and 350 bar, and observed the reduction of CO_2 to formate (HCOO-). From another perspective, our experimental work (Colín-García et al., 2011) demonstrated that olivine enhances the decomposition reaction of HCN, acting as a catalyst. This reaction is important due to the formation of some chemical species to promote further reactions.

Olivine and pyroxene also are important for the serpentinization-oxidation process responsible of the production of methane or silica. This process probably had a special importance in the origins of life on Earth (Sleep et al., 2004). Serpentinization is the reaction of olivine and pyroxene rich rocks with water producing magnetite, hydroxide, and serpentine minerals, liberating molecular hydrogen (McCollom and Seewald, 2001) with the potential to further react providing energy to other systems. According to Berndt et al. (1996) this reaction is as following:

$$\text{Olivine } (Fo_{88}) + 1.34\ H_2O = 0.25 \text{ serpentine} + 0.26 \text{ brucite} + 0.08 \text{ magnetite} + 0.08\ H_2$$

The olivine-pyroxene reactions in water are viewed as important sources for potential early chemistry on Mars (Schulte et al., 2006) and as templates for assembly and stabilization of organics in extraterrestrial and terrestrial environments (Benzerara et al., 2005). These minerals originally form from volcanism or impacts (Fairén et al., 2010; Stoeffler and Knoell, 1977), and represent dynamic and diverse mineral phases on the surface of Mars, changing depending on the evolution of the weathering conditions. As a whole, the altered Martian surface roughly had clay-type as the main minerals, especially in the Noachian characterized by a considerable water and CO_2-rich atmosphere. Olivine is a solid and its dissolution processes are of main importance for the chemistry of Mars and other planets, as stated elsewhere (Martel, 2003). Pyroxene, as in the case of olivine, is a common mineral in the Earth´s crust and the Martian surface as an abundant component of basalt. Pyroxene is of main importance in studies of extraterrestrial chemistry because it is used to trace terrestrial objects analyzing the pyroxene/olivine ratio, which gives information of the mineralogy evolutionary time (Karner et al., 2006).

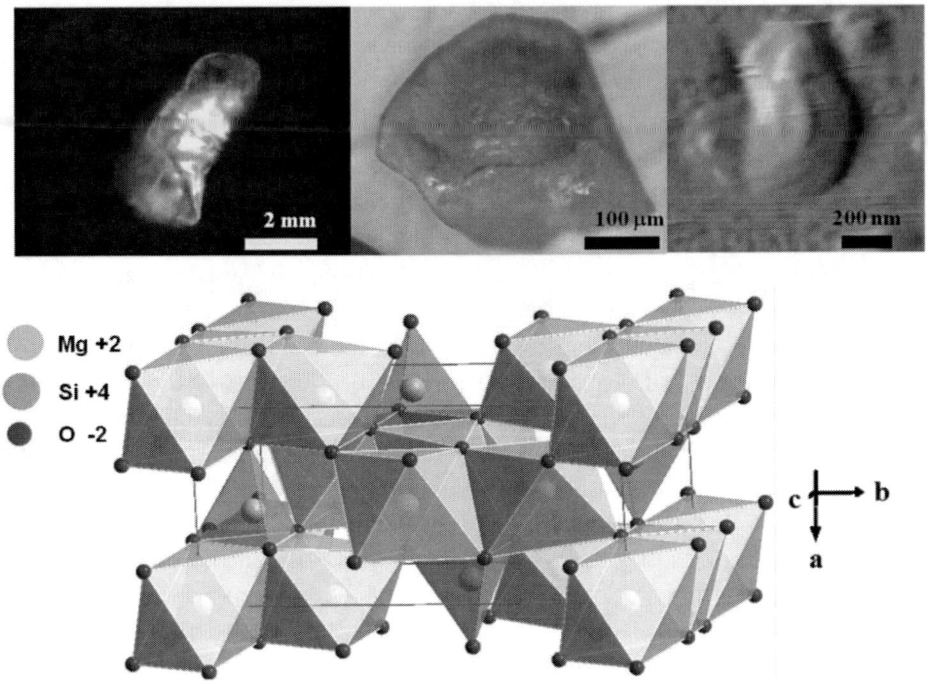

Figure 48. Olivine is one of the most important and representative minerals of Mars. Images of olivine crystals A, B (optical), C (Atomic Force Microscopy), and D structure of the crystal (Images from Maria Colín-García and Alejandro Heredia).

Opaline Silica

Opal mineral or hydrated silica ($SiO_2.nH_2O$) may have a secondary volcanic origin, silica deposits at nanometer scale, forming spheres of SiO_2 when interacting with water. On Mars, one probe found silica in association with volcanic materials, suggesting that water may have existed recently precipitating silica, as it did billions of years ago altering the other silicates such as pyroxene and olivine. This fact extends the possibility of liquid water on Mars resulting in a potential chemical environment to support life or complex chemistry. The silica deposits are present both as light-toned soils and as bedrock, which are commonly formed under active hydrothermal conditions (Skok et al., 2010; Skok et al., 2009; Squyres et al., 2008).

Carbonates

An active discussion and speculation about a warmer, wetter and higher CO_2 Mars climate exists. Such environmental conditions should have led to the formation of outcrops rich in carbonate minerals, for which evidence has been sparse. The carbonates in ancient Mars probably were produced as a result of weathering of olivine and pyroxene minerals in the presence of CO_2 during serpentinization reactions. The main CO_3^{2-} mineralogical phases from this process should be rich in magnesium and iron, which agrees with published results on certain areas of the Martian surface. The Martian calcium occurrence is of main importance due to the fact that calcium carbonates are mineralogical species exposing a CO_2 atmosphere with moisture. Anorthosites are made of more than 90% of Ca rich plagioclases (Popa et al., 2010) permitting the formation of calcium carbonates as dolomite and calcite.

Furthermore, the composition of many places studied approximates to the composition of the carbonate globules in the Martian meteorite ALH 84001 (Morris et al., 2010). The aqueous solutions without CO_3^{2-} species are amidst the main feature of ancient Martian water, although high quantities of CO_2 might be the main factor for further carbonate synthesis.

Prebiotic Chemistry on Mars

Earlier sediments on Earth have been destroyed or heavily altered by plate tectonics and the development of life; furthermore, there are no records that allow us to reconstruct the chemical pathways toward the origin of life. However, Mars may have preserved such records because it is very likely that extensive regions of this planet remain relatively undisturbed since about 3.5 Ga ago. The origin of life on Earth was surely affected by the mineralogy. Parnell, (2004) suggested that in order for chemical evolution to occur on Earth, some degree of mineral evolution was needed. It is thought that some minerals played a key role in the process: clays (Cairns-Smith, 2005), and transition metal sulfides (Russell and Hall, 1997; Wächtershäuser, 1992).The role of minerals is thus of main importance for the possible prebiotic processes that occurred on Mars. If similar chemical evolution took place on Mars, unique information would be given for prebiotic process in the Solar System through minerals and other compounds conserved in the Martian surface.

The Mars Global Surveyor has detected olivine group minerals in the surface of that planet (Hamilton and Christensen, 2005; Hoefen et al., 2003). Additionally, more recent probes, based on the thermal infrared remote sensing of low-albedo regions (presumably basalt-dominated), have also detected olivine compositions varying from Fo_{31} (iron-rich fayalite) to Fo_{91} (magnesium-rich forsterite) on the surface of Mars, which are mainly distributed along a band centered at about -30° latitude in the southern uplands (Koeppen and Hamilton, 2008). Specifically, the magnesium-rich olivine was mainly found in the rim regions of the largest basins of the red planet, and was interpreted as exposed segments of the Martian mantle. This composition of olivine is quite similar to the olivine presently found in the mantle of the Earth, which contrasts with the much higher fayalitic (iron-rich) olivine found in other regions of the Mars surface, such as the floors of outflow channels and of many craters. The floor of the Gusev crater, in particular, is lined with basalts with abundant (20-30 % of the rock) olivine of intermediate composition (McSween et al., 2006). The apparent absence of olivine in the northern lowlands (Koeppen and Hamilton, 2008) is very intriguing because, it may be the result of the alteration of olivine or its burial by fluvial and eolian sediments.

Planets evolve aftermath inner and outer processes, while different events affect their constitution their surfaces change dramatically. Minerals experience deep changes according to these transformations, but the alteration obeys a positive feedback. The environmental condition promotes the evolution of the minerals, but at the same time, the mineral composition can also affect the environment.

The assembly of molecules, their reactions, the catalysis of some reactions, and the shielding of some molecules, all these processes were suggested to have been performed by minerals, but once life emerged, biological activity could have also be able to modify the mineral surface (Figure 49).

A fundamental step in prebiotic chemistry is the formation of compounds with C-C bonds from simple carbon molecules such as CO, CO_2, and CH_4. Searching for these types of reactions on Mars can give rise to important information regarding the abiotic synthesis of organic molecules.

Figure 49. Assemble of organic molecules in layered minerals. In Earth, assembly of molecules and their stabilization might have occurred in layered minerals (inset and circle in the right image). Once life emerged the biological activity could have also be able to modify the mineral surface (photo N. Archypova).

Methane

Although not a mineral, we consider methane to be an important molecule related to life-processes. Methane could be produced through enzymatic pathways and its presence in Mars makes feasible more extended chemical processes on Mars.

The discovery of methane , the most reducing form of carbon, in the atmosphere of Mars was announced relatively early (Formisano et al., 2004), and observations show latitudinal variation ranging from about 63 parts per billion near the disk center, dropping to 52, and then 45 parts per billion northward. Several studies have looked for possible sources and sinks of methane. Since methane can only survive hundreds of years in the atmosphere before being destroyed by UV light or oxidized, there would need to be about 126 tons/year created to account for the amounts discovered. Delivery by a comet seems to be not viable, but other possible ways is that it could be of biotic origin, as on Earth, or originated by volcanism and serpentinization (Oze and Sharma, 2005).

Methane has been extensively studied because it is part of the interstellar medium and it is abundant throughout the Solar System. The interaction of methane with radiation produces more complex organic compounds. There are many studies related to the photolysis/radiolysis. Gorden and Ausloos (1967) studied the photolysis of CH_4 and of CH_4—CD_4 mixtures. The excited methane molecule dissociates to form H_2 and C_2H_4. Groth and Laudenklos (1936) found acetylene, ethylene, ethane and higher hydrocarbons. In the radiolysis, ethane is largely, although not exclusively formed. Other products are C_2H_4, C_3H_8 and nC_4H_{10}. Bennett et al. (2006) irradiated pure methane ices (CH_4) at 10 K with energetic electrons. The initially synthesized ethane, ethyl, and ethylene molecules were radiolyzed subsequently by the impinging electrons to yield the vinyl radical (C_2H_3) and acetylene (C_2H_2). These types of reactions may have importance in the pathways to complexity of simple compounds.

Radiation Effects on CO_2 as Possible Source of Organic Compounds

As it was pointed out earlier the carbon in Mars is in an oxidized form, as CO_2. This can suffer photolysis/radiolysis and could be reduced to produce organic compounds containing one or two carbon atoms (Albarran et al., 1995; Draganic et al., 1987; Draganić et al., 1991). Carbon dioxide in the atmosphere could be photolyzed in the presence of water yielding H_2CO (formaldehyde) as the main product. Nevertheless, due to its high reactivity only 1% of this formaldehyde did not react in the atmosphere. This percentage could be enough to enrich a significant amount of the Martian ocean and form more complex organic molecules.

Carbon dioxide is soluble in water, which is converted into carbonic acid (H_2CO_3) reversibly. The relative concentrations of CO_2, H_2CO_3, and the deprotonated forms HCO_3^- (bicarbonate) and CO_3^{2-} (carbonate) depend on the pH.

$$CO_2 + H_2O \rightleftharpoons H_2CO_3$$

In neutral or slightly alkaline water (pH > 6.5), the bicarbonate form predominates (>50%). In very alkaline water (pH > 10.4), the predominant (>50%) form is carbonate. If there were CO_2 dissolved in water in Mars may produce some simple compounds. In addition, CO_2 can be incorporated as carboxylic group through the formation of CO_2^- radical by radiation and the interaction with the water radiolytical products.

Draganic et al. (1987, 1991) studied the gamma irradiation of O_2-free aqueous solutions of 0.05 mol dm^{-3} ammonium bicarbonate with Co-gammas (0.0-170 kGy doses) or pulses of 10 MeV electrons. Formate, oxalate, formaldehyde and an unidentified polymer (MW14000-16000 Da) were found. The efficiency of organic synthesis within the large dose range studied is low and is explained by efficient pathways to the reformation of bicarbonate.

Albarran et al. (1987) studied experimentally the self radiolysis of calcium carbonate (solid) and determined that some more complex products were formed, among them short chain carboxylic acids (formic, oxalic, glyoxylic, acetic and glycolic), one alcohol and one aldehyde; the same authors also suggest that the co-deposition of multivalent transition metal cations, including uranium and thorium, might have given heavily irradiated solids, contributing with a large numbers of organic precursor on the early Earth. The main products are carboxylic acids, which have short half life in a high radiation field. However, some of these radiolytical products, also react under irradiation yielding more stable acids, for example, the radiolysis of acetic acid, a two carbon bearing compound, (formed also by other pathways like the photolysis of CO_2 + H_2O), produce succinic acid (a four C atoms compound) as the principal radiation product. Succinic acid is very stable under irradiation (Negrón-Mendoza and Ponnamperuma, 1976; Negrón-Mendoza and Ponnamperuma, 1982).

Scenarios for Chemical Evolution

Life is a phenomenon only known in the Earth and there are many approaches that try to explain its origin. Several environments have been proposed as possible sites for its appearance ranging from lagoon and seawater bodies to hydrothermal vents (Perry and Kolb,

2004). A parallelism between Mars and Earth has been made at times when the conditions for prebiotic processes had some similarities. This idea has been strongly reinforced by the discovery of phyllosilicates on Mars and their role for the possible occurrence of life in that planet. Among the most important solid mineral surfaces are silicates, such as olivine, which is one of the most abundant minerals on Earth and Mars, as well as carbonates, sulfides, and clays. The areas that include olivine-rich mafic and ultramafic rocks are susceptible to chemical weathering by dissolution and hydrolysis, giving rise primarily to serpentine and magnetite minerals.

When life originated, reactive minerals present on the surface of the Earth apparently participated in promoting the so-called prebiotic reactions (Hazen et al., 2008). Consequently, geological processes may have produced the energy required for the prebiotic synthesis of molecules. In fact, a thermodynamic disequilibrium is necessary to produce the chemical energy needed to maintain metabolic routes in natural environments (Janecky and Seyfried Jr, 1986; Schulte et al., 2006). Thus, geological processes most certainly contributed to the synthesis of organic molecules. Also, minerals could act as catalysts for certain chemical reactions as concentrating agents, and/or as shells that protect the molecules from external destructive energy sources (Freund et al., 1986).

On Mars, major manifestation of chemical disequilibrium is weathering of the igneous minerals to yield low-temperature alteration products, which for basalt are dominated by clays and hydroxides. Based on observations of the surface, it appears that part of the overall weathering process on Mars is the oxidation of ferrous to yield ferric minerals. For example, each mole of fayalite (Fe^{2+} olivine), which is oxidized to ferric hydroxides and ferric silicates, can provide a mole of H_2. It is conceivable that this H_2 could be coupled to CO_2 reduction to yield simple organic compounds as weathering proceeds. On the other hand, if CH_4 was introduced into an environment consisting of already-oxidized minerals, then coupled redox processes could yield organic compounds. The temperature, at which weathering reactions occur, increases with depth or heat flow, and at some elevated temperature, the processes involved would overlap those that would be considered hydrothermal.

Hydrothermal Systems as Prebiotic Sources of Organics on Mars

Early in the history of Mars there are many evidences for the presence of water in the surface together and abundant volcanic activity. In this scenario, hydrothermal systems are an unavoidable consequence of a geologically active planet which had fluid H_2O at or near its surface; hydrothermal systems would certainly have existed within the outer layers of the planet, and may have had surface manifestations as hot springs. In addition, the SNC[1] meteorites show geochemical evidence for hydrothermal alteration and isotopic evidence for exchange of water between surface and subsurface, plausibly via hydrothermal systems. Therefore, the transport properties of H_2O in hydrothermal systems lead to dynamic, rapid fluid flow. Accordingly, if magmatic processes occur on wet planets, hydrothermal systems inevitably form whilst transporting heat efficiently and rapidly away to the surface of the planet and the atmosphere. These evidences and direct chemical implications for the potential emergence of life on Mars, including the volatile budget of the planet with a subsurface

[1] The SNC meteorites (shergottite, nakhlite, and chassigny classes) comprise a group of petrologically similar specimens. The SNC meteorites are basically igneous rocks, which solidified from a cooling magma, near the surface of their parent body, probably Mars.

mineral precipitation of carbonates, sulfides and other complex solids together with a parallel similar material conditions of the Martian Noachian and Archean Earth, open the possibilities to have remains of further organic-inorganic assemblies and other higher molecular assemblies allowing towards the living chemical process on Mars.

Liquid water would circulate in the subsurface of planets fueled by the geothermal gradient and magmatism provided that water is available from sources in the crust or dissolved in the magmas. A full review of the subject by Fisk and Giovannoni (1999) lead these authors to conclude that life would survive in certain regions of the interior of Mars protected from the cold and dry conditions of the surface because underground water, carbon, nutrients, temperatures, and redox conditions may be similar to those that make life possible in the oceanic crust of the Earth. The discovery of highly opaline silica deposits in several places of the surface of Mars (Squyres et al., 2008) attests to persistent hydrothermal activity on the planet, and the possibility that life was fostered and preserved by this kind of environment, as it occurred in some terrestrial analogs of the Archean Earth. Table 4 shows secondary products that may be derived from the alteration of volcanic rocks of basaltic composition (the main rock constituent on Mars). Again, the actual identification of all the secondary minerals listed in Table 4 implies the prolonged action of hydrothermal fluids, essentially water, in the upper crustal levels of Mars and, hence, the high probability for a chemical evolution favorable for the generation of living systems (Bibring et al., 2006; Poulet et al., 2005).

The hydrothermal systems on Earth have been explored looking for evidence as potential sites for chemical evolution. The role of these environments in chemical evolution on Earth is still subjected to a very strong debate, because there are experiments in favor and against. Marshall (1994) demonstrated the amino acids synthesis in hydrothermal simulation conditions (T= 200°C) in aqueous solution. Islam et al. (2001) and Alargov et al., (2002) showed that in supercritical water flows at 300–400°C amino acids can survive for several minutes. Other authors (Miller et al., 1989) suggest that hydrothermal vents are not capable of synthesizing or to stabilize but rather to break down organic compounds and polymers. Also, Ito et al. (2006), showed that amino acids cannot be either synthesized or even survive at temperatures higher than 250 °C.

Concluding Remarks

It is widely accepted that life on Earth arose in the presence of liquid water, and for this reason water is considered as one of the conditions necessary for life on similar planets. However, the second condition is that it remains liquid long enough for allowing chemical and biotic evolution from organic compounds to complex cellular structures of the first organisms.

Having a multidisciplinary approach, here we discussed the early conditions and global processes during the Noachian history (~4.5 to 3.6 Ga) which could have allowed direct or indirectly the presence of different sources of liquid water per long enough for the generation of life. Although this time period is unknown, we consider 10 Ma that as a valid theoretical minimum for the origination of life under early Earth-like conditions.

Early planetary conditions on the history of Mars, such as a short accretion time, the presence of high amount volatiles (including water), a fast core differentiation, and the probability of a global magnetic field, suggest that Mars had similar characteristics to those found on the primitive Earth including a dense atmosphere capable of sustaining the presence of stable liquid water at its surface and sub-surface. In addition, volcanism on Mars through time has been undoubtedly a positive factor for the possible past emergence of life in that planet. Degassing of organic-making volatiles from magmas (necessary for pre-biotic chemical evolution) has occurred apparently throughout the history of Mars, and in quantities enough as to periodically regenerate dense atmospheres if previously lost.

Indeed, geomorphological, mineralogical and spectrometric evidence from Mars clearly led to the conclusion that liquid water existed on or even below the surface of that planet, but was most probably restricted to the early history of Mars. Importantly, numerous images from Mars have revealed system of gully-like structures with an apron and debris at the end, probably created by flowing water in recent geological times: namely, a few million of years ago, or perhaps much later. Since 1999, this discovery continue in debate and different hypotheses about their origin have been proposed: a) aquifers below ground warmed by energy from within Mars which could flow intermittently, b) a "rotisserie" effect at times of large obliquity when atmospheric pressure would have been high enough for water exist in its liquid phase, and/or c) melting below snowpacks where liquid water would flow beneath a protective layer of ice; these are interesting possibilities to explore throughout the Mars history since the Noachian until today. Importantly, the heat source to keep the water in its liquid phase is an open question that includes hydrothermal processes, impacts, and greenhouse gases as we discussed above.

Undoubtedly, liquid water is not enough to support life, and no organism can thrive in pure water. All pre-biotic and biotic processes require energy and organic compounds. Thereon, we discussed the high heterogeneity of environments and therefore chemistry during the Noachian and Early Hesperian. So, the chances for chemical evolution encompass a greater variety of possible reactants highly favorable for the origin of life and its survival. If life emerged on Mars, that event probably took place in the interfaces among mineral-water and the lithosphere. The triple point where the atmosphere, the hydrosphere (oceans, rivers, lagoons, or streams) and the lithosphere converged could have favored the reactions that produced more complex molecules (chemical evolution) and eventually life.

In many of the environments proposed as scenarios for the origin of life, solid surfaces are involved in the synthesis, concentration, selection, and organization of biomolecules. In this way, mineral surfaces may have facilitated a transition from geochemistry to biochemistry. The abundance and wide distribution of olivine, pyroxene, and sedimentary hematite on Earth and other large rocky bodies of the Solar System, make weathering processes of these minerals potentially important for redox reaction sites that may have provided support for life, and may have left mineralogical evidence behind as to whether or not life ever existed on Mars.

As mentioned previously, processes that may have contributed to the inventory of organic compounds, not only on the early Earth but also on early Mars, include such mineral-catalyzed reactions as water-rock reactions (for example, serpentinization) that could lead to the formation of hydrocarbons from CO_2 or the reduction of nitrogen to ammonia.

Although, several types of environments on Mars may appear as possible scenarios for chemical evolution, extant or relict Martian hydrothermal systems are, in our view, the best place where physico-chemical processes can evolved towards higher levels of chemical organization, as occurred on the Earth.

Contact between igneous rocks and liquid water at low temperatures can be the source of chemical disequilibrium that may have provided the means for organic synthesis. Minerals may have played an important role in the prebiotic process on Mars and also as shields for organics against radiation.

Multidisciplinary approaches to detect evidence of past and present prebiotic physic-chemical life-forming conditions on Mars are a necessity. Although we favor early Mars as a place where life could have emerged, we realize that much further evidence is needed to confirm or discard this idea. Following the history of Mars since its accretion, through its early atmospheric, magmatic, and magnetic stages, it is clear that many of the conditions for habitability certainly existed (e.g. abundant long-standing liquid water, nutrients, energy sources, etc.). Whether these conditions were sufficient for life's development on Mars is still an open question that would require years to come in exploration.

REFERENCES

Abramov, O., Mojzsis, S.J., 2009. Microbial habitability of the Hadean Earth during the late heavy bombardment. *Nature* 459, 419-422.

Acuña, M.H., Connerney, J.E.P., Ness, F.N., Lin, R.P., Mitchell, D., Carlson, C.W., McFadden, J., Anderson, K.A., Rème, H., Mazelle, C., Vignes, D., Wasilewski, P., Cloutier, P., 1999. Global Distribution of Crustal Magnetization Discovered by the Mars Global Surveyor MAG/ER Experiment. *Science* 284, 790-793.

Alargov, D.K., Deguchi, S., Tsuji, K., K., H., 2002. Reaction Behaviors of Glycine under Super- and Subcritical Water Conditions. *Origins of Life and Evolution of the Biosphere* 32, 1-12.

Albarrán, G., Collins, K.E., Collins, C.H., 1987. Formation of organic products in self-radiolyzed calcium carbonate. *Journal of Molecular Evolution* 25, 12-14.

Albarran, G., Negron-Mendoza, A., Ramos-Bernal, S., 1995. A plausible Route for the Synthesis of Big-organic Compounds from Inorganic Carbon on the Primitive Earth., *Chemical Evolution: Self-organization of the Macromolecules of Life.* Chela Flores, J.; Chadha, M.; Negrón, A.; Oshima, T-, pp. 49-61.

Anguita, F., Farelo, A.F., López, V., Mas, C., Muñoz-Espadas, M.J., Márquez, A., Ruiz, J., 2001. Tharsis dome, Mars: New evidence for Noachian-Hesperian thick-skin and Amazonian thin-skin tectonics. *Journal of Geophysical Research* 106, 7577-7589.

Baker, V.R., Maruyama, S., Dohm, J.M., 2007. Tharsis superplume and the geological evolution of early Mars, *Superplumes: beyond plate tectonics.* Springer, New York, pp. 507-522.

Bandfield, J.L., 2002. Global mineral distributions on Mars. *J. Geophys. Res.* 107, 5042.

Baratoux, D., Toplis, M.J., Monnereau, M., Gasnault, O., 2011. Thermal history of Mars inferred from orbital geochemistry of volcanic provinces. *Nature* 472, 338-341.

Barlow, N., 2008. *Mars: An Introduction to its Interior, Surface and Atmosphere.* Cambridge Univ Press. Cambridge, Cambridge.

Barman, A.K., Varadachari, C., Ghosh, K., 1992. Weathering of silicate minerals by organic acids: I. Nature of cation solubilisation. *Geoderma* 53, 42-63.

Bauer, S.J., Lammer, H., 2004. Planetary Aeronomy. *Atmosphere Environments in Planetary Systems*. Springer Berlin Heidelberg New York.

Bennett, C.J., Jamieson, C.S., Osamura, Y., Kaiser, R.I., 2006. Laboratory Studies on the Irradiation of Methane in Interstellar, Cometary, and Solar System Ices. *The Astrophysical Journal* 653, 792-800.

Benz, W., 2000. Low Velocity Collisions and the Growth of Planetesimals. *Space Science Reviews* 92, 279-294.

Benzerara, K., Yoon, T.H., Menguy, N., Tyliszczak, T., Brown, G.E., 2005. Nanoscale environments associated with bioweathering of a Mg-Fe-pyroxene. *Proceedings of the National Academy of Sciences of the United States of America* 102, 979-982.

Berndt, M.E., Allen, D.E., Seyfried, W.E., 1996. Reduction of CO2 during serpentinization of olivine at 300 °C and 500 bar. *Geology* 24, 351-354.

Bibring, J.-P., Langevin, Y., Mustard, J.F., Poulet, F., Arvidson, R., Gendrin, A., Gondet, B., Mangold, N., Pinet, P., Forget, F., team, T.O., Berthé, M., Gomez, C., Jouglet, D., Soufflot, A., Vincendon, M., Combes, M., Drossart, P., Encrenaz, T., Fouchet, T., Merchiorri, R., Belluci, G., Altieri, F., Formisano, V., Capaccioni, F., Cerroni, P., Coradini, A., Fonti, S., Korablev, O., Kottsov, V., Ignatiev, N., Moroz, V., Titov, D., Zasova, L., Loiseau, D., Pinet, P., Douté, S., Schmitt, B., Sotin, C., Hauber, E., Hoffmann, H., Jaumann, R., Keller, U., Arvidson, R., Duxbury, T., Forget, F., Neukum, G., 2006. Global Mineralogical and Aqueous Mars History Derived from OMEGA/Mars Express Data. *Science* 312, 400-404.

Bishop, J.L., Schelble, R.T., McKay, C.P., Brown, A.J., Perry, K.A., 2011. Carbonate rocks in the Mojave Desert as an analogue for Martian carbonates. *International Journal of Astrobiology* 10, 349-358.

Bland, P.A., Howard, L.E., Prior, D.J., Wheeler, J., Hough, R.M., Dyl, K.A., 2011. Earliest rock fabric formed in the Solar System preserved in a chondrule rim. *Nature Geoscience* 4, 244-247.

Borg, L.E., Nyquist, L.E., Wiesmann, H., Shih, C.Y., Reese, Y., 2003. The age of Dar al Gani 476 and the Differentiation History of the Martian Meteorites Inferred from their Radiogenic Isotopic Systematics. *Geochimica et Cosmochimica Acta* 67, 3519-3536.

Boynton, W.V., Feldman, W.C., Squyres, S.W., Prettyman, T.H., Brückner, J., Evans, L.G., Reedy, R.C., Starr, R., Arnold, J.R., Drake, D.M., Englert, P.A.J., Metzger, A.E., Mitrofanov, I., Trombka, J.I., d'Uston, C., Wänke, H., Gasnault, O., Hamara, D.K., Janes, D.M., Marcialis, R.L., Maurice, S., Mikheeva, I., Taylor, G.J., Tokar, R., Shinohara, C., 2002. Distribution of Hydrogen in the Near Surface of Mars: Evidence for Subsurface Ice Deposits. *Science* 297, 81-85.

Brack, A., 1998. The Molecular Origins of Life: Assembling Pieces of the Puzzle, *Origins of Life and Evolution of Biospheres.* Cambridge University Press, Springer Netherlands, p. 417.

Brearley, A.J., 2006. The action of water, in: Lauretta, D.S., McSween, H.Y. (Eds.), *Meteorites and the Early Solar System*. Univ. of Arizona, Tucson, Arizona.

Brearley, A.J., Jones, R.H., 1998. Chondritic Meteorites. *Reviews in Mineralogy and Geochemistry* 36, 3.1-3.398.

Breuer, D., Spohn, T., 2003. Early plate tectonics versus single-plate tectonics on Mars: Evidence from magnetic field history and crust evolution. *J. Geophys. Res.* 108, 5072.

Cabrol, N.A., Grin, E.A., 2010. Searching for lakes on Mars: four decades of exploration, in: Cabrol, N.A., Grin, E.A. (Eds.), *Lakes on Mars*. Elsevier, Oxford, UK, Oxford, pp. 1-30.

Cairns-Smith, A.G., 2005. Sketches for a Mineral Genetic Material. *Elements 1,* 157-161.

Canup, R.M., Agnor, C., 1998. Accretion of Terrestrial Planets and the Earth-moon System., in: Canup, R.M., Righter, K. (Ed.), *Origin of the Earth and Moon*. University of Arizona Press, Tucson, pp. 113-129.

Carr, M.H., 1986. Mars: A water-rich planet? *Icarus* 68, 187-216.

Carr, M.H., 1987. Water on Mars. *Nature* 326, 30-35.

Carr, M.H., 2006. *The surface of Mars.* Cambridge Univ. Press, Cambridge UK.

Carr, M.H., Head, J.W., 2003. Oceans on Mars: An assessment of the observational evidence and possible fate. *J. Geophys. Res.* 108, 5042.

Carter, J., Poulet, F., Bibring, J.-P., Murchie, S., 2010. Detection of Hydrated Silicates in Crustal Outcrops in the Northern Plains of Mars. *Science* 328, 1682-1686.

Cathles, L.M., Erendi, A.H.J., Barrie, T., 1997. How long can a hydrothermal system be sustained by a single intrusive event? *Economic Geology* 92, 766-771.

Chevrier, V., Poulet, F., Bibring, J.-P., 2007. Early geochemical environment of Mars as determined from thermodynamics of phyllosilicates. *Nature* 448, 60-63.

Cockell, C.S., 2006. The origin and emergence of life under impact bombardment. *Philosophical Transactions of the Royal Society B: Biological Sciences* 361, 1845-1856.

Colín-García, M., García-Martinez, J.L., Negrón-Mendoza, A., Ortega Gutiérrez, F., Ramos-Bernal, S., Sánchez-Román, M., Valdivia-Silva, J.E., 2011. Olivine: It's Role in Chemical Evolution and in the Origin of Life, in: Veress B., J., S. (Eds.), Horizons in Earth Science Research. Nova, pp. 221-244.

Connerney, J.E.P., Acuña, M.H., Wasilewski, P.J., Ness, N.F., Rème, H., Mazelle, C., Vignes, D., Lin, R.P., Mitchell, D.L., Cloutier, P.A., 1999. Magnetic Lineations in the Ancient Crust of Mars. *Science* 284, 794-798.

Craddock, R.A., Greeley, R., 2009. Minimum estimates of the amount and timing of gases released into the martian atmosphere from volcanic eruptions. *Icarus* 204, 512-526.

Dauphas, N., Pourmand, A., 2011. Hf-W-Th evidence for rapid growth of Mars and its status as a planetary embryo. *Nature* 473, 489-492.

Davies, P.C.W., 1996. The transfer of viable microorganisms between planets, in *Evolution of hydrothermal ecosystems on Earth* (and Mars?). John Wiley, Hoboken, N. J., pp. 304-314.

de Pater, I., Lissauer, J.J., 2007. *Planetary Sciences,* 2nd ed. Cambridge University Press, Cambridge.

Di Achille, G., Hynek, B.M., 2010. Ancient ocean on Mars supported by global distribution of deltas and valleys. *Nature Geosci* 3, 459-463.

Dickson, J.L., Fassett, C.I., Head, J.W., 2008. Young valley networks on Mars: *Persistent flow of water in Lyot crater, a distinctive Amazonian impact basin microenvironmen.* European Planetary Science Congress 2008.

Dohm, J.M., Anderson, R.C., Williams, J.-P., Ruiz, J., McGuire, P.C., Buczkowski, D.L., Wang, R., Scharenbroich, L., Hare, T.M., Connerney, J.E.P., Baker, V.R., Wheelock, S.J., Ferris, J.C., Miyamoto, H., 2009. Claritas rise, Mars: Pre-Tharsis magmatism? *Journal of Volcanology and Geothermal Research* 185, 139-156.

Dohm, J.M., Ferris, J.C., Baker, V.R., Anderson, R.C., Hare, T.M., Strom, R.G., Barlow, N.G., Tanaka, K.L., Klemaszewski, J.E., Scott, D.H., 2001. Ancient drainage basin of the Tharsis region, Mars: Potential source for outflow channel systems and putative oceans or paleolakes. *J. Geophys. Res.* 106, 32943-32958.

Doran, P.T., Clifford, S.M., Forman, S.L., Nyquist, L., Papanastassiou, D.A., Stewart, B.W., Sturchio, N.C., Swindle, T.D., Cerling, T., Kargel, J., McDonald, G., Nishiizumi, K., Poreda, R., Rice, J.W., Tanaka, K., 2004. Mars chronology: assessing techniques for quantifying surficial processes. *Earth-Science Reviews* 67, 313-337.

Draganic, Z.D., Negron-Mendoza, A., Navarro-Gonzalez, R., Vujosevic, S.I., 1987. The presence of polymeric material in radiolysed aqueous solutions of ammonium bicarbonate. International Journal of Radiation Applications and Instrumentation. Part C. *Radiation Physics and Chemistry* 30, 229-231.

Draganić, Z.D., Negrón-Mendoza, A., Sehested, K., Vujošević, S.I., Navarro-Gonzáles, R., Albarrán-Sanchez, M.G., Draganić, I.G., 1991. Radiolysis of aqueous solutions of ammonium bicarbonate over a large dose range. International Journal of Radiation Applications and Instrumentation. Part C. *Radiation Physics and Chemistry* 38, 317-321.

Dreibus, G., Wänke, H., 1987. Volatiles on Earth and Mars: A comparison. *Icarus* 71, 225-240.

Durand-Manterola, H.J., 2001. An estimation of the mass dragged by the solar wind from Mars's atmosphere in its geologic history. *Geofísica Internacional* 40, 315-320.

Durand-Manterola, H.J., 2003. Degassing of volatiles in the geological history of Mars. *Geofísica Internacional* 42, 157-162.

Ehlmann, B.L., Mustard, J.F., Clark, R.N., Swayze, G.A., Murchie, S.L., 2011a. Evidence for Low-grade Metamorphism, Hydrothermal Alteration, and Diagenesis on Mars from Phyllosilicate Mineral Assemblages. *Clays and Clay Minerals* 59, 359-377.

Ehlmann, B.L., Mustard, J.F., Murchie, S.L., Bibring, J.-P., Meunier, A., Fraeman, A.A., Langevin, Y., 2011b. Subsurface water and clay mineral formation during the early history of Mars. *Nature* 479, 53-60.

El Maarry, M.R., Gasnault, O., Toplis, M.J., Baratoux, D., Dohm, J.M., Newsom, H.E., Boynton, W.V., Karunatillake, S., 2009. Gamma-ray constraints on the chemical composition of the martian surface in the Tharsis region: A signature of partial melting of the mantle? *Journal of Volcanology and Geothermal Research* 185, 116-122.

Elkins-Tanton, L.T., 2008. Linked magma ocean solidification and atmospheric growth for Earth and Mars. *Earth and Planetary Science Letters* 271, 181-191.

Elkins-Tanton, L.T., Hess, P.C., Parmentier, E.M., 2005. Possible formation of ancient crust on Mars through magma ocean processes. *J. Geophys. Res.* 110, E12S01.

Fairén, A.G., Chevrier, V., Abramov, O., Marzo, G.A., Gavin, P., Davila, A.F., Tornabene, L.L., Bishop, J.L., Roush, T.L., Gross, C., Kneissl, T., Uceda, E.R., Dohm, J.M., Schulze-Makuch, D., Rodríguez, J.A.P., Amils, R., McKay, C.P., 2010. Noachian and More Recent Phyllosilicates in Impact Craters on Mars. *Proceedings of the National Academy of Sciences* 107, 12095-12100.

Farmer, J.D., Des Marais, D.J., 1999. Exploring for a record of ancient Martian life. *J. Geophys. Res.* 104, 26977-26995.

Fassett, C.I., Head, J.W., 2008. The timing of martian valley network activity: Constraints from buffered crater counting. *Icarus* 195, 61-89.

Faure, G., Mensing, T.M., 2007. *Introduction to Planetary Science: The Geological Perspective.* Springer, Dordrecht, The Netherlands.

Feldman, W.C., Prettyman, T.H., Maurice, S., Plaut, J.J., Bish, D.L., Vaniman, D.T., Mellon, M.T., Metzger, A.E., Squyres, S.W., Karunatillake, S., Boynton, W.V., Elphic, R.C., Funsten, H.O., Lawrence, D.J., Tokar, R.L., 2004. Global distribution of near-surface hydrogen on Mars. *J. Geophys. Res.* 109, E09006.

Fisk, M.R., Giovannoni, S.J., 1999. Sources of nutrients and energy for a deep biosphere on Mars. *J. Geophys. Res.* 104, 11805-11815.

Formisano, V., Atreya, S., Encrenaz, T., Ignatiev, N., Giuranna, M., 2004. Detection of Methane in the Atmosphere of Mars. *Science* 306, 1758-1761.

Freund, E., Lynch, J., Szymanski, R., 1986. Small particles in the field of industrial supported catalysts. *Ultramicroscopy* 20, 107-118.

Frey, H., 2008. Ages of very large impact basins on Mars: Implications for the late heavy bombardment in the inner solar system. *Geophys. Res. Lett.* 35, L13203.

Frey, H.V., 2004. Distribution of large visible and buried impact basins on Mars: Comparison with free-air gravity, crustal thickness, and magnetization models., in: Mackwell, S., Stansbery, E. (Ed.), *Lunar and Planetary Institute Science Conference Abstracts.*, League City, Texas.

Frey, H.V., 2006. Impact constraints on, and a chronology for, major events in early Mars history. *J. Geophys. Res.* 111, E08S91.

Gaillard, F., Scaillet, B., 2009. The sulfur content of volcanic gases on Mars. *Earth and Planetary Science Letters* 279, 34-43.

Gellert, R., Rieder, R., Brückner, J., Clark, B.C., Dreibus, G., Klingelhöfer, G., Lugmair, G., Ming, D.W., Wänke, H., Yen, A., Zipfel, J., Squyres, S.W., 2006. Alpha Particle X-Ray Spectrometer (APXS): Results from Gusev crater and calibration report. *J. Geophys. Res.* 111, E02S05.

Gomes, R., Levison, H.F., Tsiganis, K., Morbidelli, A., 2005. Origin of the cataclysmic Late Heavy Bombardment period of the terrestrial planets. *Nature* 435, 466-469.

Gooding, J.L., Keil, K., 1978. Alteration of glass as a possible source of clay minerals on Mars. *Geophys. Res. Lett.* 5, 727-730.

Gorden, R., Ausloos, P., 1967. Gas-Phase Photolysis and Radiolysis of Methane. Formation of Hydrogen and Ethylene. . *J. Chem. Phys.* 46, 4823-4835.

Gough, D.O., 1981. Solar interior structure and luminosity variations. *Solar Physics* 74, 21-34.

Gradstein, F.M., Ogg, J.G., Smith, A.G., Bleeker, W., Lourens, L.J., 2004. A new geologic time scale with special reference to Precambrian and Neogene. *Episodes* 27, 83-100.

Greeley, R., 1987. Release of Juvenile Water on Mars: Estimated Amounts and Timing Associated with Volcanism. *Science* 236, 1653-1654.

Groth, W., Laudenklos, H., 1936. Der Mechanismus des photochemischen Methanzerfalls (The mechanism of photochemical methane decomposition). *Naturwissenschaften* 24, 796-796

Grotzinger, J., Beaty, D., Dromart, G., Gupta, S., Harris, M., Hurowitz, J., Kocurek, G., McLennan, S., Milliken, R.E., Ori, G., Summer, D., 2011. The sedimentary record of Mars. *The Sedimentary Record* 9, 4-8.

Halevy, I., Zuber, M.T., Schrag, D.P., 2007. A Sulfur Dioxide Climate Feedback on Early Mars. *Science* 318, 1903-1907.

Hamilton, V.E., Christensen, P.R., 2005. Evidence for extensive, olivine-rich bedrock on Mars. Geology 33, 433-436.

Harder, H., 2000. Mantle convection and the dynamic geoid of Mars. *Geophys. Res. Lett.* 27, 301-304.

Harper, C., Nyquist, L., Bansal, B., Wiesmann, H., Shih, C., 1995. Rapid accretion and early differentiation of Mars indicated by 142Nd/144Nd in SNC meteorites. *Science* 267, 213-217.

Harrison, K.P., Chapman, M.G., 2010. Episodic Ponding and Outburst Flooding Associated with Chaotic Terrains in Valles Marineris, in: Cabrol N.A., E.A., G. (Eds.), *Lakes on Mars.* Elsevier, Amsterdam, pp. 163-194.

Hartmann, W.K., 1999. Martian cratering VI: Crater count isochrons and evidence for recent volcanism from Mars Global Surveyor. *Meteoritics and Planetary Science* 34, 167-177.

Hartmann, W.K., 2003. *A Traveler's Guide to Mars. The Mysterious Landscapes of the Red Planet.* Workman Publishing Company, Inc., New York.

Hartmann, W.K., 2005. Martian cratering 8: Isochron refinement and the chronology of Mars. *Icarus* 174, 294-320.

Hartmann, W.K., Neukum, G., 2001. Cratering Chronology and the Evolution of Mars. *Space Science Reviews* 96, 165-194.

Hazen, R.M., Papineau, D., Bleeker, W., Downs, R.T., Ferry, J.M., McCoy, T.J., Sverjensky, D.A., Yang, H., 2008. Mineral Evolution. *American Mineralogist* 93, 1693-1720.

Hoefen, T.M., Clark, R.N., Bandfield, J.L., Smith, M.D., Pearl, J.C., Christensen, P.R., 2003. Discovery of Olivine in the Nili Fossae Region of Mars. *Science* 302, 627-630.

Hynek, B.M., Phillips, R.J., 2001. Evidence for extensive denudation of the Martian highlands. *Geology* 29, 407-410.

Hynek, B.M., Phillips, R.J., Arvidson, R.E., 2003. Explosive volcanism in the Tharsis region: Global evidence in the Martian geologic record. *J. Geophys. Res.* 108, 5111.

Irvine, W.M., 1998. Extraterrestrial Organic Matter: A review. *Origins of Life and Evolution of Biospheres* 28, 365-383.

Irwin, R.P., Craddock, R.A., Howard, A.D., 2005. Interior channels in Martian valley networks: discharge and runoff production. *Geology* 33, 489-492.

Islam, M.N., Kaneko, T., K., K., 2001. Determination of Amino Acids Formed in a Supercritical Water Flow Reactor Simulating Submarine Hydrothermal Systems. *Analytical Sciences* 17, i1631-i1634.

Ito M., Gupta, L.P., H., M., H., K., 2006. Thermal Stability of Amino Acids in Seafloor Sediment in Aqueous Solution at High Temperature. *Organic Geochemistry* 37, 177-188.

Jakosky, B.M., Phillips, R.J., 2001. Mars' volatile and climate history. *Nature* 412, 237-244.

Janecky, D.R., Seyfried Jr, W.E., 1986. Hydrothermal serpentinization of peridotite within the oceanic crust: Experimental investigations of mineralogy and major element chemistry. *Geochimica et Cosmochimica Acta* 50, 1357-1378.

Johnson, C.L., Phillips, R.J., 2005. Evolution of the Tharsis region of Mars: insights from magnetic field observations. *Earth and Planetary Science Letters* 230, 241-254.

Johnson, S.S., 2008. *Mars in the late Noachian : evolution of a habitable surface environment,* Massachusetts Institute of Technology. Dept. of Earth, Atmospheric, and Planetary Sciences. Massachusetts Institute of Technology, Boston Massachusetts.

Karner, J., Papike, J.J., Shearer, C.K., 2006. Comparative planetary mineralogy: Pyroxene major- and minor-element chemistry and partitioning of vanadium between pyroxene and melt in planetary basalts. *American Mineralogist* 91, 1574-1582.

Kasting, J.F., Grinspoon, D.H., 1999. *The Faint Young Sun Problem*, in: Sonnet, C.P., Giampapa, M.S. (Eds.), The Sun in Time The University of Arizona Press, pp. 447-462.

Kleine, T., Munker, C., Mezger, K., Palme, H., 2002. Rapid accretion and early core formation on asteroids and the terrestrial planets from Hf-W chronometry. *Nature* 418, 952-955.

Knauth, L.P., Burt, D.M., 2002. Eutectic Brines on Mars: Origin and Possible Relation to Young Seepage Features. *Icarus* 158, 267-271.

Koeppen, W.C., Hamilton, V.E., 2008. Global distribution, composition, and abundance of olivine on the surface of Mars from thermal infrared data. *J. Geophys. Res.* 113, E05001.

Lahav, N., 1999. *Biogenesis: Theories of Life's Origin.* Oxford Univ. Press, New York.

Lammer, H., Kasting, J., Chassefière, E., Johnson, R., Kulikov, Y., Tian, F., 2008. Atmospheric Escape and Evolution of Terrestrial Planets and Satellites. *Space Science Reviews* 139, 399-436.

Lammer, H., Lichtenegger, H.I.M., Kolb, C., Ribas, I., Guinan, E.F., Abart, R., Bauer, S.J., 2003. Loss of water from Mars:: Implications for the oxidation of the soil. *Icarus* 165, 9-25.

Langlais, B., Purucker, M.E., Mandea, M., 2004. Crustal magnetic field of Mars. *J. Geophys. Res.* 109, E02008.

Lazcano, A., Miller, S.L., 1994. How long did it take for life to begin and evolve to cyanobacteria? *Journal of Molecular Evolution* 39, 546-554.

Lee, D.-C., Halliday, A.N., 1997. Core formation on Mars and differentiated asteroids. *Nature* 388, 854-857.

Lenardic, A., Nimmo, F., Moresi, L., 2004. Growth of the hemispheric dichotomy and the cessation of plate tectonics on Mars. *J. Geophys. Res.* 109, E02003.

Leshin, L.A., 2000. Insights into Martian water reservoirs from analyses of Martian meteorite QUE94201. *Geophys. Res. Lett.* 27, 2017-2020.

Leshin, L.A., Epstein, S., Stolper, E.M., 1996. Hydrogen isotope geochemistry of SNC meteorites. *Geochimica et Cosmochimica Acta* 60, 2635-2650.

Lillis, R.J., Manga, M., Mitchell, D.L., Lin, R.P., Acuna, M.H., 2006. Unusual magnetic signature of the Hadriaca Patera Volcano: Implications for early Mars. *Geophys. Res. Lett.* 33, L03202.

Lin, D.N.C., Papaloizou, J., 1985. On the Dynamical Origin of the Solar System, in: Black, D.C., Matthews, M.S. (Ed.), *Protostars and planets II* (A86-12626 03-90). University of Arizona Press, Tucson, AZ, pp. 981–1072.

Lundin, R., Lammer, H., Ribas, I., 2007. Planetary Magnetic Fields and Solar Forcing: Implications for Atmospheric Evolution. *Space Science Reviews* 129, 245-278.

Lundin, R., Zakharov, A., Pellinen, R., Barabasj, S.W., Borg, H., Dubinin, E.M., Hultqvist, B., Koskinen, H., Liede, I., Pissarenko, N., 1990. Aspera/Phobos measurements of the ion outflow from the MARTIAN ionosphere. *Geophys. Res. Lett.* 17, 873-876.

Luo, W., Stepinski, T.F., 2009. Computer-generated global map of valley networks on Mars. *J. Geophys. Res.* 114, E11010.

Marshall, W.L., 1994. Hydrothermal Synthesis of Amino Acids. *Geochimica et Cosmochimica Acta* 58, 2099-2106.

Martel, L.M.V., 2003. Pretty Green Mineral -- Pretty Dry Mars?, Planetary Science Research Discoveries. *Hawai'i Institute of Geophysics and Planetology,* Hawai'i, pp. 1-7.

McCollom, T.M., Seewald, J.S., 2001. A reassessment of the potential for reduction of dissolved CO2 to hydrocarbons during serpentinization of olivine. *Geochimica et Cosmochimica Acta* 65, 3769-3778.

McKay, C.P., Davis, W.L., 1991. Duration of liquid water habitats on early Mars. *Icarus* 90, 214-221.

McKay, C.P., Stoker, C.R., 1989. The early environment and its evolution on Mars: Implication for life. *Rev. Geophys.* 27, 189-214.

McSween, H.Y., Arvidson, R.E., Bell, J.F., Blaney, D., Cabrol, N.A., Christensen, P.R., Clark, B.C., Crisp, J.A., Crumpler, L.S., Des Marais, D.J., Farmer, J.D., Gellert, R., Ghosh, A., Gorevan, S., Graff, T., Grant, J., Haskin, L.A., Herkenhoff, K.E., Johnson, J.R., Jolliff, B.L., Klingelhoefer, G., Knudson, A.T., McLennan, S., Milam, K.A., Moersch, J.E., Morris, R.V., Rieder, R., Ruff, S.W., de Souza, P.A., Squyres, S.W., Wänke, H., Wang, A., Wyatt, M.B., Yen, A., Zipfel, J., 2004. Basaltic Rocks Analyzed by the Spirit Rover in Gusev Crater. *Science* 305, 842-845.

McSween, H.Y., Ruff, S.W., Morris, R.V., Bell, J.F., III, Herkenhoff, K., Gellert, R., Stockstill, K.R., Tornabene, L.L., Squyres, S.W., Crisp, J.A., Christensen, P.R., McCoy, T.J., Mittlefehldt, D.W., Schmidt, M., 2006. Alkaline volcanic rocks from the Columbia Hills, Gusev crater, Mars. *J. Geophys. Res.* 111, E09S91.

Melosh, H.J., Vickery, A.M., 1989. Impact erosion of the primordial atmosphere of Mars. *Nature* 338, 487-489.

Messenger, S., Keller, L.P., Stadermann, F.J., Walker, R.M., Zinner, E., 2003. Samples of Stars Beyond the Solar System: Silicate Grains in Interplanetary Dust. *Science* 300, 105-108.

Michalski, J.R., Niles, P.B., 2010. Deep crustal carbonate rocks exposed by meteor impact on Mars. *Nature Geosci* 3, 751-755.

Miller, S.L., Bada, J.L., N., F., 1989. What Was the Role of Submarine Hot Springs in the Origin of Life? *Origins of Life and Evolution of the Biosphere* 19, 536-537.

Milliken, R.E., Swayze, G.A., Arvidson, R.E., Bishop, J.L., Clark, R.N., Ehlmann, B.L., Green, R.O., Grotzinger, J.P., Morris, R.V., Murchie, S.L., Mustard, J.F., Weitz, C., 2008. Opaline silica in young deposits on Mars. *Geology* 36, 847-850.

Milone, E.F., Wilson, W., 2008. Solar System Astrophysics. *Planetary Atmospheres and the Outer Solar System*. Springer.

Morris, R.V., Ruff, S.W., Gellert, R., Ming, D.W., Arvidson, R.E., Clark, B.C., Golden, D.C., Siebach, K., Klingelhöfer, G., Schröder, C., Fleischer, I., Yen, A.S., Squyres, S.W., 2010. Identification of Carbonate-Rich Outcrops on Mars by the Spirit Rover. *Science* 329, 421-424.

Mustard, J.F., Murchie, S.L., Pelkey, S.M., Ehlmann, B.L., Milliken, R.E., Grant, J.A., Bibring, J.P., Poulet, F., Bishop, J., Dobrea, E.N., Roach, L., Seelos, F., Arvidson, R.E., Wiseman, S., Green, R., Hash, C., Humm, D., Malaret, E., McGovern, J.A., Seelos, K., Clancy, T., Clark, R., Marais, D.D., Izenberg, N., Knudson, A., Langevin, Y., Martin, T., McGuire, P., Morris, R., Robinson, M., Roush, T., Smith, M., Swayze, G., Taylor, H., Titus, T., Wolff, M., 2008. Hydrated silicate minerals on Mars observed by the Mars Reconnaissance Orbiter CRISM instrument. *Nature* 454, 305-309.

Mustard, J.F., Poulet, F., Gendrin, A., Bibring, J.-P., Langevin, Y., Gondet, B., Mangold, N., Bellucci, G., Altieri, F., 2005. Olivine and Pyroxene Diversity in the Crust of Mars. *Science* 307, 1594-1597.

Negrón-Mendoza, A., Ponnamperuma, C., 1976. Formation of biologically relevant carboxylic acids during the gamma irradiation of acetic acid. *Origins of Life and Evolution of Biospheres* 7, 191-196.

Negrón-Mendoza, A., Ponnamperuma, C., 1982. Role of succinic acid in chemical evolution. *Origins of Life and Evolution of Biospheres* 12, 427-431.

Neukum, G., Hiller, K., 1981. Martian Ages. *J. Geophys. Res.* 86, 3097-3121.

Nguyen, A.N., Stadermann, F.J., Zinner, E., Stroud, R.M., Alexander, C.M.O.D., Nittler, L.R., 2007. Characterization of Presolar Silicate and Oxide Grains in Primitive Carbonaceous Chondrites. *The Astrophysical Journal* 656, 1223.

Nicholson, W.L., Munakata, N., Horneck, G., Melosh, H.J., Setlow, P., 2000. Resistance of Bacillus Endospores to Extreme Terrestrial and Extraterrestrial Environments. *Microbiology and Molecular Biology Reviews* 64, 548-572.

Nikolaeva, V., Abdrakhimov, A.M., 1999. Rocks of Andesitic Composition at the Mars Pathfinder Site Reveal Chemical Analogy with Subduction-Related Andesites of Earth. *Lunar and Planetary Science Conference XXX*, 1376.

Nimmo, F., Stevenson, D.J., 2000. Influence of early plate tectonics on the thermal evolution and magnetic field of Mars. *J. Geophys. Res.* 105, 11969-11979.

Nittler, L.R., 2003. Presolar stardust in meteorites: recent advances and scientific frontiers. *Earth and Planetary Science Letters* 209, 259-273.

Nyquist, L.E., Bogard, D.D., Shih, C.Y., Greshake, A., Stöffler, D., Eugster, O., 2001. Ages and Geologic Histories of Martian Meteorites. *Space Science Reviews* 96, 105-164.

Oberbeck, V.R., Fogleman, G., 1989. Estimates of the maximum time required to originate life. *Origins of Life and Evolution of Biospheres* 19, 549-560.

Olsen, A.A., Rimstidt, J.D., 2007. Using a Mineral Lifetime Diagram to Evaluate the Persistence of Olivine on Mars. *American Mineralogist* 92, 598-602.

Owen, T., Bar-Nun, A., Kleinfeld, I., 1992. Possible cometary origin of heavy noble gases in the atmospheres of Venus, Earth and Mars. *Nature* 358, 43-46.

Oze, C., Sharma, M., 2005. Have olivine, will gas: Serpentinization and the abiogenic production of methane on Mars. *Geophysical Research Letters* 32, L10203.

Parks, G.K., 1991. *Physics of Space Plasmas.* An Introduction. Redwood City, CA; Addison-Wesley Publishing Co.

Parnell, J., 2004. Mineral Radioactivity in Sands as a Mechanism for Fixation of Organic Carbon on the Early Earth. *Origins of Life and Evolution of Biospheres* 34, 533-547.

Perry, R.S., Kolb, V.M., 2004. On the applicability of Darwinian principles to chemical evolution that led to life. *International Journal of Astrobiology* 3, 45-53.

Phillips, P.J., Zuber, M.T., Salomon, S.C., Golombek, M.P., Jakosky, B.M., Banerdt, W.B., 2001. Ancient geodynamics and global-scale hydrology on Mars. *Science* 291, 2587-2591.

Pieri, D.C., 1980. Martian Valleys: Morphology, Distribution, Age, and Origin. *Science* 210, 895-897.

Plescia, J.B., Saunders, R.S., 1979. The chronology of martian volcanoes, *Proceedings of the10th Lunar and Planetary Science Conference*, Houston, Tex, pp. 2841-2859.

Pollack, J.B., 1981. *Atmospheres of the Terrestrial Planets.*, in: Beatty, J.K., O'leary, B., Chaikin, A. (Eds.), The new Solar System Sky Publishing Corporation and Cambridge University Press, Cambridge, MA, pp. 57-70.

Popa, C., Esposito, F., Mennella, V., Colangeli, L., 2010. Occurrence of anorthosite on Mars in Xanthe Terra, *EPSC Abstracts.* 5: 589.

Poulet, F., Bibring, J.P., Mustard, J.F., Gendrin, A., Mangold, N., Langevin, Y., Arvidson, R.E., Gondet, B., Gomez, C., 2005. Phyllosilicates on Mars and implications for early martian climate. *Nature* 438, 623-627.

Rabinowitch, E., Stockmayer, W.H., 1942. Association of Ferric Ions with Chloride, Bromide and Hydroxyl Ions (A Spectroscopic Study). *Journal of the American Chemical Society* 64, 335-347.

Robinson, M.S., Mouginis-Mark, P.J., Zimbelman, J.R., Wu, S.S.C., Ablin, K.K., Howington-Kraus, A.E., 1993. Chronology, Eruption Duration, and Atmospheric Contribution of the Martian Volcano Apollinaris Patera. *Icarus* 104, 301-323.

Rubin, A.E., 2000. Petrologic, geochemical and experimental constraints on models of chondrule formation. *Earth-Science Reviews* 50, 3-27.

Russell, M.J., Hall, A.J., 1997. The emergence of life from iron monosulphide bubbles at a submarine hydrothermal redox and pH front. *Journal of the Geological Society* 154, 377-402.

Ryder, G., Koerbel, C., Mojzsis, S.J., 2000. Heavy bombardment on the Earth at 3.85 Ga: The Search for Petrographic and Geochemical Evidence, in: N., C.R., Righter, K. (Eds.), *Origin of the Earth and Moon,*. University of Arizona Press, Tucson, AZ, pp. 475– 492.

Safronov, V.S., 1954. On the Growth of Planets in the Protoplanetary *Cloud. Astronomicheskii Zhurnal* 31, 499-510.

San Pham, L.B., Karatekin, Ö., Dehant, V., 2009. Effects of Meteorite Impacts on the Atmospheric Evolution of Mars. *Astrobiology* 9, 45-54.

Schulte, M., Blake, D., Hoehler, T., McCollom, T., 2006. Serpentinization and Its Implications for Life on the Early Earth and Mars. *Astrobiology* 6, 364-376.

Schulze-Makuch, D., Dohm, J.M., Fan, C., Fairén, A.G., Rodriguez, J.A.P., Baker, V.R., Fink, W., 2007. Exploration of hydrothermal targets on Mars. *Icarus* 189, 308-324.

Segura, T.L., Toon, O.B., Colaprete, A., Zahnle, K., 2002. Environmental Effects of Large Impacts on Mars. *Science* 298, 1977-1980.

Shearer, C., Burger, P., Papike, J., Karner, J., 2008. Mineral Recorders of Martian History: Reading the Record with Microbeam Trace Element Analyses. *Microscopy and Microanalysis* 14, 526-527.

Skok, J.R., Mustard, J.F., Ehlmann, B.L., Milliken, R.E., Murchie, S.L., 2010. Silica Deposits in the Nili Patera Caldera on the Syrtis Major Volcanic Complex on Mars. *Nature Geosci* 3, 838-841.

Skok, J.R., Mustard, J.F., Murchie, S.L., 2009. Identification of Primary Noachian Crustal Blocks on Mars with CRISM Observations. 40th Lunar and Planetary Science Conference, *Lunar and Planetary Institute Science Conference Abstracts* 40, 2180.

Sleep, N.H., 1994. Martian plate tectonics. *J. Geophys. Res.* 99, 5639-5655.

Sleep, N.H., 2000. Evolution of the mode of convection within terrestrial planets. *J. Geophys. Res.* 105, 17563-17578.

Sleep, N.H., Meibom, A., Fridriksson, T., Coleman, R.G., Bird, D.K., 2004. H2-rich Fluids from Serpentinization: Geochemical and Biotic Implications. *Proceedings of the National Academy of Sciences of the United States of America* 101, 12818-12823.

Southam, G., Rothschild, L.J., Westall, F., 2007. *The Geology and Habitability of Terrestrial Planets: Fundamental Requirements for Life Geology and Habitability of Terrestrial Planets*, in: Fishbaugh, K.E., Lognonné, P., Raulin, F., Marais, D.J., Korablev, O. (Eds.). Springer New York, pp. 7-34.

Squyres, S.W., Arvidson, R.E., Ruff, S., Gellert, R., Morris, R.V., Ming, D.W., Crumpler, L., Farmer, J.D., Marais, D.J.D., Yen, A., McLennan, S.M., Calvin, W., Bell, J.F., Clark, B.C., Wang, A., McCoy, T.J., Schmidt, M.E., de Souza, P.A., 2008. Detection of Silica-Rich Deposits on Mars. *Science* 320, 1063-1067.

Staufer, U., 2011. *Atomic force microscopy on Mars – Instrumentation and experimental results*, 2nd Joint Congress of the Portuguese and Spanish Microscopy Societies 18-21 October; Vilarinho, P. Aveiro, Portugal, 2011, Aveiro, Portugal,.

Stetter, K.O., 2006. Hyperthermophiles in the history of life. Philosophical *Transactions of the Royal Society B: Biological Sciences* 361, 1837-1843.

Stoeffler, D., Knoell, H.D., 1977. Composition and Origin of Plagioclase, Pyroxene, and Olivine Clasts of Lunar Breccias 14006, 14063, 14066, 14311, 14320, and 14321, in: Lunar Science Conference, t., Houston, Tex. Proceedings. Volume 2. (A78-41551 18-91) New York, Pergamon Press, Inc. (Ed.), *Lunar Science Conference, 8th. Pergamon Press,* Houston, Tex, pp. 1849-1867.

Svetsov, V., 2007. Atmospheric erosion and replenishment induced by impacts of cosmic bodies upon the Earth and Mars. Solar System Research 41, 28-41.

Tanaka, K.L., 1986. The Stratigraphy of Mars. *J. Geophys. Res.* 91, E139-E158.

Tanaka, K.L., Isbell, N.K., Scott, D.H., Greeley, R., Guest, J.E., 1988. The resurfacing history of Mars - A synthesis of digitized, viking-based geology, *Lunar and Planetary Science Conference, 18th, Mar.* 16-20, 1987. Cambridge and New York/Houston, TX, Cambridge University Press/Lunar and Planetary Institute, Houston, TX, pp. 665-678.

Tarbuk, E.J., Lutgens, F.K., 2004. Ciencias de la Tierra. *Una introduccion a la Geologia Fisica*. Pearson-Prentice Hall., Madrid.

Tera, F., Papanastassiou, D.A., Wasserburg, G.J., 1974. Isotopic evidence for a terminal lunar cataclysm. *Earth and Planetary Science Letters* 22, 1-21.

Touboul, M., Kleine, T., Bourdon, B., Palme, H., Wieler, R., 2007. Late formation and prolonged differentiation of the Moon inferred from W isotopes in lunar metals. *Nature* 450, 1206-1209.

Valdivia-Silva, J., Negron-Mendoza, A., Ramos, S., Colin-García, M., Ortega-Gutierrez, F., García, J.L., 2011. Planetary conditions at the Hadean and Archean transitions. *Possible Scenarios for the Origin of life*. Nova Science Publishers, Inc, New York.

Wächtershäuser, G., 1992. Groundworks for an evolutionary biochemistry: The iron-sulphur world. *Progress in Biophysics and Molecular Biology* 58, 85-201.

Wanke, H., Dreibus, G., Wright, I.P., 1994. Chemistry and Accretion History of Mars [and Discussion]. Philosophical Transactions of the Royal Society of London. *Series A: Physical and Engineering Sciences* 349, 285-293.

Wetherill, G.W., 1990. Formation of the Earth. *Annual Review of Earth and Planetary Sciences* 18, 205-256.

Wilde, S.A., Valley, J.W., Peck, W.H., Graham, C.M., 2001. Evidence from detrital zircons for the existence of continental crust and oceans on the Earth 4.4[thinsp]Gyr ago. *Nature* 409, 175-178.

Wilson, L., Scott, E.D., Head, J.W., III, 2001. Evidence for episodicity in the magma supply to the large Tharsis volcanoes. *J. Geophys. Res.* 106, 1423-1433.

Wood, B.J., Halliday, A.N., 2005. Cooling of the Earth and core formation after the giant impact. *Nature* 437, 1345-1348.

Wood, B.J., Walter, M.J., Wade, J., 2006. Accretion of the Earth and segregation of its core. *Nature* 441, 825-833.

Wyatt, M.B., McSween, H.Y., 2002. Spectral evidence for weathered basalt as an alternative to andesite in the northern lowlands of Mars. Nature 417, 263-266.

Zimbelman, J.R., Salomon, S.C., Sharpton, V.L., 1991. The evolution of volcanism, tectonics, and volatiles on Mars-an overview of recent progress. *Proceedings, Lunar and Planetary Science* 21, 613-626.

In: Mars
Editor: Alberto G. Fairén

ISBN: 978-1-62618-102-1

Chapter 7

LIFE IN THE DRY VALLEYS, ANTARCTICA: INDICATORS FOR THE EXISTENCE OF LIFE ON MARS?

***Kris M. Hart**[*1]**, James M. Dohm**[†2] **and William C. Mahaney**[‡3]*

[1]School of Chemical Sciences, Dublin City University, Dublin, Ireland
[2]Department of Hydrology and Water Resources, University of Arizona, Tucson, US
[3]Quaternary Surveys, Thornhill, Ontario, Canada

ABSTRACT

The origin and evolution of life in Antarctica may be considered to offer parallels to possible biogenesis on Mars through correlative threads of which are explored herein. While the Antarctic analogues described here cover a time span of at most the last 50 million years, similar lines of evolution on Mars in cold dry climatic conditions, albeit with possible ameliorative periods, covers a much longer interval of billions of years. With similar mixes of ultrabasic to silicic rocks on both planets, weathered regoliths (paleosols) on Mars are likely to have produced abiotic structures capable of supporting life now and/or in the distant past. Moreover, Martian paleosols could reveal a mineral and chemical codex, one with pedostratigraphic profiles of much greater complexity than presently hypothesized by many researchers, and along with this a diverse and complex array of microbial communities, perhaps with much different evolutionary pathways than on Earth.

* Corresponding Author: email: hart.kris1@gmail.com
† email: jmd@hwr.arizona.edu
‡ arkose@rogers.com

INTRODUCTION

The planet Mars has been the subject of speculation for its potential as a habitat capable of supporting life (e.g., Schulze-Makuch et al., 2008; Fairén et al., 2010, Dohm et al., 2011) for hundreds and perhaps thousands of years. *In situ* investigation for extraterrestrial life (be it extinct or extant) on the Red Planet began with the Viking Lander missions during the 1970s (Klein, 1999), yielding ambiguous experimental results largely interpreted to be a biological chemical reactions (Klein, 1978), though a biological explanation could not be ruled out (Levin and Straat, 1981). The instrument payload of the Landers detected water vapor, water-ice, and most of the chemical elements essential for terrestrial life in surface/near-surface rock materials and in the atmosphere at both landing sites (Mazur et al., 1978). Arguably no organic molecules other than those apparently derived from terrestrial contaminants were found by the on-board gas chromatography–mass spectrometer (GC-MS), though this is currently disputed (Navarro-González et al., 2010). The labelled release experiment for metabolic activity, however, delivered positive signals, which initiated a controversy over the possible life-detection methodology of the Viking Mission. The presence of strong unknown oxidants in the regolith has been proposed as a possible mechanism for the destruction of putative organics and a contributor to false-positive results (Levin and Straat, 1981; McKay et al., 1998a; Navarro-González et al., 2006). Corroborating this possibility, the Phoenix mission detected perchlorate in surface/near-surface rock materials, which may have been the oxidant that influenced the Viking measurements (Hecht et al., 2009). Insufficient sensitivity of the GC-MS could be another reason for not finding the organics, as the detection limit of the GC-MS was between 10^2 and 10^6 bacteria per gram of soil analyzed (Levin and Straat, 1981; McKay et al., 1998a; Glavin et al., 2001). Insufficient thermal isolation of the instrument is yet another explanation for why there could have been false-positive results (Miller et al., 2001). As there are unknowns for each option, the potential presence of life cannot be ruled out and thus the debate continues (e.g., Levin and Straat, 1981; Miller et al., 2001; Navarro-González et al., 2006; Biemann, 2007; Houtkooper and Schulze-Makuch, 2007). Subsequent missions have paved the way for assessing prime targets of elevated life potential (Dohm et al., 2004, 2011), such as those surface and near-surface environs that exhibit present/past water and energy conditions (Hock et al., 2007; Furfaro et al., 2008), and have yielded information that will likely be folded into future smart robotic reconnaissance missions (Fink et al., 2005).

Investigations, partly based on Earth analogues, since Viking have resulted in a greater understanding of the paleoclimate and geology of Mars. An understanding of the recent and ancient geologic and hydrologic pasts of Mars is essential to making implicit assumptions as to the emergence of life, as well as identifying and characterizing primary targets likely to yield traces of its existence (e.g., Dohm et al., 2004, 2011). Mars has had an active and varied geologic history similar in many respects to that of Earth (Baker et al., 2007). The first several hundred million years of Mars' evolution, or so, are generally believed to be warm and wet (Baker et al., 1991, 2007; Carr, 1996; Baker, 2001), while recent work suggests that cold and wet conditions may have prevailed (Fairén, 2010). Distinct networks of valleys dissect the Martian surface suggesting significant hydrological activity, especially highlighted during the Noachian Period or more than 3.7 billion years ago based on the estimated absolute ages determined through impact crater investigations (Hartmann and Neukum, 2001) (e.g.,

Scott et al., 1995). Though surface conditions between Earth and Mars may have varied significantly during this relatively early stage of evolution, on Earth, self-replicating, proto-organisms are estimated to have emerged between approximately 4.0 - 4.2 billion years ago for deep marine hydrothermal abiogenesis and 3.7 - 4.0 billion years ago for terrestrial abiogenesis (Maher and Stevenson, 1988).

Later during the Hesperian Period (~ 3.8 - ~ 3.0 billion years ago, based on Hartmann and Neukum (2001)) the Martian climate became generally much colder and drier. As temperatures persistently decreased, the scarcity of liquid water was inevitable (however, this does not mean the complete absence of liquid water during this period). The Martian hydrosphere developed into a thick layering of ice giving rise to the second age of Mars being referred to as "Snowball Mars" (Fairén et al., 2010). This semipermanent, ice-rich permafrost constituted almost all of the surface water including the "Oceanus Borealis". Exceptions are episodic events where volcano-ice interactions occurred, providing short-lived outbursts of liquid water from surrounding ice features and underlying ground water (Baker et al., 1991, 2007; Dohm et al., 2001a, 2009; Baker, 2006). The prolonged existence of the ice-rich permafrost layer is purported to have been about 1 - 2 km thick at the equatorial regions and 5 - 6 km thick at the poles (Kuzmin, 2005). The global climate change on Mars was caused by decreasing atmospheric pressure to below the triple point of water (McKay et al., 1992) attributed to the disappearance of the planetary magnetic field (Chassefière and Leblanc, 2004), including the shutdown of a hypothesized phase of plate tectonism not necessarily Earth-like (Baker et al., 2007), catastrophic impact (Brain and Jakosky, 1998), and gradual erosion of the atmosphere by solar wind (Jokosky et al., 1994).

The hyperarid Mars observed today is extremely cold and exposed to a full spectrum of solar radiation. Cold, arid, and radiated surface conditions have dominated during the Amazonian Period (the last 3 billion years to present, based on Hartmann and Neukum et al., 2001), and present extreme conditions to what we perceive as viable life forms. Mean surface temperatures and pressures of -60°C and ~ 8 kPa, respectively, indicate that liquid water at the surface is rare and highly transient (Fairén et al., 2009). The potential existence of strong surface oxidants (Atreya et al., 2006), in addition to the constant bombardment by solar generated UV radiation (between 190 - 300 nm) is yet another constraint to the survival of organisms on/in the current Martian cryosphere. Thus, if microbes exist in surface soils/paleosols, they will populate profiles that look somewhat like similar pedons in the Antarctic (Mahaney et al., 2001; Hart et al., 2011).

Studies of the microbiological component of Earth's biome have exponentially grown over the last decade, with environmental microbiologists adopting more invasive techniques to study microbe abundance and diversity. Until the late 1990's, nearly all information about prokaryotes was based upon measurements of isolated species (< 5000), a number that represents only ~ 0.1% of the estimated total diversity of prokaryotes in the Earth's biome (Curtis et al., 2002; Madsen, 2005). Traditional microbiology has depended upon laboratory cultivation, where a sub-sample is propagated over varying time spans and conditions to allow the biomass morphology to be observed (microscopy) and biochemical characterisation or phylogenetic analysis to be carried out on the cultivars. The development of molecular biology on the conserved regions of nucleic acids has allowed for the characterization of complex microbial communities without the need for cultivation and the biases such methods bring (DeLong and Pace, 2001).

The benefits of culture-independent molecular biology when applied to environmental samples has been to extend the known diversity of prokaryotes from 12 divisions of bacteria, as proposed by Woese, (1987), to 52 (Rappé and Giovannoni, 2003; Schloss and Handelsman, 2004), with new discoveries being made daily. Another recent development in modern microbiology is the cryptogenic technique, *Fluorescence in Situ Hybridization* (FISH), where microorganisms can be observed *in situ* at the genus level to reveal abundances, location, and community structure (Pernthaler et al., 2001). FISH is a staining technique (of sorts) that uses rRNA-fluorescent probes that attach to conserved regions of the cells chromosome. The technique holds great value when determining community order from various biomes, but is limited by the light gathering capacity of the particular microscope used. In fact, technological limitations are still one of the biggest obstacles to studying microbial terrestrial ecosystems. Culturing methods still have merit in environmental microbiology (e.g. nutritional studies using isotopic tracers), especially for testing stress factors (e.g. oxygen, pH, oxidative damage, desiccation, etc.) and response to stimuli by prokaryotes (e.g. light, carbon sources, nutrients, etc.). The application of modern molecular techniques to environmental samples, that demonstrate the limits life can exist, are of great significance to forming hypotheses of extraterrestrial life within the Mars-containing solar system. However, Madsen (1998) presents compelling arguments that demonstrate the limits of what laboratory-based experiments can show in regard to environmental applicability.

Prokaryotes are the most successful and long-lived group of organisms known, with an evolutionary history stretching back at least 3.5 billion years on Earth (Madsen, 1998) and likely longer on Mars if existent. Their ubiquitous distribution, relatively high rates of growth and metabolic activity, and unparalleled enzymatic diversity and physiological responsiveness place them as the chief recycling agents for the elements within our biosphere (Madsen, 2005). Exobiology studies are currently limited to Earth-based analogues of our planetary neighbours where scientists probe the limits of what is understood to be 'life' (Dohm et al., 2011). Extremophilic organisms, especially microorganisms, have been documented to exist in the most supposedly harsh conditions such as volcanic fumaroles (Stetter, 2005; Costello et al., 2009), high alpine slopes/peaks (Spalding, 1979; Swan, 1992; Mahaney, 1990), hyperarid environments (Warren-Rhodes et al., 2006, 2007a,b; Hock et al., 2007; Navarro-González et al., 2010), and cold deserts (Mahaney et al., 2001; de la Torre et al., 2003; Hart et al., 2011). A good example of one such barren landscape is the McMurdo Dry Valleys (herein called the Dry Valleys, see Figure 50), where cold desert soils (Cryosols) are among the driest and coldest on Earth (Campbell and Claridge, 1987; Campbell et al., 1997; Burkins et al., 2001). The Dry valleys of South Victoria Land, Antarctica, are one of the largest ice-free areas of continental Antarctica and compare favorably with those expected for the Martian crust (Mahaney et al., 2001, 2009; Mahaney and Dohm, 2010). This review will mainly focus on the current knowledge of the Antarctic Dry Valleys ecosystem and its relevance to Martian habitability. More importantly, perhaps, is the type of Martian environment which might be conducive to biogenesis, now or in the distant past. If life developed early on Mars, much earlier than on Earth, could it have been sustained when a thinner atmosphere and cold and dry conditions largely prevailed especially highlighted during the Amazonian Period? Would evolution have run apace with developing colder and drier conditions and would new species correlate with organisms known to exist in Antarctic Cryosols? Cryosolic organisms in the Antarctic have had 30 million years or so to develop. On Mars, similar organisms could have had as much time and, more likely, much greater time (e.g., several billions of years) to build an evolutionary genetic system capable of dealing with extreme desiccation and extremely cold temperatures equator to pole.

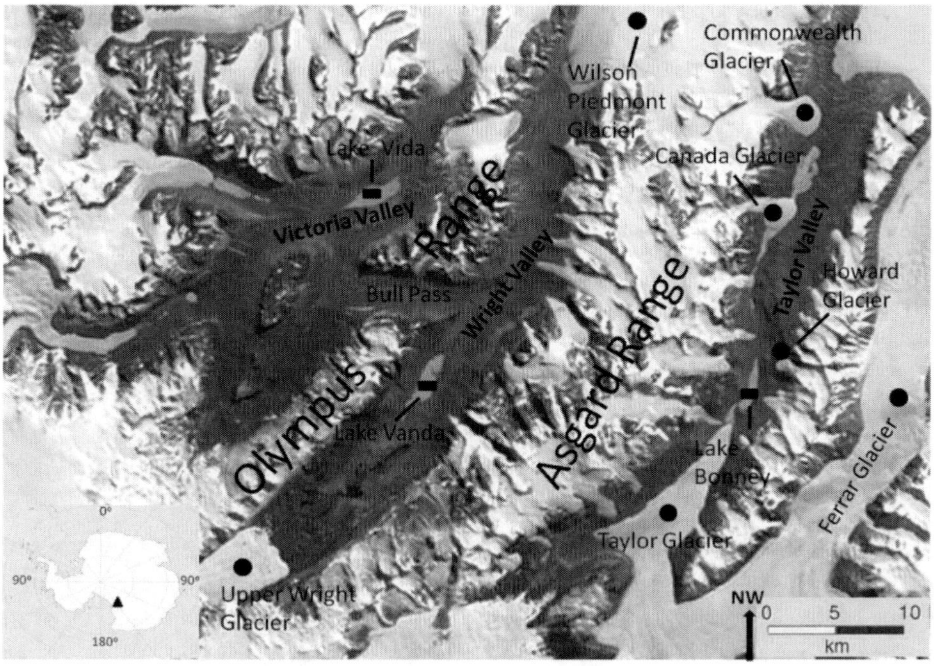

Figure 50. The McMurdo Dry Valleys (77° S, 161° E), Antarctica, and some points of interest, such as glaciers and periodic lakes. Modified image courtesy of: http://earthobservatory.nasa.gov/IOTD/view.php?id=2140. Inset; continent of Antarctica showing approximate geographical position of the McMurdo Dry Valleys (▲). Image courtesy of US Geological survey; available at: http://lima.usgs.gov/.

CRYOSOLS OF THE DRY VALLEYS, ANTARCTICA

Environmental Context and Constraints to Life

The Antarctic continent is mostly covered with ice sheets, representing one of the most physically and chemically extreme habitats. The extant microbial communities are mostly uncultured and taxonomically distinct, with a high reliance upon abiotic processes. Ice-free areas only account for ~ 0.4% of the land area, though much of these regions have been ice-free since the last glacial maximum (Bockheim, 2008). The Cryosols (cold desert soils) within the Dry Valleys of the Antarctic are resident in one of the coldest (-30°C mean temperature) and driest (< 10 cm precipitation y^{-1}) on Earth (Campbell et al., 1997; Burkins et al., 2001), comparing favorably with those expected in the Martian crust, and hence making it a suitable Martian analogue (Mahaney et al., 2001, 2009; Mahaney and Dohm, 2010). Potassium/argon dates acquired from lava flows indicate that the Dry Valleys have been in existence for > 4 million years (Prentice et al., 1998). The ice plateau that dominates most of the continent influences the climatic conditions of the Dry Valleys. Katabatic winds travel from the South Pole, across and off the plateau, and into the valleys throughout the

Transantarctic Mountains, bringing cold and dry air at high velocity. This leads to ablation of snow and ice along the valley floor which, further compounded by low precipitation, results in some of the driest soils on Earth (Hopkins et al., 2006). Further, light regimes, extreme chemical gradients, and elevated ultraviolet-A and -B solar radiation further compound the inhospitality of this Antarctic region (Doran et al., 2002a). Microorganisms existing under these cold, highly saline, nutrient-poor, and water-limited conditions are still known to proliferate (Mahaney et al., 2001, 2012; Hart et al., 2011), dominated chiefly by phototrophic and heterotrophic species.

While the physical, chemical, and mineralogical properties of Cryosols in the Antarctic Dry Valleys have been worked out and their geochronological relationships established, little is known about their microbial composition (Mahaney et al., 2001). Additionally, the Fe- and salt-rich horizons of the Dry Valley Cryosols bear some resemblance to that of the surface and near-surface rock materials of Mars (Yen et al., 2005; Lane et al., 2004), providing some of the mineralogical conditions conducive to microbial growth. Antarctic Cryosols are devoid of water except when rare snowfall melts during the short summer season, and minor amounts of meltwater seep through the surface pavement to the Fe-rich and salt-rich horizons below (Figure 51). Near the Taylor Glacier, Cryosols are formed in a mix of dolerite, quartz sandstone, and granite, all constituting transported regolith left behind in high alpine valleys, episodically invaded by ice during the Early to Middle Miocene. Chemical weathering (chiefly oxidation), possibly chelation with forms of organic carbon, and physical processes such as salt crystal growth amount to increments (or deposits) of about 1-2 cm/weathering thickness/million yr^1 (Mahaney et al., 2009). Sources of organic carbon within the area are limited to fresh input from various microorganisms (e.g., lichen, algae (Finegold et al., 1990) and cyanobacteria (Wood et al., 2008)) and ancient fossilized organic matter. Matsumoto et al. (1987), for example, postulated that the erosion and transport of sandstones from the Beacon Supergroup into the Cryosols carried with them hydrocarbons (steranes and visual kerogen) and silicified wood fragments, providing a source of nutrients for microorganisms in the substrate. The majority of organic materials are assumed to be contributed by vascular plants and, to a lesser extent, microorganisms, but input from sedimentary sources (Beacon Supergroup) is also a possibility as per above.

Figure 51. Leading edge of the seasonal Von Guerard stream, McMurdo Dry Valleys, Antarctica. The stream is one of many in the area usually associated with glacial melt water or rare snowfall events. Surrounding desert pavement consisting mostly from a mix of dolerite, quartz sandstone, and granite, all constituting transported regolith left behind in high alpine valleys. Photo courtesy of Chris Jaros (INSTAAR).

Though the above sources are possible, aeolian processes are likely to be the major transport system of nutrients to the Dry Valleys (Swan, 1992; Lancaster, 2002) via local and/or (to a lesser degree) deposition from surrounding continents such as South America and Australia (Deckker et al., 2010 and references there in). Lancaster (2002) attributes the flux of deposition to low sediment supply and, to a lesser degree, wind speeds. Wind speeds and elevation of the test sites have been particular factors in deposition rates showing that the valley floors experienced much higher aeolian flux compared to all other locations. The influence of wind-borne transportation in remote locations such as Antarctica is extremely important for delivering limiting nutrients, such as inorganic N and P. Deuerling (2010) claims that upon dissolution of inorganic nutrients, their counterpart organic nutrients increase in glacial and lake leachates due to the influence of native biota.

Endolithic Communities – Life in the Rocks

The endolithic environment can be defined as the pore space or cracks within rocks that provide habitation for microorganisms and act as an interface between biology and geology (Walker and Pace, 2007a). These endolithic ecosystems, which are predominantly photosynthesis-based, inhabit exposed rocks at depths usually ranging from the upper few millimeters to centimeters. Knowledge of these relatively newly discovered niches (Friedmann et al., 1967) has been greatly accelerated and enhanced by the application of molecular techniques, revealing endolithic communities to be relatively depauperate in comparison to other terrestrial microbiological communities (de la Torre et al., 2003; Walker and Pace, 2007a; Walker and Pace, 2007b; Omelon, 2008). Low-diversity communities are extremely useful for astrobiologists, geobiologists, and microbial ecologists, as they are robust, ubiquitous, and appear to occur with some degree of uniformity (Friedmann and Ocampo-Friedmann, 1984).

Endolithic microbes have been reported in both rock outcrops and strewn rock materials of relatively hot and cold deserts (Friedmann, 1980; de la Torre et al., 2003; Warren-Rhodes et al., 2006, 2007a,b; Hock et al., 2007; Rios del los et al., 2007; Omelon, 2008; Navarro-González et al., 2010), high altitude/alpine locales (Walker and Pace, 2007b; Horath and Bachofen, 2009), geothermal environments (Glamoclija, et al., 2004; Walker et al., 2005; Gaylarde et al., 2006), and cliff faces (Gerrath et al., 2000; Casamatta et al., 2002). The discovery of endolithic communities in the Dry Valleys of Antarctica stimulated great interest as soil horizons were once believed to be sterile (Friedmann and Ocampo, 1976). The existence of community structures of extant microorganisms living under extreme conditions has led to intensive reconnaissance of physical and chemical environments supporting such growth. The application that such research has had in regards to polar deserts is presented briefly here, but for more detailed overviews please see Walker and Pace (2007a) and Omelon (2008).

The exposed rock faces of relatively hot and cold deserts may represent a final niche, where relatively complex communities can exist under extreme conditions or where specialist microorganisms express adaptive strategies allowing them to proliferate in zones devoid of excessive inter-species competition. Endolithic environments most likely act as buffers between microorganisms and the external conditions. For instance, intense solar radiation, extreme temperature fluctuations, desiccation, and strong winds are factors that would most likely inhibit surface growth on the rock face (epilithic growth) (Friedman, 1982). The type of rock best suited to house such communities is poorly understood (Walker and Pace, 2007b), but is thought to be influenced by mineral composition, pore structure, translucence, and surface permeability. External factors are most likely to be as important such as water availability, pH, exposure, and delivery of nutrient sources (e.g., aeolian dust deposition) (Omelon et al., 2007).

Water is the primary constituent for life, as is currently understood, and therefore, it is also one of the most important factors in sustaining endolithic ecosystems. In hot deserts, water availability and abundance mostly occurs as fog or dew that accumulates upon the surfaces of soils or rocks, but rapidly evaporates as daytime temperatures rise (Nicholson, 2011), perhaps optimally expressed in the Atacama Desert (Warren-Rhodes et al., 2006, 2007a, b; Hock et al., 2007). For polar deserts, a moisture gradient is more commonly observed but with the Dry Valleys experiencing much lower annual rain and snow fall events than the high Arctic (Gooseff et al., 2003; Omelon et al., 2003). The pore spaces within the rock protect the below-surface environment from desiccation and deflect most of the polar katabatic winds. The organisms that make up the pore-space community have adapted themselves to frequent and prolonged desiccation events (Potts, 1999), and though some cyanobacteria appear to have no adaptive mechanism to continue to photosynthesize under low water conditions, they can resume photoautotrophic activities rapidly after re-wetting (e.g., members of the genus *Chroococcidiopsis* (Billi, 2010)). The production of water-retention molecules such as trehalose and sucrose (Wharton, 2002) is commonly observed from the cultivatable fraction of endolithic communities (Friedmann et al., 1993). As liquid water mostly enters the pore spaces of rocks via percolation, these events are reliant on rare snowfall events in the Dry Valleys, and the melting of the snow after solar heating (Friedmann and McKay, 1985). Though liquid water is only available when temperatures rise to > 0°C, it may also be likely that mineral assemblages such as sulphates and phyllosilicates provide liquid water molecules to endolithic communities at temperatures below 0°C after chemical dissolution of the surrounding rock (Barker and Banfield, 1998). The production of extracellular polymeric substances (EPS) by community members retain moisture for prolonged periods allowing endolithic primary production to continue after external moisture has frozen or evaporated (Potts, 2001; Mager and Thomas, 2011).

The influence of light on a photosynthetically driven system such as those in the Dry Valleys is a vital aspect. The translucence of the host rock is a strong influence on the depth of penetration of the community, especially the ability to attenuate photosynthetically active radiation (PAR; ~ 400 - 700 nm). Light penetration into the rock-face decreases between 70 - 90% for every millimetre travelled into the surface, though this distance increases ten-fold under damp conditions due to increased refractive index (Nienow et al., 1988). Rock color, mineralogy, and internal structure are likely to be strong physical factors that limit light penetration (Walker and Pace. 2007b). The endolithic microbial communities expand as stratified horizons according to community position, with photosynthesizers dominant at the upper layers (made obvious by the layers of pigmented molecules), though present throughout the colony (see Figure 1 from Walker et al., 2005; Wong et al., 2010). Photo-protective molecules throughout the vertical community, such as scytonemin and mycosporine-like (Wynn-Williams et al., 1999), are likely present for cellular defense against excessive UV exposure.

Temperature regimes in the Dry Valleys and other polar regions severely limit metabolic activity as temperatures can be on average -20°C, with air temperatures scarcely rising above 0°C (Doran et al., 2002a). Nonetheless, Antarctic communities rely on solar heating to melt snow and ice and allow for photosynthesis reactions to take place (Walker and Pace, 2007a). The thermal properties of rock, where heat capacity allows for the absorbance of solar heat, can provide an internal microclimate up to 20°C warmer than the external environment (Cockell et al., 2003; Omelon et al., 2006). The heat capacity of rocks in cold climates also helps to mitigate the effects of rapid temperature changes by slowing freeze-thaw cycles sufficiently to allow endolithic communities to regulate intra-cellular ice formation (Omelon, 2008). According to Siebert et al. (1996), many isolates of Antarctic endoliths are psychrophilic with optimal growth rates between 0°C and 20°C. Temperature can have a deciding factor on the fate of colonisation of a particular site, with factors such as orientation and compositional colour determining if colonization is possible or not (Omelon, 2008).

Access to nutrients is a limitation experienced by endolithic communities by the very nature of their protected environment. A rock substrate would be considered oligotrophic (Walker and Pace, 2007a), and cultivation studies of endolithic community members supports this hypothesis (Siebert et al., 1996) where even the addition of nutrients to the system does not stimulate growth (Johnston and Vestle, 1989, 1991). Siebert et al. (1996) suggest that it is likely that the host rock provides some nutrients, but as a wide variety of rock types are colonised, external weathered material or dust particles provide the majority of essential nutrients (Omelon et al., 2007).

The presence of slow-growing and long-lived endolithic communities in polar deserts aptly exemplify life at the extremes where microorganisms can colonize an area despite the harsh environment. The age of some Antarctic endolithic communities has been placed at 10^3 - 10^4 years using ^{14}C dating of the bulk carbon (Sun and Friedmann, 1999), and the colonization of newly exposed rock faces taking only a few years to be established, though this is currently poorly understood (Walker and Pace, 2007a). In general, the organisms found in these niches include primary producers such as cyanobacteria and algae, consumers (fungi), and opportunistic decomposers (fungi and heterotrophic bacteria). Anaerobic bacteria, archaea, and chemoautotrophic bacteria do not appear to represent significant contributions to the microbial diversity of endolithic communities (Cowan and Tow, 2004; Pointing et al., 2009). Studies from diverse locations such as the Antarctic Dry Valleys (de la Torre et al., 2003), Rocky Mountains (Walker and Pace, 2007b), Yellowstone Park (Walker et al., 2005), Central Tibet (Wong et al., 2010), and Shark Bay, western Australia (Papineau et al., 2005), show that endolithic communities are among the least complex known on the terrestrial surface (Walker and Pace, 2007a; Omelon, 2008). The ability of cyanobacteria and fungi to protect themselves from high doses of radiation (mostly UV) and to survive repeated bouts of desiccation for prolonged periods has provided freedom to speculate on their astrobiological significance.

The techniques being developed to further probe the limits of life have been extensively utilized in polar regions because they currently represent the best analogues to extraterrestrial environments such as Mars (Mahaney et al., 2001, 2009). The recent advances in geobiology have allowed for a wider understanding of the interactions between minerals and microbes (Madsen, 2005). The presence of biosignatures in terrestrial rocky habitats, no matter how tentative, could be employed to probe the pre-existence of such communities through studies of mineral modification and/or precipitation (Johnston and Vestal, 1989; Banfield et al.,

2001). The search for biosignatures in sheltered areas such as lithic environments holds great promise when one considers that the familiar chemical basis of terrestrial life on Earth may not apply elsewhere in the solar system and these structures may act as refuges for extinct and extant life.

Edaphic Communities – Life in the Dry, Salty, and Cold Soils

Until recently, little was known about the microbial diversity of Antarctic soils. The upper surface soil, so called "active zone", extends between a few mm to ~ 45 cm below the surface (Bockheim et al., 2007; Marinova et al., 2012). The extreme nature of the near-surface soils of the Dry Valleys, characterized by low temperatures, desiccation, high soil salinity, low nutrient input, high (summer) UV radiation levels, and strongly abrasive katabatic winds, all contribute to the appearance of a sterile to low diversity of taxa. Culture-based studies and microscopy observations have revealed sterile (Horowitz et al., 1972) to very low species diversity (Morelli et al., 1972; Johnson et al., 1978; Hart et al., 2011) indicating rather than the more cosmopolitan populations found across the world (Aislabie et al., 2006). The majority of the bacteria identified using these classical techniques belong to the genera *Corynebacterium, Arthrobacter, Micrococcus, Brevibacteriumm, Bacillus, Pseudomonas,* and *Flavobacterium* (Johnson et al., 1978; Gounot, 1991). More recent studies employing molecular biology have shown the Dry Valley soil community to be much more diverse and widespread throughout the immediate area (Cowan et al., 2002; Aislabie et al., 2006; Smith et al., 2006; Mahaney et al., 2012). The structure of these biological communities is still relatively simple (in comparison to more temperate soils) with only few discernable trophic levels and limited invertebrate taxa (Cary et al., 2010).

The floors of the Dry Valleys are comprised of dry permafrost with a polygonal surface terrain, containing a desert pavement of dolerite, quartz sandstone, and granite surrounded by mineral-enriched soils (Mahaney et al., 2001). The majority of the parent material is believed to be sourced from moraine deposits, weathered gneiss, schist, and outwash sediments (Beyer et al., 2001). A study carried out by Bockheim and Hall (2002a) suggests that 70% of the Dry Valleys soils are classified as Anhyorthels, which reflects the minimal cryoturbation in Antarctica. Organic carbon concentrations of Dry Valley soils are low (< 0.01 - 0.03%; Mahaney et al., 2001; Hart et al., 2011) and less than most other desert soils due to the lack of higher vegetation (Boyd et al., 1970; Blume et al., 2002).

The surfaces of Antarctic soils are subject to regular and large variations in temperature with mean annual temperatures ranging from -30°C to -15°C (Doran et al., 2002b). The austral summer brings a short (~ December to ~ February) period where temperatures exceed 0°C allowing for the availability of liquid water (McKay and Friedmann, 1985; Doran et al., 2002b). Daily multiple freeze-thaw cycles are known to occur during summer months when solar heating between cloud cover can create temperature fluctuations of 20°C or more (Aislabie et al., 2006). During winter months, temperatures plummet to between -40°C and -60°C (Doran et al., 2002a). The surfaces of the mineral-enriched soils of the Dry Valleys are exceptionally dry, containing typically < 1 - 2% water content (Bockheim, 1997; Cary et al., 2010). The deposition of snow in the Dry Valleys is generally rare and much of the water equivalent (< 10 cm yr^{-1}) is removed before penetrating soils due to sublimation (Witherow et al., 2006). The low moisture content of the constant and prevailing katabatic winds descending from the polar ice sheet into the open Dry Valleys is another source of rapid evaporation (Nylen et al., 2004). Antarctica, however, contains 37% of the world's permafrost (most of this lies beneath the East Antarctic ice sheet) and much of the soils of the Dry Valley contain buried ice or ice-cemented soil (Bockheim and Hall, 2002). This permafrost layer is often found within 30 cm of the soil surface (Bockheim, 2002) and can influence overriding microbiota by providing a humidity gradient between shallow wet soils and the dry surface (Hagedorn et al., 2007). The high incidence of evaporation and low surface run-off from the mineral-enriched soils has led to a large build-up of aerosol-deposited salts over geological time (Mahaney et al., 2001). Highly elevated salt levels in the soils provide additional constraints to life-forms as the bioavailability of water becomes further restricted.

For approximately six months of the year, the Dry Valleys receive solar radiation but elevated in UVA (320 - 400 nm) and UVB (280 - 320 nm; UVC is outside the range of the solar radiation field at the Earth's surface). The seasonal ozone depletion over Antarctica (caused by the anti-cyclone phenomenon observed at both poles (Lamb et al., 1955)), results in higher incidences of UVB radiation (Jackson et al., 2011 and references therein). This is particularly pertinent to soil-dwelling bacteria, especially cyanobacteria, where exposure to solar radiation is required and possible UV toxicity is difficult to avoid (Billi, 2010). There are large differences between individual species of cyanobacteria and their response to UV damage. For example, the Antarctic cyanobacterium, *Phormidium murrayi,* showed inhibited growth for 5 days under UVB bombardment, whereas another species isolated from the Dry Valleys (*Oscillatoria priestleyi*) remained inactive under the same conditions (Vincent and Quesada, 1994). Photoautotrophic edaphic communities in arid sandy soils are important as they aggregate sand particles and form thin semi-permeable crusts. These fragile crusts trap aeolian dust particles and reduce evapotranspiration (Verrecchia et al., 1995). However, cyanobacteria have been shown to play a minor role in Dry Valley soil ecosystems with the majority of bacteria belonging to heterotrophic species (Smith et al., 2006; Wood et al., 2008; Pointing et al., 2009). Pointing et al. (2009) observed *Chroococcidiopsis*-like and oscillatorian cyanobacterial morphotypes upon rock surfaces (sandstone and quartz, respectively), along with copious amounts of extracellular polymeric substances (EPS; i.e. biofilm) surrounding the colonized site. The production of EPS by microbial communities is believed to provide protection from many environmental stresses (Roberson and Firestone, 1992; McKew et al., 2011). Using (nucleic acid based) molecular analysis, cyanobacterial signatures were absent from the surrounding soil with only the detection of green, non-sulphur chloroflexi phototrophs in low abundance (Pointing et al., 2009). Cyanobacteria are well distributed throughout the ice-free areas of Antarctica, showing a high degree of diversity among geographical locations (Namsaraev et al., 2010), though the overall diversity is low (Wood et al., 2008; Pointing et al., 2009) in comparison to more temperate soils. Stable isotope evidence from the organic component of Dry Valley soils indicate depleted $\delta\ ^{13}C$ enrichment, indicative of autotrophic fixation (Hopkins et al., 2009) as they fractionate carbon sources according to isotope (preferring the energetically less demanding ^{12}C). In the absence of higher plants and lichens, it is plausible that cyanobacterial fixation of CO_2 and/or N_2 provides sufficient nutritional input to support a limited community. Studies of the mineral-enriched soils of the Dry Valleys identified the presence of *Chroococcales*, *Nostocales* and *Oscillatoriales* (Cavacini, 2001; Broady, 2005; Smith et al., 2006; Wood et al., 2008), but larger numbers of, as yet unidentified cyanobacteria (according to the low-similarity scores of existing 16S rDNA clone libraries), suggest that up > 15 species could be present in a sample (Wood et al., 2008).

The paucity of ice-free Antarctic soils also relates to domain diversity (Woese et al., 1990) where archaeal and eukaryal DNA sequences are completely absent (Pointing et al., 2009) to extremely low (Cary et al., 2010). Many soil types of the Dry Valleys lack higher order species such as metazoans, though nematodes are well documented across the region exploiting brief periods when liquid water becomes available (Poage et al., 2008). Protozoa are present in limited water film communities, generally dominated by flagellates and amoebae. Bamforth et al. (2005) found viable soil protozoa in 100% of the surveyed locations (n = 12) but could not confirm *in situ* activity. Symbiotic associations such as lichens are documented for the Dry Valleys consisting of both mycobionts and phycobionts existing on rock faces (Pointing et al., 2009), and several lichen-forming species of fungi (e.g. *Caloplaca*) have been detected in the upper soil horizons (Fell et al., 2006). Lichens have been reported to form a major fraction of the extractable organic matter from Dry Valley soils, though the immediate origin remains elusive (Hart et al., 2011). Saprophytic fungi are present being mainly represented by two major groups: Ascomycetes and Basidiomycetes (Fell et al., 2006; Pointing et al., 2009). Algal sequences are low in both biomass and diversity for Dry Valley soils consisting of only Chlorophyta (*Chlamydomonas*, *Chlorococcum*, and *Acrosiphonia*) and Stramenophiles (*Hantzschia, Paulellia, Heterococcus* and *Chrysolepidomona*) (Fell et al., 2006; Pointing et al., 2009). Algal mats are commonly known to accumulate at small ponds or within wet soils that form during the austral summer. Free-living algae are most likely to exist under quartz stones (hypolithic) where some protection from the elements is afforded (Broady, 1981).

The lack of archaeal presence in Dry Valley soils is worthy of note, as this microbial domain is ubiquitous throughout most terrestrial soils (Bates et al., 2010, and references therein) and even found in other niches within the Dry Valleys (Cary et al., 2010). Many Archaea species are known to be extremophiles, oligothrophic and autotrophic (Berg et al., 2010; Zhang et al., 2010), capable of flourishing under varying conditions. Not many studies have focused attention on retrieving sequences or cultivating archaea from the Dry Valley soils (Pointing et al., 2009). General absence from studies that have attempted their isolation using phylogenetic methods suggests that archaea are unable to tolerate the extreme xeric environment (Rothrock and Garcia-Pichel, 2005) or other approaches require development.

Opportunistic heterotrophs appear to dominate Dry Valley soil populations consisting of several phyla. 14 divisions of phyla have been identified to date with the dominant groups (> 3%) being Acidobacteria, Alphaproteobacteria, Betaproteo-bacteria, Gammaproteobacteria, Cyanobacteria, Gemmatimonas, Bacteroidetes, Actinobacteria and Deinococcus and Thermus, according to Cary et al. (2010). Dominant phyla were determined through several phylogenetic studies (de la Torre et al., 2003; Aislabie et al., 2006; Smith et al., 2006; Aislabie et al., 2008; Niederberger et al., 2008), possibly providing a better understanding of the soil diversity for this isolated habitat. Considering both the extreme nature of the habitat and the theoretically conservative trophic system (Cary et al., 2010), the diversity is surprisingly high (Smith et al., 2006; Mahaney et al., 2012). Among the most common isolates to be identified at the species level were Actinobacteria such as *Arthrobacter* spp., *Brevibacterium* spp., and *Corynebacterium* spp., including *Micrococcus*, *Norcardia*, and *Streptomyces*. Other common isolates are from the genera's *Bacillus*, *Flavobacterium,* and *Acinetobacter* spp. (Johnson et al., 1978; Liu et al., 2000; Hart et al., 2011). The Dry Valleys are largely characterized according to the varying environmental conditions, and the direct effect this has on the soil microbial community. As such, varying levels of microbial diversity are observed depending on sample location with some species clustering in certain locations, such as the drier soils of Wright Valley (Aislabie et al., 2006), where *Deinococcus* and *Rubrobacter* dominate.

The presence of heterotrophic microbes in soils supports the assumption that viable trophic structures exist in Dry Valley soils. Further, the isolation of psychrophilic and psychrotrophic indigenous species such as *Psychrobacter* (Shivaji et al., 2005) and *Arthrobacter cryotolerans* (Ganzert et al., 2011a) indicates *in situ* activity.

A large number of reported sequences extracted from Dry Valley soils are poorly represented within clone libraries (e.g., Mahaney et al., 2012). This occurs when no cultured and characterized organism with a similar genetic identity is present for comparison leading to the isolate remaining unknown.

The high proportion of unknown sequences of Antarctic isolates indicates novel taxa with no known close relative. Despite the global dispersion of terrestrial species to the Antarctic continent from atmospheric transportation (Pearce et al., 2009), geographical isolation and adaption to environmental factors have led to the development of distinct soil communities fundamentally different than those of other soil types across the globe (Niederberger et al., 2008).

Permafrost Communities – Life in the Deeper Frozen Soils

Permafrost soil is soil or sediment that is constantly at $\leq 0^{o}C$ for two or more years (Harris et al., 1988). Permafrost is strictly a thermal phenomenon not dependent on the surrounding composition of the ground. Permafrost in the Northern Hemisphere has been examined in several regions (e.g., Alaska, Canadian and Siberian Arctic, Greenland, and the alpine mountainous regions).

The permafrost of the Northern Hemisphere is generally both 'wet' in terms of its water content (as compact massive ice or soil/ice cement) (Romanovsky et al., 2010) and represent huge stores of organic carbon (Tarnocai et al., 2009). Northern Hemisphere polar soils are overlaid with wet soils for at least part of the year after summer snowmelts leading to liquid infiltration of the subsurface and recharge of the ice-layer. In the southern hemisphere, the soils of the high-elevation Dry Valleys of Antarctica are extremely xeric for much of the year. Glacial melt water and occasional snowfall are rapidly sublimated from surface soils leading to highly desiccated epipedons (Hagedorn et al., 2007; McKay, 2009). According to Bockheim et al. (2007), 43% of Dry Valley soil is dry frozen. The remaining permafrost consists of ice-cemented soil (55%) and ground/buried ice (2%) with higher-elevation soils being predominantly dry frozen. The penetration depth of ice-cemented soil approaches 1000 m, but the maximum depth of dry permafrost is unknown. Current evidence suggests dry permafrost is limited to a few meters (McGinnis, 1981; Bockheim and Tarnocai, 1998; Bockheim et al., 2007) with some reports conservatively estimating it to be < 1 m (McKay et al., 1998b; McKay, 2009; Marinova et al., 2011). The image in Fig. 52 shows dry permafrost conditions for the particular sampling site with the soil remaining ice-free to a depth of ~ 45 cm.

Figure 52. Depth profile of dry permafrost (University Valley, Antarctica). The ground is dry and ice-free extending to a depth of ~ 45 cm. Photo courtesy of Margarita Marinova (BAER institute and NASA Ames Research Center).

Permafrost has been shown in several studies to harbor viable biota, surviving extended geological periods (Steven et al., 2006; Gilichinsky et al., 2007; Gilichinsky et al., 2008 and references therein), though only ice-bearing soils have been thoroughly investigated. The viable biological component of permafrost in the Dry Valleys has not been extensively investigated to date, though Gilichinsky et al. (2007) showed that aerobic and anaerobic bacteria (Proteobacteria [α, β, and γ subdivisions] and Gram-positive), fungi, yeasts, and algae were all readily cultured under various temperature regimes using various culturing techniques. Arctic permafrost can contain significant populations of microorganisms with direct microscopy counts, determining 10^6 - 10^7 cells / g dry weight of soil (Gilichinsky et al., 2007). When compared to the Arctic permafrost, microbial populations of Antarctic permafrost microbial populations are considered to be of an order of magnitude lower, with population estimates of 10^5 - 10^6 g dry soil of soil, though estimates vary upwards depending on soil type (Cowan et al., 2002; Steven et al., 2006; Gilichinsky et al., 2007). Studies of the microbial diversity of Antarctica are numerous, but to date, have centered mostly on the more active surface layers. According to Vorobyova et al. (1997), *Arthrobacter*, *Bacillus,* and *Streptomyces* are the most predominant bacterial genera's in Antarctic permafrost with most species being psychrotrophic. In addition, Nitrospina, green non-sulphur bacteria,

Fibrobacter, Acidobacterium, and various methanogens have also been recovered from Antarctic ice-bearing paleosols (Gilichinsky et al., 2007). Using culture-based studies, a large proportion of isolates remain unidentified (> 40%), despite comparison to DNA-based clone libraries. Comparisons of the surface/active-layer clones to the deeper, subsurface-permafrost clones have shown high similarity indicating that the identified bacterial species may have inhabited Antarctica for as long as the age of the permafrost.

Using light and electron microscope observational studies, the identification of microbial life within permafrost has been achieved for Antarctic permafrost soils. The presence of intact, partially degraded, and empty cells have been made for several microbial types including eukaryotes (Steven et al., 2006; Gilichinsky et al., 2007). Examining the cell ultrastructure for both aerobic and anaerobic communities, Gilichinsky et al. (2007) determined that immediately after thawing, cell wall integrity was maintained with no signs of cell lysis. This data corroborates with the findings of several other studies of Arctic permafrost communities (e.g., Soina and Vorobyova, 1994). Cell types show characteristics of both Gram-positive and Gram-negative bacteria with cell morphologies generally described as cocci or rods. The characteristics of intact cells typically include coatings with capsular material, thickened cell walls, and the cytoplasm containing numerous aggregates (Dmitriev et al., 2000; Sonia et al., 2004; Gilichinsky et al., 2007). The cells which are identified in Antarctic permafrost are generally small, ranging in size between 0.5 - 1.0 μm in diameter (Gilichinsky et al., 2007). The small size (< 1 μm) of viable microbial cells within the Antarctic permafrost, may be a physiological response to cryogenic stress, and corroborates well with their northern counterparts (Steven et al., 2006).

A recent culture-independent molecular study of extractable 16s rRNA has shown potentially high bacterial diversity (Mahaney et al., 2012). Over 50% of cloned species were unidentified and likely to represent novel genera, with no obvious dominance of the identifiable genera. Viable bacterial DNA identified in the samples indicates that microbes were present and possibly active to some degree in the Antarctic permafrost at the time of data acquisition, at least to a depth of ~ 10 cm. The range of bacteria is consistent with the observations of other researchers (e.g. Peeters et al., 2011; Ganzert et al., 2011b; Harding et al., 2011), and the very low levels of organic carbon (< 0.01 - 0.03%) and extreme aridity demonstrate that oligotrophic conditions existed at these sites during sample acquisition (Mahaney et al., 2001, 2009, 2012). The extracted DNA from this particular study may have come from dormant microbes existing in a freeze-dried state or encased within spores, though psycrotolerant microbes have been demonstrated to be active at very low temperatures (Liu et al., 2000; White et al., 2000; Aislabie et al., 2000; Ruberto et al., 2005). If indeed the microbes from this site were inactive, it has already been demonstrated that both Arctic and Antarctic microorganisms can readily be re-animated under cultivation conditions (Soina et al., 2004; Hart et al., 2011). The correlation of some possibly salt-tolerant species from the Genebank library (Tripathi et al., 2002; Mehnaz et al., 2010; Tkavc et al., 2011) with the isolated sequences in this study is highly relevant. Mahaney et al. (2001, 2009) showed that the Dry Valley soils contain significant concentrations of salts (due to soil desiccation), and therefore, the relevance that many of the sequences match those of halophiles; thus, hints at the likelihood of specialized communities being adapted for survival at these sites when in the presence of liquid water.

Microbial *in situ* activity within permanently cryogenic environments such as permafrost are of high relevance to astrobiological investigations, as it is recognised that most solar bodies, theoretically capable of supporting life, are cryogenic in comparison to Earths biosphere. The majority of bacteria isolated from permafrost samples are mesophiles capable of surviving multiple freeze-thaw cycles (described by Friedman as "a community of survivors" (Friedmann, 1994)), but generally cultivated at temperatures > 4°C (Soina et al., 2004). Several species of bacteria and yeasts have been reported to be capable of sustaining normal activities (excluding reproduction) at temperatures as low as -18°C (Bowman et al., 1997). Table 9 provides a brief overview of the currently known sub-zero temperature limits for some diverse microbial species. There are few reported investigations using culture-independent techniques to describe permafrost communities (Steven et al., 2006), and with the exception of a limited number of stable isotope labelling studies (Rivkina et al., 2000; Gilichinsky et al., 2003), almost none exist that describe *in situ* activity. Studies employing culture independent and stable isotope probing (SIP) of recently melted permafrost have, however, showed that communities respond rapidly to upward temperature mobility (Martineau et al., 2010; Mackelprang et al., 2011) and represent a significant impact to terrestrial biogeochemical systems.

Table 9. Current lower limits for microbial growth at sub-zero temperatures. N/A = not available

Species	Lowest Temperate Growth (°C)	Isolated Environment	Reference
Bacteria			
Psychrobacter glacincola	-18.0	Antarctic Sea-Ice	Bowman et al., 1997
Psychromonas ingrahamii	-12.0	Alaskan Sea-Ice	Breezee et al., 2004
Colwellia psychrerythraea	-14.5	Arctic Sea-Ice	Huston, 2003
Psychrobacter urativorans	-10.0	Antarctic Soil	Bowman et al., 1996
Psychrobacter cryopegella	-10.0	Siberian Permafrost	Bakermans and Nealson, 2004
Fungi and Yeasts			
Rhodotorula glutinis	-18.0	Food Storage Facility	Collins and Buick, 1989
Mrakia nivalis	-7.0	N/A	Larkin and Stokes, 1969
Humicola marvinii	-2.5	N/A	Larkin and Stokes, 1969
Algae			
Nitzschia frigida	-8.0	Greenland Sea-Ice	Aletsee and Jahnke, 1992
Thalassosira antarctica	-6.0	Greenland Sea-Ice	Aletsee and Jahnke, 1992
Archaea			
Methanococcoides burtonii	-2.0	Antarctic Saline Lake	Reid et al., 2006
Halorubrum lacusprofundi	-1.0	N/A	Reid et al., 2006

There is a clear lack of data regarding *in situ* microbial communities of the Antarctic permafrost. Studies have generally used culturing techniques that represent only ~ 0.1 % of the viable population (Gilichinsky et al., 2008). In addition to a lack of studies in the existing literature targeting the Antarctic dry permafrost, with the exception of the preliminary study by Mahaney et al. (2012), sampling of the deeper sub-surface, which could yield distinctive biological information, has yet to be performed.

The enduring presence of microorganisms in ancient permafrost samples suggests that a DNA repair mechanism is capable of stabilising microbial systems over millions of years, which at the very least is efficient at neutralizing the cumulative effect of cosmic background radiation that doses Arctic and Antarctic sites for at least 6 months of each solar year (Gilichinsky et al., 2003). A large number of molecular adaptations in Antarctic paleosols are clear evidence that many cellular processes contribute to a microorganism's capacity for growth at low temperature. Microbial metabolic activity has been recorded for temperatures as low as -28°C in frozen soils (Gilichinsky et al., 2007), both in Arctic and Antarctic permafrost, largely due to nearly 5% of the pore space being unfrozen brine; this access to water prevents cells from developing ice crystals, and so the viable cells present are in an overcooled state and in equilibrium with surrounding unfrozen briney-water films (Gilichinsky et al., 2007). Active communities of psychrophilic bacteria thrive within liquid brine inclusions in frozen environments at temperatures as low as -23°C in Siberia, Greenland, the Arctic (Rivkina et al., 2000; Bakermans et al., 2003; Junge et al., 2004; Tung et al., 2005, 2006), and Antarctica (Priscu et al., 1999; Mahaney et al., 2001; De Angelis et al., 2005; Gilichinsky et al., 2007; Laybourn-Parry, 2009 and references therein). These microbial communities derive energy from the metabolism of dissolved constituents in the brine (including organic carbon, nitrogen, iron, and sulphur compounds) (Grasby et al., 2003; Tung et al., 2006; Mikucki et al., 2009), possibly serving as analogues to search for iron-sulphur based ecosystems living at extremely low temperatures in dilute acidic subterranean brines or within tiny brine pockets in crustal ice on Mars.

Martian Analogues Resembling Earths Cold Deserts

More and more evidence is pointing to a Martian landscape that has been sculpted by hydrological cycling-related processes, which includes the development of seas, fluvial valleys, ice sheets, glaciers, rock glaciers, and polygonal-patterned ground (Scott et al., 1995; Baker, 2001; Fairén et al., 2003; Mahaney et al., 2007; Baker et al., 2007; Dohm et al., 2009a). Such activity is expected to have generated Cryosols in parts of Mars, especially given advanced ages that Martian profiles are expected to exhibit, and with greater age, higher grades of weathering compared with Cryosols on Earth. Structurally-controlled basins within the extremely ancient geologic terrains of Mars, such as Terra Sirenum and Terra Cimmeria (Dohm et al., 2002, 2005; Baker et al., 2007; Tanaka et al., 2012, in preparation), for example, may contain paleosols that record geologic, hydrologic, and climatic activity extending as far back as the Early Noachian epoch (Anderson et al., 2012) or greater than 3.95 billion years ago, estimated absolute ages of which are based on the impact crater chronology of Hartmann and Neukum (2001). Indeed, pedostratigraphic profiles of much greater complexity than the ones described by Mahaney et al. (2001) may exist in these old surfaces and record paleoclimatic oscillations ranging from warmer/wetter to colder/drier over much of Martian history.

Complex basement structural fabrics, high impact crater densities, highly degraded promontories, several of which have been interpreted to be volcanic constructs (Scott and Tanaka, 1986; Hodges and Moore, 1994; Dohm and Tanaka, 1999; Dohm et al., 2001a), and distinct banded magnetic anomalies (Acuña et al., 1999; Connerney et al., 1999), collectively point to such extremely ancient terrains that have not been destroyed by magmatic-driven activity, such as exemplified at Tharsis (Dohm et al., 2001b, 2007) and giant impact crater events such as Hellas and Argyre (Scott and Tanaka, 1986; Greeley and Guest, 1987; Crown et al., 1992; Tanaka et al., 2012).

Basins and structurally controlled valleys resulting from the giant impact events, Hellas and Argyre, may also be optimal caches of far-reaching Cryosols; a new impact crater database indicates that the Argyre and Hellas impact events occurred nearly 3.93 and 4.13 billion years ago, respectively (Robbins and Hynek, 2012). These features display evidence for lacustrine, glacial, periglacial, and rock glacial activity (e.g., Kargel and Strom, 1994; Head and Pratt, 2001; Kargel, 2004; Dohm et al., 2009a), likely resulting from endogenic-driven hydrological cycling largely related to the growth of Tharsis (Baker et al., 1991, 2007; Dohm et al., 2001b, 2007, 2009a; Fairén et al., 2003), though obliquity has also contributed to environmental change (e.g., Laskar et al., 2002), but to a much lesser extent. During the dynamic phases of Tharsis activity, which would include major transfers of ground water and changing environmental conditions in and nearby Tharsis (Dohm et al., 2001a, b, 2007, 2009a, 2011), outflow channel development, and ponding of the floodwaters in the northern plains (e.g., Baker et al., 1991, 2007; Fairén et al., 2003), aqueous conditions away from Tharsis such as within and surrounding the ancient giant impact basins, would have included snow fall and the growth of glaciers and the development of periglacial landscapes. Such activity is not necessarily distinct from that of the Dry Valleys in that the Dry Valleys have recorded resurfacing events related to climatic change at regional to global scales. Resurfacing events include the transgression of ice sheets that intermingled with the Dry Valley infill materials such as demonstrated at New Mountain (Mahaney et al., 2001). Fog, which likely develops in the giant impact basins (Neumann et al., 2003 and references therein), is significant to life in the hyperarid Atacama Desert; fog sources from the Pacific Ocean through the coastal range along structurally controlled gaps (e.g., Warren-Rhodes et al., 2007a,b; Hock et al., 2007).

Taking into account the composition and age of Antarctic Cryosols discussed above and extrapolating to Mars, one might consider that paleosols would have similar properties multiplied many times over. Certainly, the mineralogy of Mars may differ from that of the Earth (e.g., Taylor et al., 2006), but rock types of intermediate to felsic compositions such as andesites and granites, respectively, cannot be ruled out, as the primary rocks in the extremely ancient terrains such as Terra Cimmeria and Terra Sirenum and the mountain ranges of Thaumasia highlands and Coprates rise, have yet to be identified through orbital and *in situ* reconnaissance (Dohm et al., 2009b). These terrains, which includes Coprates rise, a near replicate of the Wind River Mountain Range of Wyoming in morphologic expression but twice the size and marked by impact craters (Figures 53 and 54), record extremely ancient dynamic crustal/lithospheric conditions during the embryonic (first several hundred million years) stage of Mars' evolution (Dohm et al., 2001a, b, 2002; Fairén et al., 2002; Baker et al., 2007). Further, considering that pedogenesis in the Antarctic amounts to 1 cm/depth/10^6 yr, what depth might be expected on Mars? If parts of the ancient terrains have been under a cold dry climate for more than 3 billion years, for example, soil depths could be expected to be greater, perhaps reaching to several meters or more, exhibiting deeper profiles than shown in Figure 52. Other normal differentiation of terrestrial paleosols, horizon by horizon, might be infinitely more complex on Mars than on Earth, especially if obliquity changes and dynamic environmental conditions related to major phases of Tharsis development have led to periods of climatic amelioration where permafrost could melt out and meteoric water find its way down-profile. There may, in fact, be many episodes of climatic amelioration over 3 billion years giving rise to deep and complex pedostratigraphic sections with no end of buried pedons, somewhat similar to sections described in the New Mountain area of Antarctica (Mahaney et al., 2001).

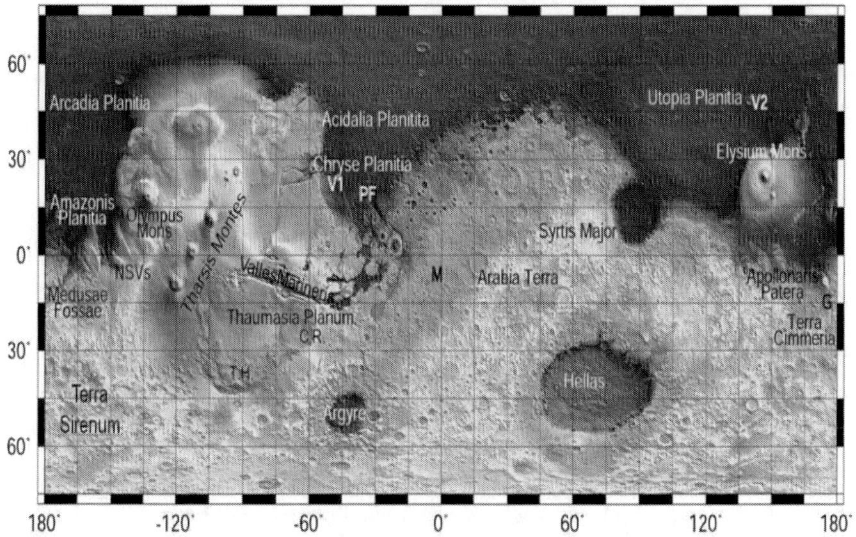

Figure 53. Mars Orbiter Laser Altimeter topographic map showing features of interest, including the ancient Thaumasia highlands (T.H.) and Coprates rise (C.R.) mountain ranges.

Discussion

Cold environments are common on Earth, with a large proportion of the terrestrial surface permanently frozen or at temperatures below the optimum growth requirement for microbes (< 10 - 15°C; Chin et al., 2010). Every continent on Earth has large representative areas where mesophilic growth optima are rarely attained. The deep oceans (< 500 m) and surface waters of the temperate and polar regions typically range from 12°C downward. This is highly significant as low-temperature environments still contain high microbial diversity, where cold-tolerant microorganisms can grow at ≤ 4°C (Zhang et al., 2009). Psychrophiles are microorganisms that can grow and reproduce at temperatures at or below 0°C, and they are found in various locations ranging from the deepest ocean ridges to the stratosphere (Summit and Baross, 2001; Sattler et al., 2001). Environments with low temperatures and high salt concentrations such as the deep oceans and the Antarctic Dry Valleys, depress the freezing point of water to -1.8°C. As liquid water levels decrease (freezing or evaporation), the concentration of salts increase creating brine pockets where the freezing point of water can be as low as -20°C (Margesin et al., 2008). The growth of bacteria has been measured within such environments to demonstrate that metabolism is not constrained to external temperature (Bakermans and Nealson, 2004). For this reason, many currently known psychrophiles are also thought to be halophiles (Pampelotto, 2010).

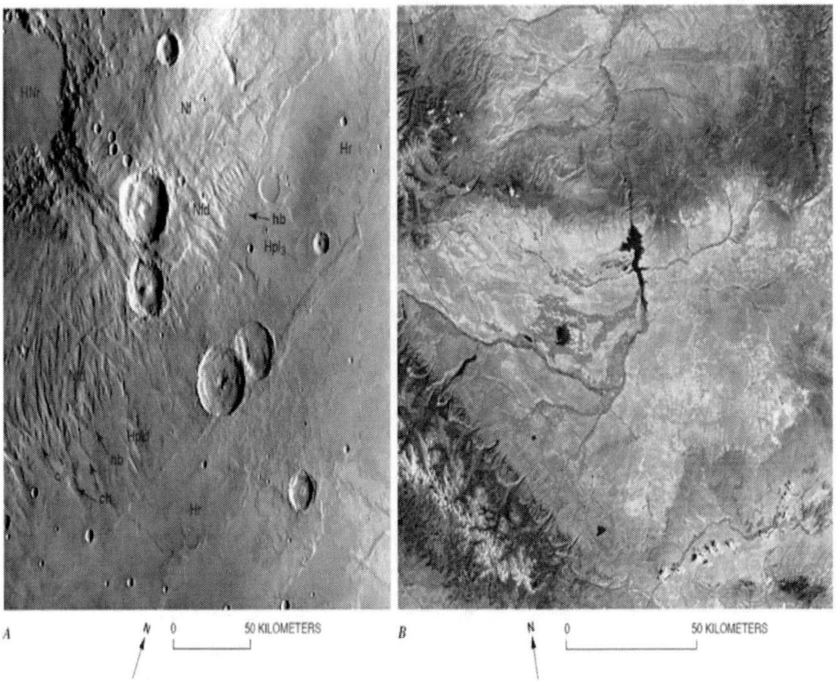

Figure 54. A: Based on Dohm et al. (2001a), cuestas (c) and hogbacks (hb), which are caused by tectonic tilting and differential erosion, and channels (ch) of the southeast part of Coprates Rise. Symbols c1, Nf, Nfd, Hpl3, Hr, and Hpld denote material of degraded craters, older fractured material, fractured and dissected material, smooth unit, younger ridged plains material, and younger dissected material. (Viking image 610A44; resolution 224 m/pixel) B: Wind River Mountains of Wyoming that display similar morphologic features to Coprates rise. (Landsat TM image; plate 97 (Short et al., 1976)).

Desert environments are classified as areas with severe shortages of moisture, lacking perennial and integrated systems of drainage. Deserts occupy approximately one third of the terrestrial land surface (Shmida, 1985) and occur on every continent ranging from hot (e.g., Great Victoria Desert, Australia; the Sahara Desert, North Africa) to cold (e.g., Dry Valleys, Antarctica; Gobi Desert, China) deserts. Higher organisms such as plants and insects can tolerate the harsh conditions of most deserts but some hyperarid deserts appear sterile to the casual observer. In such xeric locations such as the Dry Valleys, Antarctica, high alpine peaks, and the Atacama Desert, Chile, microorganisms have been found to be largely ubiquitous (Swan, 1992; Warren-Rhodes et al., 2006; Gilichinsky et al., 2007; Walker and Pace, 2007b; Pointing et al., 2009; Schmidt et al., 2010), though activity at the trophic level requires further investigation to demonstrate the extent of biogeochemical impact.

The Dry Valleys of Antarctica and the highest mountain peaks are extreme by our understanding of what the limitations to life may be in order to exist (or even proliferate?), and environmental constraints such as low temperatures are not as 'extreme' for microorganisms, unlike high temperatures (Gounot, 1991). The extreme Martian conditions at its most hospitable are quite similar to the Dry Valleys, Antarctica (Reid et al., 2006). Considering the large diversity of microbial species discovered to date along with accumulating evidence of *in situ* activity within various niches, including those at sub-zero temperatures and with rare inputs of moisture, the capacity for extant life on Mars is theoretically feasible.

The summits of mountains and high alpine ridges (≥ 4000 m) also represent some of the most extreme of all terrestrial environments (Mahaney, 1990; Swan, 1992; Mahaney et al., 2008; Costello et al., 2009). These locations are exposed to extreme swings in temperature (experiencing several fluctuations across the freezing point each day), high wind speeds, low atmospheric precipitation, high UV exposure, and lower atmospheric pressures than found nearer sea level (Schmidt et al., 2010). Limited ecosystems still exist, surrounding rock ramparts, where shallow peat soils accumulate or locations where rock debris may amass (Bardgett and Leemans, 1996). These niches can support higher vegetation such as alpine grasses and dwarf-shrub heaths depending on ridge elevation and nutrient availability (Swan, 1992). Primary productivity for higher plants is almost entirely dependent on nutrient release from microorganisms, which in turn is constrained by low temperatures and xeric conditions (Bardgett and Leemans, 1996). According to Swan (1992), the known upper elevation limit for higher eukaryotic species is > 7000 m, whereupon only bacteria, fungi, and yeasts are generally found above this zone. The deposition of wind-born (aeolian) essential nutrients and substrates, which become lodged in crevices and clefts of rock faces, is most likely the source for most heterotrophic microbial growth (Swan, 1992; Schmidt et al., 2009). Such environments are known as allobiospheres where photosynthesis is depressed by the environmental extremes, and the active inhabitants depend on their sustenance from imported material (Edwards, 1988). However, Schmidt et al. (2009, 2010) have reported that photosynthetic microbes are present in the dry, barren soils of the high Andes and the high Himalayas. Using a culture-independent study, the dominant algal clades described by the research group for the high Himalayas were highly similar to those reported for the Dry Valleys, Antarctica, indicating a link between these environmentally similar but geographically distant systems (Schmidt et al., 2010) as predicted by the Baas-Becking hypothesis (Martiny et al., 2006 and references there in). A comprehensive study of culturable bacteria and fungi was carried out by Michael Boyer working with the Quaternary Project on Mt. Kenya (1976-1987), detailed in Mahaney (1990). Three different profile distributions of microorganisms were discovered, that showed a slow decline from the surface zone of maximum biological activity, disjunct where distributions disappeared with depth and resumed lower down in the profile and bimodal where decline in numbers approached a nil count but then increased. Extremely ancient Martian mountain ranges, such as Coprates rise and Thaumasia highlands, may also record fluvial, glacial, and rock glacier activity (Dohm et al., 2001a; Baker et al., 2007; Rossi et al., 2011), paleosol (soil) development, and possible fossilized and extant life, similar to the Wind River Mountain Range (Mahaney, 1987).

The enumeration of extant microbes and culture-dependent studies have shown that microbial biomass is significant at peaks > 5000 m, but the question remains whether these residents are dormant or active at such altitudes of mountainous terrains. Are the organisms deposited from aeolian processes capable of active growth and reproduction under such extreme conditions? Or, do they form a layer of biologically dormant life that can be readily re-animated upon culturing? Each geochemical setting provides its own set of varying resources that can be exploited by microorganisms owing to their small size, high surface area, high rates of metabolic activity, genetic transfer, potentially high growth rate, and vast enzymatic versatility (Madsen, 2005). The topic of microbial biogeography has received much interest with some authors proposing that local populations develop according to the surrounding environment (as would higher organisms) (Martiny et al., 2006), consistent with Dohm et al. (2011) who emphasize that biology, hydrology, and geology are interlinked on Earth such that certain types of life are often associated with specific conditions, including rock type, pressure, temperature, and chemistry. This is more commonly referred to as the 'Baas-Becking hypothesis' which postulates that everything is everywhere but the environment selects (Martiny et al., 2006). Others propose that because microorganisms have such wide dispersal owing to their small size and large populations, that the local environmental conditions permit the development of distinctive microbial populations (Martiny et al., 2006; Namsaraev et al., 2010).

Applying modern phylogenetic techniques to such samples may provide a better understanding of the microbial diversity but little information on activity, unless coupled with isotopic tracer studies (e.g. Dumont and Murrell, 2005). Although, to date, very view studies have used culture-independent molecular techniques to investigate the microbial ecology of high mountain peaks (Schmidt et al., 2008, 2010). The logistical difficulties and inaccessibility of high alpine areas, coupled with sample storage integrity issues, are obstacles that must be overcome in the near future to provide us with a clearer picture of the *in situ* microbial activity of these 'barren' environments, and hence advance our knowledge of extremophiles.

Crytoendolithic microorganisms have developed strategies and genetic modifications to protect themselves from desiccation, high doses of UV radiation, oligotrophic conditions, and extreme swings in temperature suggesting that microorganisms are capable of developing strategies that can survive Martian surface conditions. The development of microbial colonies consisting of several trophic levels existing within the protected surface of rocks suggests that life can develop strategies to survive almost any terrestrial environment. The well documented cyanobacteria, *Chroococcidiopsis*, is commonly located in both hot and cold deserts, where it is capable of slow but continuous growth under the most xeric conditions and exposure to high doses of UV radiation (Billi, 2010). These versatile organisms colonize many habitats such as microscopic fissures (chasmoendoliths) or structural cavities (cryptoendoliths) of rocks, forming biofilms between the stone-soil interface upon the surface of desert pavements (hypoliths), or growing within brine pockets and halite deposits (Wierzchos et al., 2006). It is speculated that desiccation and UV tolerance of *Chroococcidiopsis* is dependent on the ability to avoid and/or repair lethal damage to DNA (Billi et al., 2000; Billi, 2010). The ability of this organism to proliferate in such extreme environments has not gone unnoticed, and several studies investigating the limits of life to exist under extraterrestrial conditions have been carried out with some success (Cockell et al., 2005; Grilli-Caiola and Billi, 2007).

Mars has high alpine regions similar to Earth and on a somewhat greater scale. Consider Coprates Rise (Mars) and its above noted resemblance to the Wind River Mountains of Wyoming (Figure 54), a terrestrial tilted fault block mountain with an extensively studied glacial history, but with little known about populations of bacteria and fauna (Mahaney, 1978, 1984). In this particular case, as a terrestrial analogue to Mars, little is known about the variation in microorganisms in soils, only that different distributions results from variations in organic carbon and influx of ground water and less about their origin. The situation is similar in other high alpine areas such as the high Andes (Mahaney et al., 2008) where occasional fungi and bacteria have been observed in soils.

Whether sufficient carbon exists on Mars is an open question, as in turn, whether ^{14}C might be used on Mars. In the future, accelerator mass spectrometry (AMS) may well be able to date minute quantities of organic carbon on the red planet, but this residue would probably come from the remains of fossil bacteria and fungi. Could microbes on Mars have evolved with a different metabolic and physiologic system, perhaps one not based on carbon?

SUMMARY

Although there are no places on Earth that can fully be compared to Mars today, there is strong evidence to suggest that during the long periods between environmental and climatic adjustments, conditions may have been sufficiently stable for life to have emerged. Within the confines of the Earth's biome, life proliferates within almost every niche it is able to establish. From hyperarid deserts and volcanic fumaroles to anthropogenic nuclear waste dumps, microorganisms thrive and continue to recycle the elements around them in an efficient manner. Also, due to the evolutionary adaptability demonstrated on Earth, if existent, Martian organisms may have developed in environments in which there is no terrestrial analog, as well as traveled down evolutionary pathways distinct from that of Earth. The pace of new discoveries that demonstrate microbially mediated processes under diverse and complex systems is intensifying. The application of modern molecular biology to terrestrial environmental samples has not only shown the bewildering diversity and activity of microbes, but also their ability to survive under 'extreme' conditions.

The Dry Valleys, Antarctica, present us with an optimal location for testing hypotheses about the limits to which life can survive in hostile locations. Studies targeting endolithic communities have provided detailed accounts of the active microbial interactions within a relatively stable ecosystem. The protection afforded to these abiotic/biotic interactive structures is highly relevant to assessing the limits to which life may adapt to extreme conditions, especially when theorizing the onset of catastrophic climate change.

Sterility of the Antarctic continental mineral soils has been successfully disproven using various culture-dependent and -independent methods, but it still remains unclear how much microbial activity is taking place during the brief summer periods or indeed throughout the entire Polar year. The majority of microorganisms isolated using DNA-based techniques remain unidentified owing to their novelty, and many genera's may display properties currently unknown to science. Few experiments have been published that demonstrate the *in situ* activity of soil and permafrost microbial communities. The application of modern molecular biology (e.g., SIP and real-time quantitative PCR (qPCR)) to soils of the Dry Valleys, while under specific experimental conditions, could be extremely useful in further clarifying the habitability of such locations by microorganisms. Further, using analytical and instrumental methods commonly applied in environmental chemistry laboratories, the detection of biosignatures under extreme and/or long-term conditions must be investigated. The long periods of relative stability experienced at the southern pole may provide opportunities for the scientific community to develop more refined techniques to detect extremely small quantities of organic material and/or biologically derived minerals. Such studies will have a direct bearing on future exobiological studies of Mars, which include determination of prime targets of life-habitability potential and ultimately whether similar microbes exist on the Red Planet.

References

Acuña MH., Connerney JEP., Ness NF., Lin RP., Mitchell D., Carlson CW., McFadden J., Anderson KA., Reme H., Mazelle C., Vignes D., Wasilewski P., Cloutier P. (1999) Global distribution of crustal magnetization discovered by the Mars Global Surveyor MAG/ER experiment. Science 284, 790–793.

Aislabie J., Foght J., Saul D. (2000) Aromatic hydrocarbon degrading bacteria from soil near Scott Base, Antarctica. Polar Biology 23, 183–188.

Aislabie JM., Chhour K-L., Saul DJ., Miyauchi S., Ayton J., Paetzold RF., Balks MR. (2006) Dominant bacteria in soils of Marble Point and Wright Valley, Victoria Land, Antarctica. Soil Biology and Biochemistry 38, 3041-3056.

Aislabie JM., Jordan S., Barker GM. (2008) Relation between soil classification and bacterial diversity in soils of the Ross Sea region, Antarctica. Geoderma 144, 9-20.

Aletsee L., Jahnke J. (1992) Growth and productivity of the psychrophilic marine diatoms *Thalassiosira antarctica* Comber and *Nitzchia frigida* Grunow in batch cultures at temperatures below the freezing-point of sea-water. Polar Biology 11, 643–647.

Anderson RC., Dohm JM., Robbins S., Hynek B., Andrews-Hanna J. (2012) Terra Sirenum: window into pre-Tharsis and Tharsis phases of Mars evolution. Lunar and Planetary Science Conference XXXXIII, #2803 (abstract) [CD-ROM].

Atreya SK., Wong A-S., Renno NO., Farrell WM., Delory GT., Sentman DD., Cummer SA., Marshall JR., Rafkin SCR., Catling DC. (2006) Oxidant enhancement in Martian dust devils and storms: implications for life and habitability. Astrobiology 6(3), 439-450.

Baker VR. (2001) Water and the Martian landscape. Nature 412, 228–236.

Baker VR., Strom RG., Gulick VC., Kargel JS., Komatsu G., Kale VS. (1991) Ancient oceans, ice sheets and the hydrological cycle on Mars. Nature 352, 589–594.

Baker VR. (2006) Geomorphological evidence for water on Mars. Elements 2(3), 139-143.

Baker VR., Maruyama S., Dohm JM. (2007) Tharsis superplume and the geological evolution of early Mars. In: Yuen DA., Maruyama S., Karato S-I., Windley BF. (Eds.), Superplumes: Beyond Plate Tectonics. London, Springer, pp. 507-523.

Bakermans C., Tsapin AI., Souza-Egipsy V., Gilichinsky DA., Nealson KH., (2003) Reproduction and metabolism at -10°C of bacteria isolated from Siberian permafrost: Environmental Microbiology 5, 321-326.

Bakermans C., Nealson KH. (2004) Relationship of critical temperature to macromolecular synthesis and growth yield in *Psychrobacter cyropegella*. Journal of Bacteriology 186(8), 2340-2345.

Bamforth SS., Wall DH., Virginia RA. (2005) Distribution and diversity of soil protozoa in the McMurdo Dry Valleys, Antarctica. Polar Biology 25, 756-762.

Banfield JF., Moreau JW., Chan CS., Welch SA., Little B. (2001) Mineral biosignatures and the search for life on Mars. Astrobiology 1(4), 447-465.

Bardgett R., Leemans D. (1996) Soil microbial activity on exposed mountain ridges in Snowdonia (Eryri), North Wales. Soil Biology and Biochemistry 28(10-11), 1533-1536.

Barker WW., Banfield JF. (1998) Zones of chemical and physical interaction at interfaces between microbial communities and minerals: a model. Geomicrobiology 15, 223-244.

Bates ST., Berg-Lyons D., Caporaso JG., Walters WA., Knight R., Fierer N. (2010) Examining the global distribution of dominant archaeal populations in soil. The ISME Journal 5(5), 908-917.

Berg IA., Kockelkorn D., Ramos-Vera WH., Say RF., Zarzycki J., Hügler M., Alber BE., Fuchs G. (2010) Autotrophic carbon fixation in archaea. Nature Reviews 8, 447-460.

Beyer L., White DM., Bölter M. (2001) Soil organic matter composition, transformation, and microbial colonisation of Gelic Podzols in the coastal region of East Antarctica. Australian Journal of Soil Research 39, 543-563.

Biemann K. (2007) On the ability of the Viking gas chromatograph–mass spectrometer to detect organic matter. Proceedings of the National Academy of Sciences of the USA 104, 10310–10313.

Billi D. (2010) Genetic tools for desiccation- and radiation-tolerant cyanobacteria of the genus, *Chroococidiopis*. Current Research, Technology and Education Topics in Applied Microbiology and Microbial Biotechnology 2(2), 1517-1521.

Billi D., Friedmann EI., Hofer KG., Grilli Caiola M., Ocampo-Friedmann R. (2000) Ionizing-radiation resistance in the desiccation-tolerant cyanobacterium *Chroococcidiopsis*. Applied and Environmental Microbiology 66, 1489-1492.

Blume H., Kuhn D., Bölter M. (2002) Soils and soilscapes In: Beyer L., Bölter M. (Eds.), Geoecology of Antarctic Ice-Free Coastal Landscapes. Ecological Studies. Springer-Verlag, Berlin, pp. 91-113.

Bockheim JG. (1997) Properties and classification of cold desert soils from Antarctica. Soil Science Society of America Journal 61, 224-231.

Bockheim JG. (2002) Landform and soil development in the McMurdo Dry Valleys, Antarctica: a regional synthesis. Arctic, Antarctic and Alpine Research 34, 308-317.

Bockheim JG. (2008) Functional diversity of soils along environmental gradients in the Ross Sea region, Antarctica. Geoderma 144, 32-42.

Bockheim JG., Tarnocai C. (1998) Nature, occurrence and origin of dry permafrost. Permafrost. Seventh International Conference (Proceedings), Yellowknife, Canada, 23rd-27th June 1998. No. 55.

Bockheim JG., Hall KJ. (2002) Permafrost active-layer dynamics and perglacial environments of continental Antarctica. South African Journal of Science 98, 82-90.

Bockheim JG., Campbell IB., McLeod M. (2007) Permafrost distribution and active-layer depths in the McMurdo Dry Valleys, Antarctica. Permafrost and Periglacial Processes 18, 217-227.

Bowman JP., Cavanagh J., Austin JJ., Sanderson K. (1996) Novel *Psychrobacter* species from Antarctic ornithogenic soils. International Journal of Systematic Bacteriology 46, 841–848.

Bowman JP., Nichols DS., McMeekin TA. (1997) *Psycrobacter glacincola* sp. nov., a halotolerant, psychrophilic bacterium isolated from Antarctic sea ice. Systematic Applied Microbiology 20, 209-215.

Boyd WL., Rothenberg I., Boyd JW. (1970) Soil microorganisms at Paradise Harbour, Antarctica. Ecology 51, 1040-1045.

Brain DA., Jakosky BM. (1998) Atmospheric loss since the onset of the Martian geologic record: combined role of impact erosion and sputtering. Journal of Geophysical Research 103, 22,689-22,694.

Breezee J., Cady N., Staley JT. (2004) Subfreezing growth of the sea ice bacterium "*Psychromonas ingrahamii*". Microbial Ecology 47, 300-304.

Broady PA. (1981) The ecology of sublithic terrestrial algae at the Vestfold Hills, Antarctica. British Phycological Journal 16, 231-240.

Broady PA. (2005) The distribution of terrestrial and hydro-terrestrial algal associations at three contrasting locations in southern Victoria Land, Antarctica. Archives for Hydrobiological Supplement, Algological Studies 118, 95-112.

Burkins MB., Virginia RA., Wall DH. (2001) Organic carbon cycling in Taylor Valley, Antarctica: quantifying soil reservoirs and soil respiration. Global Change Biology 7, 113–125.

Campbell IB., Claridge GGC. (1987) Antarctica: Soils Weathering Processes and Environment. Elsevier, Amsterdam. pp. 368.

Campbell IB., Claridge GGC., Balks MR., Campbell DI. (1997). Moisture content in soils of the McMurdo Sound and Dry Valley region of Antarctica. In: Lyons WB., Howard-Williams C., Hawes HI. (Eds.), Ecosystem Processes in Antarctic Ice-Free Landscapes. Balkema, Rotterdam, pp. 61-76.

Carr MH. (1996) Water on Mars. Oxford University Press, New York.

Cary SC., McDonald IR., Barrett JE., Cowan DA. (2010) On the rocks: the microbiology of Antarctic Dry Valley soils. Nature Reviews – Microbiology 8, 129-138.

Casamatta DA, Verb RG, Beaver JR, Vis ML. (2002) An investigation of the cryptobiotic community from sandstone cliffs in southeast Ohio. International Journal of Plant Science 163, 837-845.

Cavacini P. (2001) Soil algae from Northern Victoria Land (Antarctica). Polar Bioscience 14, 45-60.

Chassefière E., Leblanc F. (2004) Mars atmospheric escape and evolution; interaction with the solar wind. Planetary and Space Science 52(11), 1039-1058.

Chin JP., Megaw J., Magill CL., Nowotarski K., Williams JP., Bhaganna P., Linton M., Patterson MF., Underwood GJC., Mswaka AY., Hallsworth JE. (2010) Solutes determine the temperature windows for microbial survival and growth. Proceedings of the National Academy of Sciences of the USA 107(17), 7835-7840.

Cockell CS., McKay CP., Omelon C. (2003) Polar endoliths – an anti-correlation of climate extremes and microbial diversity. International Journal of Astrobiology 1, 305-310.

Cockell CS., Schuerger AC., Billi D., Friedmann EI., Panitz C. (2005) Effects of a Simulated Martian UV Flux on the cyanobacterium, *Chroococcidiopsis* sp. 029. Astrobiology 5, 127-140.

Collins MA., Buick RK. (1989) Effect of temperature on the spoilage of stored peas by *Rhodotorula glutinis*. Food Microbiology 6(3), 135-141.

Connerney JEP., Acuña MH., Wasilewski PJ., Kletetschka G., Ness NF., Rème H., Lin RP., Mitchell DL. (1999) The global magnetic field of Mars and implications for crustal evolution. Science 284, 790-793.

Costello EK., Halloy SRP., Reed SC., Sowell P., Schmidt SK. (2009) Fumarole-supported islands of biodiversity within a hyperarid high-elevation landscape on Socompa Volcano, Puna de Atacama, Andes. Applied and Environmental Microbiology 75(3), 735-747.

Cowan DA., Russell N., Mamais A., Sheppard DM. (2002) Antarctic Dry Valley mineral soils contain unexpectedly high levels of microbial biomass. Extremophiles 6, 431-436.

Cowan DA., Tow LA. (2004) Endangered Antarctic environments. Annual Review of Microbiology 58, 649-690.

Cowan DA., Russell NJ., Mamais A., Sheppard DM. (2002) Antarctic Dry Valley mineral soils unexpectedly high levels of microbial biomass. Extremophiles 6, 431-436.

Crown DA., Price KH., Greeley R. (1992) Geologic evolution of the east rim of the Hellas basin, Mars. Icarus 100, 1-25.

Curtis TP., Sloan WT., Scannell JW. (2002) Estimating prokaryotic diversity and its limits. Proceedings of the National Academy of Sciences of the USA 99(16), 10494-10499.

De Angelis M., Steffensen JP., Legrand M., Clausen H., Hammer C. (2005) Primary aerosol (sea salt and soil dust) deposited in Greenland ice during the last climatic cycle: comparison with east Antarctic records. Journal of Geophysical Research 102, 26681-26698.

Deckker de P., Norman M., Goodwin ID., Wain A., Gingele FX. (2010) Lead isotopic evidence for an Australian source of aeolian dust to Antarctica at times over the last 170,000 years. Palaeogeography, Palaeoclimatology, Palaeoecology 285, 205-223.

de la Torre JR., Goebel BM., Friedmann EI., Pace NR. (2003) Microbial diversity of cryptoendolithic communities from the McMurdo Dry Valleys, Antarctica. Applied and Environmental Microbiology 69(7), 3858-3867.

DeLong EF., Pace NR. (2001) Environmental diversity of bacteria and archaea. Systematic Biology 50(4), 470-478.

Deuerling KM. (2010) Aeolian Sediments of the McMurdo Dry Valleys, Antarctica. Masters Dissertation, Ohio State University, Columbus, Ohio, USA.

Dmitriev VV., Suzina NE., Rusakova TG., Gilichinsky DA., Duda VI. (2000) Ultrastructural characteristics of natural forms of microorganisms isolated from permafrost grounds of eastern Siberia by the method of low-temperature fractionation. Doklady Biological Science 378, 304–306.

Dohm JM., Tanaka KL. (1999) Geology of the Thaumasia region, Mars: plateau development, valley, and magmatic evolution. Planetary and Space Science 47, 411-431.

Dohm JM., Tanaka KL., Hare TM., (2001a) Geologic Map of the Thaumasia Region of Mars: USGS Misc. Inv. Ser. Map I-2650U.S. Geological Survey Map I-2650, scale 1:5,000,000.

Dohm JM., Ferris JC., Baker VR., Anderson RC., Hare TM., Strom RG., Barlow NG., Tanaka KL., Klemaszewski JE., Scott DH. (2001b) Ancient drainage basin of the Tharsis region, Mars: potential source for outflow channel systems and putative oceans or paleolakes. Journal of Geophysical Research 106, 32,942-32,958.

Dohm JM., Maruyama S., Baker VR., Anderson RC., Ferris JC., Hare TM. (2002) Plate tectonism on early Mars: diverse geological and geophysical evidence. Houston, Texas, Lunar and Planetary Institute, Lunar and Planetary Science Conference XXXIII, abstract 1639 (CD-ROM).

Dohm JM., Ferris JC., Barlow NG., Baker VR., Mahaney WC., Anderson RC., Hare TM. (2004) The Northwestern Slope Valleys (NSVs) region, Mars: a prime candidate site for the future exploration of Mars. Planetary and Space Sciences 52, 189-198.

Dohm JM., Kerry K., Keller JM., Baker VR., Maruyama S., Anderson RC., Ferris JC., Hare TM. (2005) Mars geological province designations for the interpretation of GRS data. Lunar and Planetary Science Conference XXXVI, #1567 (abstract) [CD-ROM].

Dohm JM., Maruyama S., Baker VR., Anderson RC. (2007) Traits and evolution of the Tharsis superplume, Mars. In: Yuen DA., Maruyama S., Karato S-I., Windley BF. (Eds.), Superplumes: Beyond Plate Tectonics: London, Springer, pp. 523-537.

Dohm JM., Baker V., Boynton W., Fairén A., Ferris J., Mahaney WC., Finch M., Furfaro R., Hare T., Janes D., Kargel J., Karunatillake S., Keller J., Kerry K., Kim K., Komatsu G., Schulze-Makuch D., Marinangeli L., Ori G., Ruiz J. (2009a) GRS evidence and the possibility of paleooceans on Mars. Journal of Planetary and Space Sciences 57(5-6), 664-684.

Dohm JM., Anderson RC., Williams J-P., Ruiz J., McGuire PC., Buczkowski DL., Wang R., Scharenbroich L., Hare TM., Connerney JEP., Baker VR., Wheelock SJ., Ferris JC., Miyamoto H. (2009b) Claritas rise: Pre-Tharsis magmatism. Journal of Volcanology and Geothermal Research 185, 139-156.

Dohm JM., Miyamoto H., Ori GG., Fairén AG., Davila AF., Komatsu G., Mahaney WC., Williams J-P., Joye SB., Di Achille G., Oehler DZ., Marzo GA., Schulze-Makuch D., Acocella V., Glamoclija M., Pondrelli M., Boston P., Hart KM., Anderson RC., Baker VR., Fink W., Kelleher BP., Furfaro R., Gross C., Hare TM., Frazer AR., Ip F., Allen CCR., Kim KJ., Maruyama S., McGuire PC., Netoff D., Parnell J., Wendt L., Wheelock SJ., Steele A., Hancock RGV., Havics RA., Costa P., Krinsley D. (2011) An inventory of potentially habitable environments on Mars: geological and biological perspectives. In: Gerry B., Bleacher J. (Eds.), Analogs for Planetary Exploration, GSA Special Paper SPE483, Geological Society of America, Boulder, Colorado, pp. 317-347.

Doran PT., McKay CP., Clow GD., Dana GL., Fountain AG., Nylen T., Lyons WB. (2002a) Valley floor climate observations from the McMurdo dry valleys, Antarctica, 1986-2000. Journal of Geophysical Research 107, doi:10.1029/2001JD002045.

Doran PT., Priscu JC., Lyons WB., Walsh JE., Fountain AG., McKnight DM., Moorhead DL., Virginia RA., Wall DH., Clow GD., Fritsen CH., McKay CP., Parsons AN. (2002b) Antarctic climate cooling and terrestrial ecosystem response. Nature 415, 517-520.

Dumont MG., Murrell CJ. (2005) Stable isotope probing – linking microbial identity to function. Nature Reviews Microbiology 3, 499-504.

Edwards JS. (1988) Life in the allobiosphere. Trends in Ecology and Evolution 3(5), 111-114.

Fairén AG. (2010) A cold and wet Mars. Icarus 208, 165-175.

Fairén AG., Ruiz J., Anguita F. (2002) An origin for the linear magnetic anomalies on Mars through accretion of terranes: implications for dynamo timing. Icarus 160, 220-223, doi:10.1006/icar.2002.6942.

Fairén AG., Dohm JM., Baker VR., de Pablo MA., Ruiz J., Ferris JC., Anderson RC. (2003) Episodic flood inundations of the northern plains of Mars. Icarus 165, 53-67.

Fairén AG., Schulze-Makuch D., Rodríguez AP., Fink W., Davila A., Uceda ER., Furfaro R., Amils R., McKay CP. (2009) Evidence for Amazonian acidic liquid water on Mars - a reinterpretation of MER mission results. Planetary and Space Science 57, 276-287.

Fairén AG., Davilla AF., Lim D., Bramall N., Bonaccorsi R., Zavaleta J., Uceda ER., Stoker C., Wierzchos J., Dohm JM., Amils R., Andersen D., McKay CP. (2010) Astrobiology through the ages of Mars: the study of terrestrial analogues to understanding the habitability of Mars. Astrobiology 10(8), 821-843.

Fell JW., Scorzetti G., Connell L., Craig S. (2006) Biodiversity of micro-eukaryotes in Antarctic Dry Valley soils with < 5% soil moisture. Soil Biology and Biochemistry 38, 3107-3119.

Finegold L., Singer MA., Federle TW., Vestal JR. (1990) Composition and thermal properties of membrane lipids in cryptoendolithic lichen microbiota from Antarctica. Applied and Environmental Microbiology 56, 1191-1194.

Fink W., Dohm JM., Tarbell MA., Hare TM., Baker VR. (2005) Next-generation robotic planetary reconnaissance missions: a paradigm shift. Planetary and Space Sciences 53, 1419-1426.

Friedmann EI. (1980) Endolithic microbial life in hot and cold deserts. Origins of Life and Evolution of Biospheres 10(3), 223-235.

Friedmann EI. (1982) Endolithic microorganisms in the Antarctic cold desert. Science 215, 1045-1053.

Friedmann EI. (1994) Permafrost as microbial habitat. In: Gilichinsky DA. (Ed.), Viable Microorganisms in Permafrost. Russian Academy of Sciences, pp. 21-26.

Friedman EI., Ocampo R. (1976) Endolithic blue-green algae in the dry valleys: primary producers in the Antarctic desert ecosystem. Science 193, 1274-1279.

Friedmann EI., Ocampo-Friedmann R. (1984) Endolithic microorganisms in extreme dry environments: analysis of a lithobiontic microbial habitat, In: Klug MJ., Reddy CA. (Eds.), Current Perspectives in Microbial Ecology. American Society of Microbiology, Washington, pp. 177-185.

Friedmann EI., McKay CP. (1985) A method for continuous monitoring of snow: application to the cryptoendolithic microbial community of Antarctica. Antarctic Journal of the United States 20, 179-181.

Friedmann I., Lipkin Y., Ocampo-Paus R. (1967) Desert algae of the Negev (Israel). Phycologia 6, 185-200.

Friedmann EI., Kappen L., Meyer MA., Nienow JA. (1993) Long-term productivity in the cryptoendolithic microbial community of the Ross Desert, Antarctica. Microbial Ecology 25, 51-69.

Furfaro R., Dohm JM., Fink W., Kargel J., Schulze-Makuch D., Fairén, AG., Palmero-Rodriguez A., Baker VR., Ferré, PT., Hare TM., Tarbell MA., Miyamoto H., Komatsu G. (2008) The search for life beyond Earth through fuzzy expert systems. Planetary and Space Sciences 56, 448-472.

Ganzert L., Bajerski F., Mangelsdorf K., Lipski A., Wagner D. (2011a) *Arthrobacter livingstonensis* sp. nov. and *Arthrobacter cryotolerans* sp. nov., salt-tolerant and psychrotolerant species from Antarctic soil. International Journal of Systematic and Evolutionary Microbiology 61, 979-984.

Ganzert L., Lipski A., Hubberten H-W., Wagner D. (2011b) The impact of different soil parameters on the community structure of dominant bacteria from nine different soils located on Livingston Island, South Shetland Archipelago, Antarctica. FEMS Microbiological Ecology 76, 476-491.

Gaylarde PM., Jungblut A-D., Gaylarde CC., Neilan BA. (2006) Endolithic phototrophs from an active geothermal region in New Zealand. Geomicrobiology Journal 23, 579-587.

Gerrath JF., Gerrath JA., Matthes U., Larson DW. (2000) Endolithic algae and cyanobacteria from cliffs of the Niagara Escarpment, Ontario, Canada. Canadian Journal of Botany 78(6), 807-815.

Gilichinsky DA., Rivkina E., Shcherbakova V., Laurinavichuis K., Tiedje JM. (2003) Supercooled water brines within permafrost – an unknown ecological niche for microorganisms: a model for astrobiology. Astrobiology 3, 331-341.

Gilichinsky DA., Wilson GS., Friedmann EI., McKay CP., Sletten RS., Rivkina EM., Vishnivetskaya TA., Erokhina LG., Ivanushkina NE., Kochkina GA., Shcherbakova VA., Soina VS., Spirina EV., Vorobyova EA., Fyodorov-Davydov DG., Hallet B., Ozerskaya SM., Sorokovikov VA., Laurinavichyus KS., Shatilovich AV., Chanton JP., Ostroumov VE., Tiedje JM. (2007) Microbial populations in Antarctic permafrost: biodiversity, state, age, and implication for astrobiology. Astrobiology 7(2), 275-311.

Gilichinsky DA., Vishnivetskaya T., Petrova M., Spirina E., Mamykin V., Rivkina E. (2008) Bacteria in Permafrost. In: Margesin R., Schinner F., Marx J-C., Gerday C. (Eds.), Psychrophiles – From Biodiversity to Biotechnology. Springer-Verlag, Berlin. pp, 83-102.

Glamoclija M., Garrel L., Berthon J., López-García P. (2004) Biosignatures and bacterial diversity in hydrothermal deposits of Solfatara Crater, Italy. Geomicrobiology Journal 21, 529-541

Glavin DP., Schubert M., Botta O., Kminek G., Bada JL. (2001) Detecting pyrolysis product from bacteria on Mars. Earth and Planetary Science Letters 185, 1-5.

Gooseff MN., Barrett JE., Doran PT., Fountain AG., Lyons WB., Parsons AN., Porazinska DL., Virginia RA., Wall DH. (2003) Snow-patch influence on soil biogeochemical processes and invertebrate distribution in the McMurdo Dry Valleys, Antarctica. Arctic, Antarctic, and Alpine Research 35(1), 91-99.

Gounot A-M. (1991) Bacterial life at low temperature: physiological aspects and biotechnological implications. Journal of Applied Bacteriology 71, 386-397.

Grasby SE., Allen CC., Longazo TG., Lisle JT., Griffin DW., Beauchamp B. (2003) Supraglacial sulfur springs and associated biological activity in the Canadian high Arctic - signs of life beneath the ice. Astrobiology 3, 583-596.

Greeley R., Guest JE. (1987) Geologic map of the eastern equatorial region of Mars. USGS Misc. Inv. Ser. Map I-1802B (1:15,000,000).

Grilli-Caiola M., Billi D. (2007) *Chroococcidiopsis* from desert to Mars. In: Seckbach J. (Ed.), Algae and Cyanobacteria in Extreme Environments. Vol 11, COLE book series. Berlin, Springer, pp. 553-568.

Hagedorn B., Sletten RS., Hallet B. (2007) Sublimation and ice condensation in hyperarid soils: modeling results using field data from Victoria Valley, Antarctica. Journal of Geophysical Research 112, F03017, doi:10.1029/2006JF000580.

Harding T., Jungblut AD., Lovejoy C., Vincent WF. (2011) Microbes in high Arctic snow and implications for the cold biosphere. Applied and Environmental Microbiology 77(10), 3234-3243.

Harris SA., French HM., Heginbottom GH., Ladanyi B., Sego DC., van Everdingen RO. (1988) Glossary of Permafrost and Related Ground-Ice Terms. Technical Memorandum No. 142. National Research Council of Canada.

Hart KM., Szpak MT., Mahaney WC., Dohm JD., Jordan SF., Frazer AR., Allen CCR., Kelleher BP. (2011) A bacterial enrichment study and overview of the extractable lipids from paleosols in the Dry Valleys, Antarctica: Implications for future Mars reconnaissance. Astrobiology 11(4), 303-321.

Hartmann WK., Neukum G. (2001) Cratering chronology and the evolution of Mars. Space Science Reviews 96, 165-194.

Head JW., Pratt S. (2001) Extensive Hesperian-aged south polar ice sheet on Mars: Evidence for massive melting and retreat, and lateral flow and ponding of meltwater. Journal of Geophysical Research 106, 12,275-12,299.

Hecht MH., Kounaves SP., Quinn RC., West SJ., Young SMM., Ming DW., Catling DC., Clark BC., Boynton WV., Hoffman J., DeFlores LP., Gospodinova K., Kapit J., Smith PH. (2009) Detection of perchlorate and the soluble chemistry of Martian soil at the Phoenix landing site. Science 325, 64-67.

Hock AN., Cabrol NA., Dohm JM., Piatek J., Warren-Rhodes K., Weinstein S., Wettergreen DS., Grin EA., Moersch J., Cockell CS., Coppin P., Ernst L., Fisher G., Hardgrove C., Marinangeli L., Minkley E., Ori GG., Waggoner A., Wyatt M., Smith T., Thompson D., Wagner M., Jonak D., Stubbs K., Thomas G., Pudenz E., Glasgow J. (2007) Life in the Atacama: a scoring system for habitability and the robotic exploration for life. Journal of Geophysical Research 112, G04S08.

Hodges CA., Moore HJ. (1994) Atlas of Volcanic Landforms on Mars: U.S. Geological Survey Professional Paper 1534, pp. 194.

Hopkins DW., Sparrow AD., Novis PM., Gregorich EG., Elberling B., Greenfield LG. (2006) Controls on the distribution of productivity and organic resources in Antarctic Dry Valley soils. Proceedings of the Royal Society B 273, 2687-2695.

Hopkins DW., Sparrow AD., Gregorich EG., Elberling B., Novis P., Greenfield LG. (2009) Isotopic evidence for the provenance and turnover of organic carbon by soil microorganisms in the Antarctic dry valleys. Environmental Microbiology 11, 597-608.

Horath T., Bachofen R. (2009) Molecular characterization of a endolithic microbial community in dolomite rock in the central Alps. Microbial Ecology 58, 290-306.

Horowitz NH., Cameron RE., Hubbard JS. (1972) Microbiology of the Dry Valleys of Antarctica. Science 176(4032), 242-245.

Houtkooper JM., Schulze-Makuch D. (2007) A possible biogenic origin for hydrogen peroxide on Mars: the Viking results reinterpreted. International Journal of Astrobiology 6, 147-152.

Huston AL. (2003) Bacterial adaptation to the cold: in situ activities of extracellular enzymes in the North Water polynya and characterization of a cold-active aminopeptidase from *Colwellia psychrerythraea* strain 34H, Doctoral Dissertation, School of Oceanography, University of Washington, pp. 153.

Jackson DR., Orsolini YJ., Engelsen O. (2011) Low-ozone events in the southern polar summer as indicated by Met Office ozone analyses. Journal of Geophysical Research 116, D06302, doi:10.1029/2010JD014858.

Jakosky BM., Pepin RO., Johnson RE., Fox JL. (1994) Mars atmospheric loss and isotopic fractionation by solar-wind-induced sputtering and photochemical escape. Icarus 111, 271-288.

Johnson RM., Madden JM., Swafford JR. (1978) Taxonomy of Antarctic bacteria from soils and air primarily of the McMurdo Station and Victoria Land Fry Valleys regions. In: Parker B. (Ed.), Terrestrial Biology III, Antarctic Research Series 30. AGU, Washington, pp. 35-64.

Johnston CG., Vestal JR. (1989) Does iron inhibit cryptoendolithic communities? Antarctic Journal of the United States 21(5), 225-226.

Johnston CG., Vestal JR. (1991) Photosynthetic carbon incorporation and turnover in Antarctic cryptoendolithic microbial communities: are they the slowest growing communities on Earth? Applied and Environmental Microbiology 57, 2308-2311.

Junge K., Eicken H., Deming JW. (2004) Bacterial activity at -2 to -20°C in Arctic wintertime sea ice. Applied and Environmental Microbiology 70, 550-557.

Kargel JS. (2004) Mars: A Warmer Wetter Planet. Praxis-Springer, pp. 557.

Kargel JS., Strom RG. (1992) Ancient glaciation on Mars. Geology 20, 3-7.

Klein HP. (1978) The Viking biological experiments on Mars. Icarus 34, 666-674.

Klein HP. (1999) Did Viking discover life on Mars? Origins of Life and Evolution of Biospheres 29(6), 625-631.

Kuzmin RO. (2005) Ground ice in the Martian regolith. In: Tokano T. (Ed.), Water on Mars and Life, Advances in Astrobiology and Biogeophysics, Springer, Heidelberg, pp. 155-189.

Lamb HH., Commander GP., Britton RN. (1955) General atmospheric circulation and weather variations in the Antarctic. The Geographical Journal 121(3), 334-349.

Lancaster N. (2002) Flux of eolian sediment in the McMurdo Dry Valleys, Antarctica: a preliminary assessment. Arctic, Antarctic and Alpine Research 34(3), 318-323.

Lane MD., Darby-Dyar M., Bishop JL. (2004) Spectroscopic evidence for hydrous iron sulfate in the Martian soil. Geophysical Research Letters 31, doi:10.1029/2004GL021231.

Laskar J., Levarard B., Mustard JF. (2002) Orbital forcing of the Martian polar layered deposits. Nature 419, 375-377.

Laybourn-Parry J. (2009) No place too cold. Science 324, 1521-1522.

Levin GV., Straat PA. (1981) A search for a nonbiological explanation of the Viking Labeled Release life detection experiment. Icarus 45, 494-516.

Liu H., Xu Y., Ma Y., Zhou P. (2000) Characterization of *Micrococcus antarcticus* sp. nov., a psychrophilic bacterium from Antarctica. International Journal of Systematic and Evolutionary Microbiology 50, 715-719.

Mackelprang R., Waldrop MP., DeAngelis KM., David MM., Chavarria KL., Blazewicz SJ., Rubin EM., Jansson JK. (2011) Metagenomic analysis of a permafrost microbial community reveals a rapid response to thaw. Nature 480, 368-371.

Madsen EL. (1998) Epistemology of environmental microbiology. Environmental Science and Technology 32, 429-439.

Madsen EL. (2005) Identifying microorganisms responsible for ecologically significant biogeochemical processes. Nature 3, 439-446.

Mager DM., Thomas AD. (2011) Extracellular polysaccharides from cyanobacterial soil crusts: A review of their role in dryland soil processes. Journal of Arid Environments 75, 91-97.

Mahaney WC. (1978) Late-Quaternary Soil-Stratigraphy in the Wind River Mountains, Western Wyoming, In: Mahaney WC. (Ed.), Quaternary Soils, Geo-Abstracts, Norwich, UK, pp. 223-264.

Mahaney WC. (1984) Evaluation of dating methods used to assign ages in the Wind River and Teton Ranges, Western Wyoming. In: Mahaney WC. (Ed.), Quaternary Dating Methods, Vol. 7 - Developments in Palaeontology and Stratigraphy. Elsevier, Holland, pp. 355-374.

Mahaney WC. (1990) Ice on the Equator. Wm Caxton Press, Ellison Bay, Wisconsin. pp. 386.

Mahaney WC., Dohm J. (2010) Life on Mars? Microbes in Mars-like Antarctic environments. Journal of Cosmology 5, 951-958.

Mahaney W., Miyamoto H., Dohm JM., Baker VR., Cabrol NA., Grin EA., Berman DC. (2007) Rock glaciers on Mars: Earth-based clues to Mars' recent paleoclimatic history. Planetary and Space Science 55, 181-192.

Mahaney WC., Milner MW., Kalm V., Dirszowsky RW., Hancock RGV., Beukens RP. (2008) Evidence for a Younger Dryas glacial advance in the Andes of northwestern Venezuela. Geomorphology 96, 199-211.

Mahaney WC., Dohm JM., Baker VR., Newsom HE., Malloch D., Hancock RGV., Campbell I., Sheppard D., Milner MW. (2001) Morphogenesis of Antarctic paleosols: Martian analogue. Icarus 154, 113-130.

Mahaney WC., Dohm J., Kapran B., Hancock RGV., Milner MW. (2009) Secondary Fe and Al in Antarctic paleo-sols: correlation to Mars with prospect for the presence of life. Icarus 203, 320-330.

Mahaney WC., Hart KM., O'Reilly SS., Allen CCR., Dohm JM., Hancock RGV., Kelleher BP., Milner MW. (2012) Coleoptera and microbe biomass in Antarctic Dry Valley paleosols adjacent to inland ice: implications for Mars. Planetary and Space Science 60, 386-398.

Maher KA., Stevenson DJ. (1988) Impact frustration of the origin of life. Nature 331, 612-614.

Margesin R., Schinner F., Marx JC., Gerday C. (2008) Psychrophiles: from Biodiversity to Biotechnolgy. Springer, New York.

Marinova MM., McKay CP., Heldmann JL., Davila AF., Andersen DT., Jackson WA., Lacelle D., Paulson G., Pollard WH., Zacny K. (2011) Dry soils: the highlands of the Antarctic Dry Valleys and the defining environmental conditions. EPSC Abstracts Vol. 6, EPSC-DPS2011-1319. EPSC-DPS Joint Meeting 2011, 2-7 October 2011, Nantes, France.

Marinova MM., McKay CP., Pollard WH., Heldmann JL., Davila AF., Andersen DT., Jackson WA., Lacelle D., Paulson G., Zacny K. Distribution of depth to ice-cemented soils in the high-elevation Quartermain Mountains, Dry Valleys of Antarctica. Antarctic Science, (*submitted*).

Martineau C., Whyte LG., Greer CW. (2010) Stable isotope probing analysis of the diversity and activity of methanotrophic bacteria in soils from the Canadian high Arctic. Applied and Environmental Microbiology 76(17), 5773-5784.

Martiny JBH., Bohannan BJM., Brown JH., Colwell RK., Fuhrman JA., Green JL., Horner-Devine MC., Kane M., Adams Krumins J., Kuske CR., Morin PJ., Naeem S., Øvreås L., Reysenbach A-L., Smith VH., Staley JT. (2006) Microbial biogeography: putting microorganisms on the map. Nature Reviews Microbiology 4, 102-112.

Matsumoto GI., Machihara T., Suzuki N., Funaki M., Watanuki K. (1987) Steranes and triterpanes in the Beacon Supergroup samples from southern Victoria Land in Antarctica. Geochimica Cosmochimca Acta 51(10), 2663-2671.

Mazur P., Barghoorn ES., Halvorson HO., Jukes TH., Kaplan IR., Margulis L. (1978) Biological implications of the Viking Mission to Mars. Space Science Reviews 22, 3-34.

McGinnis LD. (1981) Dry Valley drilling project. In: Antarctic Research Series Vol. 33, AGU, Washington. pp. 465.

McKay CP. (2009) Snow recurrence sets the depth of dry permafrost at high elevations in the McMurdo Dry Valleys of Antarctica. Antarctic Science 21(1), 89-94.

McKay CP., Friedmann EI. (1985) The cryptoendolithic microbial environment in the Antarctic cold desert: temperature variations in nature. Polar Biology 4, 19-25.

McKay CP., Friedmann EI., Wharton RA., Davies WL. (1992) History of water on Mars: a biological perspective. Advances in Space Research 12, 231-238.

McKay CP., Grunthaner FJ., Lane AL., Herring M., Bartman RK., Ksendzov A., Manning JL., Williams RM., Ricco AJ., Butler MA., Murray BC., Quinn RC., Zent AP., Klein HP., Levin GV. (1998a) The Mars oxidant experiment (MOx) for Mars '96. Planetary and Space Sciences 46, 769-777.

McKay CP., Mellon MT., Friedmann EI. (1998b) Soil temperatures and stability of ice-cemented ground in the McMurdo Dry Valleys, Antarctica. Antarctic Science 10(1), 31-38.

McKew BA., Taylor JD., McGenity TJ., Underwood GJC. (2011) Resistance and resilience of benthic biofilm communities from a temperate saltmarsh to desiccation and rewetting. The ISME Journal 5, 30-41.

Mehnaz S., Baig DN., Lazarovits G. (2010) Genetic and phenotypic diversity of plant growth promoting *Rhizobacteria* isolated from sugarcane plants growing in Pakistan. Journal of Microbial Biotechnology 20(12), 1614-1623.

Mikucki JA., Pearson A., Johnston DT., Turchyn AV., Farquhar J., Schrag P., Anbar AD., Priscu JC., Lee PA. (2009) A contemporary microbially maintained subglacial ferrous "ocean". Science 324, 397-400.

Miller JD., Straat PA., Levin GV. (2001) Periodic analysis of the Viking Lander Labeled Release Experiment: instruments, methods, and missions for astrobiology. Proceedings of SPIE 4495, 96-107.

Morelli FA., Cameron RE., Gensel DR., Randall LP. (1972) Monitoring of Antarctic Dry Valley drilling sites. Antarctic Journal of the US 7, 92-94.

Namsaraev Z., Mano M-J., Fernandez R., Wilmotte A. (2010) Biogeography of terrestrial cyanobacteria from Antarctic ice-free areas. Annals of Glaciology 51(56), 171-177.

Navarro-González R., Navaro KF., de la Rosa J., Iñiguez E., Molina P., Miranda LD., Morales P., Cienfuegos E., Coll P., Raulin F., Amils R., McKay CP. (2006) The limitations on organic detection in Mars-like soils by thermal volatilization- gas chromatography-MS and their implications for the Viking results. Proceedings of the National Academy of Sciences of the USA 103, 16089-16094.

Navarro-González R., Vargas E., de la Rosa J., McKay CP. (2010) Pyrolysis of Atacama soils with added perchlorates: implications for the Viking results [abstract 5072]. In: Astrobiology Science Conference 2010: Evolution and Life: Surviving Catastrophes and Extremes on Earth and Beyond, Lunar and Planetary Institute, Houston.

Neumann GA., Smith DE., Zuber MT. (2003) Two Mars years of clouds detected by the Mars Orbiter Laser Altimeter. Journal of Geophysical Research 108, doi:10.1029/2002JE001849.

Nicholson SE. (2011) Dryland Climatology. Cambridge University Press, Cambridge.

Niederberger TD., McDonald IR., Hacker AL., Soo RM., Barrett JE., Wall DH., Cary C. (2008) Microbial community composition in soils of Northern Victoria Land, Antarctica. Environmental Microbiology 10, 1713-1724.

Nienow JA., McKay CR., Friedmann EI. (1988) The cryptoendolithic microbial environment in the Ross Desert of Antarctica: light in the photosynthetically active region. Microbial Ecology 16(3), 271-289.

Nylen TH., Fountain AG., Doran PT. (2004) Climatology of the katabatic winds in the McMurdo dry valleys, southern Victoria Land, Antarctica. Journal of Geophysical Research 109, D03114, doi:10.1029/2003JD003937.

Omelon CR. (2008) Endolithic microbial communities in polar desert habitats. Geomicrobiology 25, 404-414.

Omelon CR., Ferris FG., Pollard WH., Andersen DT., Whyte L. (2003) High Arctic cryptoendolithic microorganisms in a polar desert environment In: *Phillips* M., *Springman* SM., *Arenson* LU. (*Eds*.), Proceedings of the 8th International Conference on Permafrost, 21-25 July 2003, Zurich, Switzerland.

Omelon CR., Pollard WH., Ferris FG. (2006) Environmental controls on microbial colonization of high Arctic cryptoendolithic habitats. Polar Biology 30, 19-29.

Omelon CR., Pollard WH., Ferris FG. (2007) Inorganic species distribution and microbial diversity within high Arctic cryptoendolithic habitats. Microbial Ecology 54(4), 740-752.

Pampelotto PH. (2010) Resistance of microorganisms to extreme environmental conditions and its contribution to astrobiology. Sustainability 2, 1602-1623.

Papineau D., Walker JJ., Mojzsis SJ., Pace NR. (2005) Composition and structure of microbial communities from stromatolites of Hamelin Pool in shark bay, Western Australia. Applied and Environmental Microbiology 71(8), 4822-4832.

Pearce DA., Bridge PD., Hughes KA., Sattler B., Psenner R., Russell NJ. (2009) Microorganisms in the atmosphere over Antarctica. FEMS Microbial Ecology 69, 143-157.

Peeters K., Ertz D., Willems A. (2011) Culturable bacterial diversity at the Princess Elisabeth Station (Utsteinen, Sør Rondane Mountains, East Antarctica) harbours many new taxa. Systematic and Applied Microbiology 34, 360-367.

Pernthaler J., Glöckner FO., Schönhuber W., Amann R. (2001) Fluorescence *in situ* hybridization. In: Paul J. (Ed.), Methods in Microbiology: Marine Microbiology, vol. 30. Academic Press, London, pp. 207-226.

Poage MA., Barrett JE., Virginia RA., Wall DH. (2008) The influence of soil geochemistry on nematode distribution, McMurdo Dry Valleys, Antarctica. Arctic, Antarctic and Alpine Research 40(1), 119-128.

Pointing SB., Chan Y., Lacap DC., Lau MCY., Jurgens JA., Farrell RL. (2009) Highly specialized microbial diversity in hyper-arid polar desert. Proceedings of the National Academy of Sciences of the USA 106, 19964-19969.

Potts M. (1999) Mechanisms of desiccation tolerance in cyanobacteria. European Journal of Phycology 34, 319-328.

Prentice ML., Kleman J., Stroeven AP. (1998) The composite glacial erosional landscape of the Northern McMurdo Dry Valleys: implications for Antarctic Tertiary glacial history. In: Priscu JC. (Ed.), Antarctic Research Series, Vol. 72: Ecosystem Processes in a Polar Desert: The McMurdo Dry Valleys, Antarctica. American Geophysical Union, Washington, pp. 1-38.

Priscu JC., Adams EE., Lyons WB., Voytek MA., Mogk DW., Brown RL., McKay CP., Takacs CD., Welch KA., Wolf CF., Kirshtein JD., Avci R. (1999) Geomicrobiology of subglacial ice above Lake Vostok, Antarctica. Science 286, 2141-2144.

Rappé MS., Giovannoi SJ. (2003) The uncultured majority. Annual Review of Microbiology 57, 369-394.

Reid IN., Sparks WB., Lubow S., McGrath M., Livio M., Valenti J., sowers KR., Shukla HD., MacAuley S., Miller T., Suvanasuthi R., Belas R., Colman A., Robb FT., DasSarma P., Müller JA., Coker JA., Cavicchioli R., Chen F., DasSarma. (2006) Terrestrial models for extraterrestrial life: methanogens and halophiles at Martian temperatures. International Journal of Astrobiology doi:10.1017/S1473550406002916.

Rios de los A., Wierzchos J., Grube M., Sancho LG., Ascaso C. (2007) Community structure and micro-scale distribution in endolithic microbial desert ecosystems. Geophysical Research Abstracts 9, 06711.

Rivkina EM., Friedmann EI., McKay CP., Gilichinsky D. (2000) Metabolic activity of permafrost bacteria below the freezing point. Applied and Environmental Microbiology 66, 3230-3233.

Roberson EB., Firestone MK. (1992) Relationship between desiccation and exopolysaccharide production in a soil *Pseudomonas sp*. Applied and Environmental Microbiology 58(4), 1284-1291.

Robbins SJ., Hynek BM. (2012) Impact history of large bollides at Mars: implications for the late-heavy bombardment and isochron uncertainties. Lunar and Planetary Science Conference XXXXIII, #1649 (abstract) [CD-ROM].

Romanovsky VE., Smith SL., Christiansen HH. (2010) Permafrost thermal state in the polar northern hemisphere during the International Polar Year 2007-2009: a synthesis. Permafrost and Periglacial Processes 21, 106-116.

Rothrock MJ, Jr., Garcia-Pichel F. (2005) Microbial diversity of benthic mats along a tidal desiccation gradient. Environmental Microbiology 7, 593-601.

Ruberto LAM., Vazquez S., Lobalbo A., MacCormack WP. (2005) Psychrotolerant hydrocarbon-degrading *Rhodococcus* strains isolated from polluted Antarctic soils. Antarctic Science 17, 47-56.

Sattler B., Puxbaum H., Psenner R. (2001) Bacterial growth in supercooled cloud droplets. Geophysical Research Letters 28(2), 239-242.

Schloss PD., Handelsman J. (2004) Status of the microbial census. Microbiology and Molecular Biology Reviews 68(4), 686-691.

Schmidt SK., Reed SC., Nemergut DR., Grandy AS., Cleveland CC., Weintraub MN., Hill AW., Costello EK., Meyer AF., Neff JC., Martin AM. (2008) The earliest stages of ecosystem succession in high-elevation (5000 meters above sea level), recently deglaciated soils. Proceedings of the Royal Society B 275, 2793-2802.

Schmidt SK., Nemergut DR., Miller AE., Freeman KR., King AJ., Seimon A. (2009) Microbial activity and diversity during extreme freeze-thaw cycles in perigalcial soils, Cordillera Vilcanota, Peru. Extremophiles 13, 807-816.

Schmidt SK., Lynch RC., King AJ., Karki D., Robeson MS., Nagy L., Williams MW., Mitter MS., Freeman KR. (2010) Phylogeography of microbial phototrophs in the dry valleys of the high Himalayas and Antarctica. Proceedings of the Royal Society B 278, 702-708.

Schulze-Makuch D., Fairén AG., Davila AF. (2008) The case for life on Mars. International Journal of Astrobiology 7(2), 117-141.

Scott DH., Tanaka KL. (1986) Geologic Map of the Western Equatorial Region of Mars: U.S. Geological Survey Miscellaneous Investigations Series Map I-1802-A, scale 1:15,000,000.

Scott DH., Dohm JM., Rice JW. Jr., (1995) Map of Mars showing channels and possible paleolake basins, *USGS Misc. Inv. Ser.* Map I-2461 (1:30,000,000).

Shivaji S., Reddy GSN., Suresh K., Gupta P., Chintalapati S., Schumann P., Stackebrandt E., Matsumoto GI. (2005) *Psychrobacter vallis* sp. nov. and *Psychrobacter aquaticus* sp. nov., from Antarctica. *International Journal of Systematic and Environmental Microbiology* 55, 757-762.

Shmida A. (1985) Biogeography of the desert floras of the world. In: Evenari M., Noy-Meir I., Goodall DW. (Eds.), Hot Deserts - *Ecosystems of the World, Vol. 12a*, Elsevier, Amsterdam, pp. 23-77.

Short N., Lowrnan Jr., PD., Freden SC., Finch Jr., WA. (1976) Mission to Earth: Landsat Views of the World. *NASA* SP-360, Washington, D.C., p. 459.

Siebert J., Hirsch P., Hoffmann B., Gliesche CG., Peissl K., Jendrach M. (1996) Cryptoendolithic microorganisms from Antarctic sandstone of Linnaeus Terrace (Asgard range): diversity, properties and interactions. *Biodiversity and Conservation* 5, 1337-1363.

Smith JJ., Tow LA., Stafford W., Cary C., Cowan DA. (2006) Bacterial diversity in three different Antarctic cold desert mineral soils. *Microbial Ecology* 51, 413-421.

Soina V., Vorobyova E. (1994) Preservation of microbial cell structures in the permafrost. In: Gilichinsky D. (Ed.), *Viable Microorganisms in Permafrost. Pushchino Research Center, Russian Academy of Sciences*, Pushchino, pp. 37-48.

Soina V., Mulyukin AL., Demkina EV., Vorobyova EA., El-Registan GI. (2004) The structure of resting bacterial populations in soil and subsoil permafrost. *Astrobiology* 4, 345-358.

Spalding JB. (1979) The aeolian ecology of White Mountain peak, California: windblown insect fauna. *Arctic and Alpine Research* 11(1), 83-94.

Stetter KO. (2005) Volcanoes, hydrothermal venting, and the origin of life. In: Martí J., Ernst GGJ. (Eds.), *Volcanoes and the Environment*. Cambridge University Press, Cambridge, pp. 175-206.

Steven B., Léveillé R., Pollard WH., Whyte LG. (2006) Microbial ecology and biodiversity in permafrost. *Extremophiles* 10, 259-267.

Summit M., Baross JA. (2001) A novel microbial habitat in the mid-ocean ridge seafloor. *Proceedings of the National Academy of Sciences of the USA* 98(5), 2158-2163.

Sun HJ., Friedmann EI. (1999) Growth on geological time scales in the Antarctic cryptoendolithic microbial community. *Geomicrobiology* 16, 193-202.

Swan LW. (1992) The aeolian biome. *Bioscience* 42(4), 262-270.

Tarnocai C., Canadell JG., Schuur EAG., Kuhry P., Mazhitova G., Zimov S. (2009) Soil organic carbon pools in the northern circumpolar permafrost region. *Global Biogeochemical Cycles 23*, GB2023, doi:10.1029/2008GB003327.

Tanaka KL., Dohm JM., Fortezzo CM., Irwin RP III., Kolb EJ., Skinner JA. Jr., Hare TM., Platz T., Michael G., Robbins S. (2012) The geology of Mars: what the new global map shows. *Lunar and Planetary Science Conference XXXXIII,* #2702 (abstract) [CD-ROM].

Tanaka KL., Dohm JM., Fortezzo CM., Irwin RP III., Kolb EJ., Skinner Jr. JA., Hare TM., Platz T., Michael G., Robbin S. *New global geologic map of Mars,* (in preparation).

Taylor GJ., Stopar JD., Boynton WV., Karunatillake S., Keller JM., Brückner J., Wänke H., Dreibus G., Kerry KE., Reedy RC., Evans LG., Starr RD., Martel LMV., Squyres SW., Gasnault O., Maurice S., d'Uston C., Englert P., Dohm JM., Baker VR., Hamara D., Janes D., Sprague AL., Kim KJ., Drake DM., McLennan SM., Hahn BC. (2006) Variations in K/Th on Mars. *Journal of Geophysical Research* 111, E03S06, doi:10.1029/2006JE002676.

Tkavc R., Ausec L., Oren A., Gunde-Cimerman N. (2011) Bacteria associated with *Artemia* spp. along the salinity gradient of the solar salterns at Eilat (Israel). *FEMS Microbial Biology* 77, 321-321.

Tripathi AK., Verma SC., Ron EZ. (2002) Molecular characterization of a salt-tolerant bacterial community in the rice rhizosphere. *Research in Microbiology* 153, 579-584.

Tung HC., Bramall NE., Price PB. (2005) Microbial origin of excess methane in glacial ice and implications for life on Mars. *Proceedings of the National Academy of Sciences of the USA* 102, 18292-18296.

Tung HC., Price PB., Bramall NE., Vrdoljak G. (2006) Microorganisms metabolizing on clay grains in 3-km-deep Greenland basal ice. *Astrobiology* 6, 69-86.

Verrecchia E., Yair A., Kidron GJ., Verrecchia K. (1995) Physical properties of the psammophile crytpogamic crust and their consequences to the water regime of sandy soils, north-western Negev Desert, Israel. *Journal of Arid Environments* 29, 427-437.

Vincent WF., Quesada A. (1994) Ultraviolet radiation effects on cyanobacteria: Implications for Antarctic microbial ecosystems. In: Weiler CS., Penhale PA. (Eds.), *Ultraviolet Radiation in Antarctica and Biological Effects - Antarctic Research Series Vol. 62*. pp. 111-124.

Vorobyova E., Soina V., Gorlenko M., Minkovskaya N., Zalinova N., Mamukelashvili A., Gilichinsky DA., Rivkina E., Vishnivetskaya T. (1997) The deep cold biosphere: facts and hypothesis. *FEMS Microbiological Review* 20, 277-290.

Walker JJ., Spear JR., Pace NR. (2005) Geobiology of a microbial endolithic community in the Yellowstone geothermal environment. *Nature* 434, 1011-1014.

Walker JJ., Pace NR. (2007a) Endolithic microbial ecosystems. *Annual Review of Microbiology* 61, 331-347.

Walker JJ., Pace NR. (2007b) Phylogenetic composition of Rocky Mountain endolithic microbial ecosystems. *Applied and Environmental Microbiology* 73(11), 3497-3504.

Warren-Rhodes KA., Rhodes KL., Pointing SB., Ewing SA., Lacap DC., Gómez-Silva B., Amundson R., Friedmann EI., McKay CP. (2006) Hypolithic cyanobacteria, dry limit of photosynthesis, and microbial ecology in the hyperarid Atacama Desert. *Microbial Ecology* 52(3), 389-398.

Warren-Rhodes K., Weinstein S., Dohm JM., Piatek J., Minkley E., Hock A., Pane D., Ernst LA., Fisher G., Emani S., Waggoner AS., Cabrol NA., Wettergreen DS., Apostolopoulos D., Coppin P., Grin E., Diaz C., Moersch J., Ori GG., Smith T., Stubbs K., Thomas G., Wagner M., Wyatt M. (2007a) Searching for microbial life remotely: Satellite-to-rover habitat mapping in the Atacama Desert, Chile. *Journal of Geophysical Research* 112, G04S05, doi:10.1029/2006JG000283.

Warren-Rhodes K., Weinstein S., Piatek JL., Dohm JM., Hock A., Minkley E., Pane D., Ernst LA., Fisher G., Emani S., Waggoner AS., Cabrol NA., Wettergreen DS., Grin E., Coppin P., Diaz C., Moersch J., Ori GG., Smith T., Stubbs K., Thomas G., Wagner M., Wyatt M., Boyle L Ng. (2007b) Robotic ecological mapping: habitats and the search for life in the Atacama Desert. *Journal of Geophysical Research* 112, G04S06, doi:10.1029/2006JG000301.

Wharton DA. (2002) Life at the Limits: Organisms in Extreme Environments. Cambridge University Press, Cambridge.

White PL., Wynn-Williams DD., Russell NJ. (2000) Diversity of thermal responses of lipid composition in the membranes of the dominant culturable members of an Antarctic fellfield soil bacterial community. *Antarctic Science* 12, 386-393.

Wierzchos J., Ascaso C., McKay CP. (2006) Endolithic cyanobacteria in halite rocks from the hyperarid core of the Atacama Desert. *Astrobiology* 6, 415-122.

Witherow RA., Lyons WB., Bertler NAN., Welch KA., Mayewski PA., Sneed SB., Nylen T., Handley MJ., Fountain A. (2006) The aeolian flux of calcium, chloride and nitrate to the McMurdo Dry Valleys landscape: evidence from snow pit analysis. *Antarctic Science* 18(4), 497-505.

Woese CR. (1987) Bacterial evolution. Microbiological Reviews 51(2), 221-271.

Woese CR., Kandler O., Wheelis ML. (1990) Towards a natural system of organisms: proposal for the domains archaea, bacteria and eucarya. *Proceedings of the National Academy of Sciences of the USA* 87, 4576-4579.

Wong FKY., Lau MCY., Lacap DC., Aitchison JC., Cowan DA., Pointing SB. (2010) Endolithic microbial colonization of limestone in a high-altitude arid environment. *Environmental Microbiology* 59, 686-699.

Wood SA., Rueckert A., Cowan DA., Cary SC. (2008) Sources of edaphic cyanobacterial diversity in the Dry Valleys of Eastern Antarctica. *The ISME Journal* 2, 308-320.

Wynn-Williams DD., Edwards HGM., Garcia-Pichel F. (1999) Functional biomolecules of Antarctic stromatolitic and endolithic cyanobacterial communities. *European Journal of Phycology* 34, 381-391.

Yen AS., Gellert R., Schröder C., Morris RV., Bell III JF., Knudson AT., Clark BC., Ming DW., Crisp JA., Arvidson RE., Blaney D., Brückner J., Christensen PR., DesMarais DJ., de Souza Jr PA. Economou TE., Ghosh A., Hahn BC., Herkenhoff KE., Haskin LA., Hurowitz JA., Joliff BL., Johnson JR. Klingelhöfer GS., Madsen MB., McLennan SM., McSween HY., Richter L., Rieder R., Rodionov D., Soderblom L., Squyres SW., Tosca NJ., Wang A., Wyatt M., Zipfel J. (2005) An integrated view of the chemistry and mineralogy of Martian soils. *Nature* 436, 49-54.

Zhang D-C., Liu H-C., Xin Y-H., Yu Y., Zhou P-J., Zhou Y-G. (2009) *Planomicrobium glaciei* sp. nov., a psychrotolerant bacterium isolated from a glacier. International *Journal of Systematic and Evolutionary Microbiology* 59, 1387-1390.

Zhang L-M., Offre PR., He J-Z., Verhamme DT., Nicol GW., Prosser JI. (2010) Autotrophic ammonia oxidation by soil thaumarchaea. *Proceedings of the National Academy of Sciences of the USA* 107(40), 17240-17245.

In: Mars
Editor: Alberto G. Fairén
ISBN: 978-1-62618-102-1

Chapter 8

SAMPLING POSSIBLE SUBSURFACE NICHES ON MARS: TARGETING ASTROBIOLOGICAL MISSIONS

Akos Kereszturi*
Konkoly Astronomical Institute, Research Centre for Astronomy and Earth Sciences, Hungarian Academy of Sciences, Budapest, Hungary
Hungarian Astronomical Association, Polaris Observatory
Karoly Nagy Astronomical Foundation, Budapest, Hungary

ABSTRACT

Subsurface sampling will be important in the robotic exploration of Mars in the future. In this chapter theoretical argumentation is summarized on the selection of ideal sites, including parameters that were defined for the earlier four candidate landing sites of Mars Science Laboratory. To identify ideal locations for subsurface analysis the dust and dune coverage, the cap layer distribution, and location of outcrops could be determined from above. While imaging information is enough for this, but few data is available on the rock hardness on Mars, and more results would be necessary to put the locations into global context and to understand the role of secondary cratering in age estimations. More laboratory work is also necessary to improve our knowledge on the extraction and preservation of organic materials.

Using remote images at the four earlier candidate landing sites of Mars Science Laboratory, at Eberswalde and Holden crater only 3-9% of exposed strata is present and individual outcrops are 100-400 m in diameter. For the Gale crater and Mawrth Valles regions, these parameters were 46-35% of exposed strata, and average outcrop diameter of ~300 m. At the first two sites elongated outcrops were present in larger number, while in the second group the large outcrops were farther away from each other. The analysis suggests that for missions aimed at subsurface sampling different exploration strategy would be ideal at different terrains, and the target terrain's characteristics could be taken into account during the planning phase.

* Akos Kereszturi. E-mail: kereszturiakos@gmail.com.

INTRODUCTION

This chapter outlines the background knowledge that is useful in the selection of landing sites aimed at the subsurface analysis of Mars. This research is a continuation and expansion of the author's earlier work on the analysis of crater exposed outcrops in Terra Meridiani (Kereszturi 2012a). Subsurface sampling is important for astrobiology, as the harsh surface conditions on the planet might have altered any possible biosignature, while buried ancient, water related shallow subsurface structures can hold these signatures and should be accessible with carefully planned sampling attempts. Some possible subsurface settings are summarized in Figure 55, where interesting layers of fine grained materials with long exposure to liquid water are indicated.

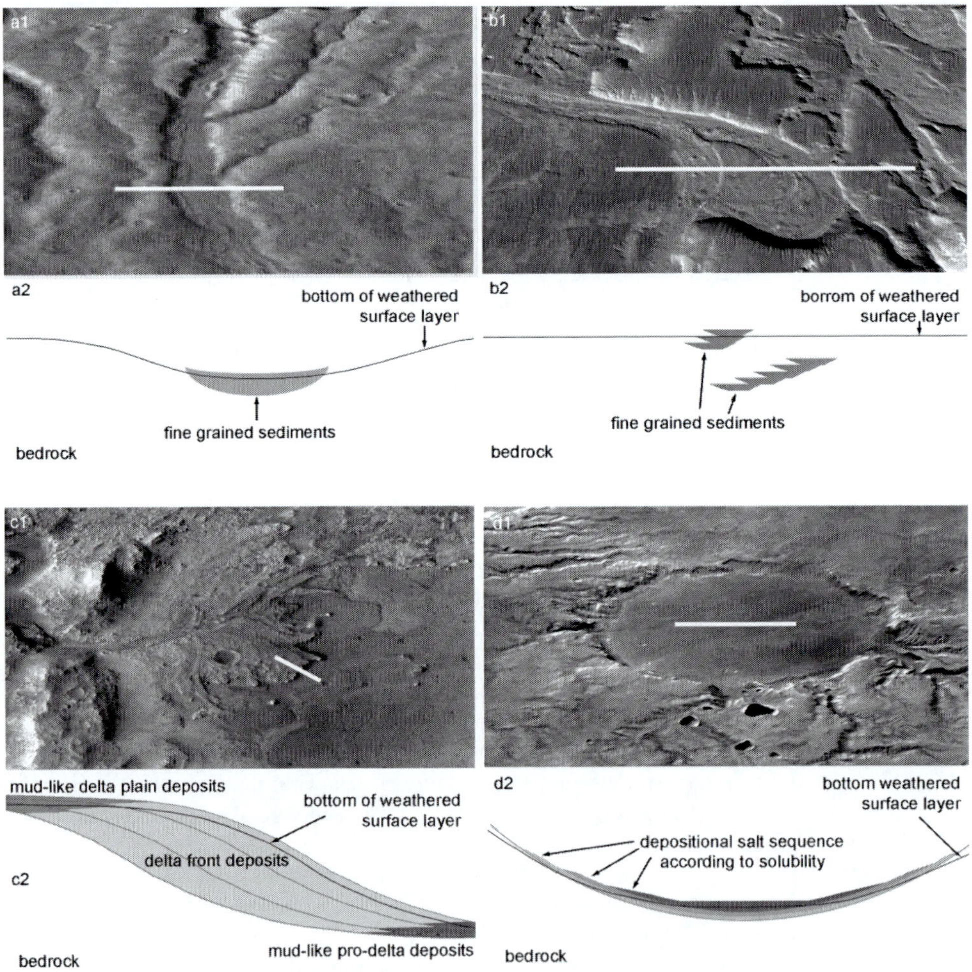

Figure 55. Images (a1, b1, c1, d1) and hypothetic cross sections (a2, b2, c2, d2), where the cross-profile's location is marked with white line, and important subsurface layers with gray colour. a) 12 km wide section of Warrego Vallis tributary system, b) a cut-off meander in the distributary fan of Eberswald crater (2 km wide), c) delta in Jezero-crater with a hypothetic foreset dominated (Edmonds et al. 2011) cross-profile, d) ancient crater lake with sequence of salt deposits. Useful subsurface targets could be acquired only below the weathered surface layer (marked with black lines).

Since the Viking missions our knowledge and experience in landing site selection has improved greatly, several publications (Arvidson et al. 2008, Golombek et al. 2009, Bridges et al. 2003) summarize the process of how to find the ideal landing site that satisfies both the requirements of scientists and engineers.

The design of missions and detectors are mostly aimed at to understand the possibility of past and present life on Mars, and the search strategy is also optimized for this goal (Westall et al. 2000). In this chapter some important points are reviewed in order to compare different landing sites regarding the access and astrobiological potential of the shallow subsurface zone there, a topic that fuses the scientific background with exploration issues.

The approach in this chapter gives example for the framework to collect information for site selection, to identify parameters, to compare possible sites and to realize useful laboratory tests. Landing site selection beyond the design of the spacecraft is important to realize vehicle traverses and acquisition of samples (Moore et al. 1989). Previous landing site selections beside the engineering parameters (that are relevant for all landers, in both surface and subsurface missions), focused mostly on surface and not subsurface conditions (Golombek et al. 1997) from scientific point of view, as it is the natural result of the scientific aims. Despite the focus on the surface, there were several attempts already beforehand (Anderson et al. 2003, Grant et al. 2004) or after (Golombek et al. 1999) the mission to estimate some parameters that would be relevant to shallow subsurface related scientific issues and sample acquisition. The findings from these attempts are summarized below.

Basic types of surface materials and rock coverage values were determined for the two Viking landers, and classified drift, crusty to cloddy, blocky and rock materials at the landing sites. In the case of Viking Lander-2 the relative deep penetration and deformation of the surface material around of the footpad suggested low density materials (Moore et al. 1989). The top cm thin layer of the surface covering material was also important above all regarding the rock abundance, dune and dust coverage, which influence the safety and traffic ability on the surface (Golombek et al. 2003). The subsurface was partly also taken into account in the case of Phoenix lander (Arvidson et al. 2007) where water ice was expected several cm below the surface based on GRS (Gamma Ray Spectrometer onboard Mars Odyssey spacecraft) and thermal inertia computations (Bandfield 2007). During the site selection of the unfortunately failed Beagle-2 mission, the shallow subsurface was analyzed regarding the presence, distribution, and rough coarse size of sedimentary material there (Bridges et al. 2003), where the MOLE equipment (Gibson et al. 2003) should have done its subsurface analysis.

Parameters Relevant to the Subsurface

Subsurface sampling is important to analyze past conditions on Mars by in-situ methods (Navarathinam et al. 2011). The attempt to fit the landing site selection process to the aim of subsurface sampling is relatively new in planetary science as most missions (except Phoenix with its very shallow subsurface access) were aimed at surface analysis.

Minerals and rocks observed on and close to the surface do not necessarily represent the conditions that were present during their formation as weathering altered them later. Samples acquired from below the surface are still not absolutely relevant for environmental reconstruction because atmospheric oxidants, along with energetic solar particles and galactic cosmic rays, can penetrate to certain depth. Based on theoretical considerations even subsurface samples may show surface weathering effects as impacts mix the regolith over geological timescales (Westall et al. 2000, Gose and Morris 1977). These effects make need for more focused consideration on the targeting of subsurface sampling. The framework outlined in this paper may assist with this issue.

The topmost layer of the Martian regolith is often composed of cemented fine particles. The possibility of analyzing this layer is outside the scope of this chapter, even though it is interesting from an engineering point of view and holds important information on the surface processes and conditions of the present and recent past (ElShafie et al. 2010, Manthri et al. 2007).

Four example target sites were analyzed according to the factors listed below, which were considered as possible landing sites of the MSL mission (Golombek et al. 2011) at the beginning of 2011. Although this mission is aimed surface analysis, these sites serve as good examples to give insight how subsurface access could be estimated and compared for different locations. The four landing sites were characterized by the following parameters:

- thickness and areal coverage of *cap layer*, to identify where are important subsurface structures closer than 2 m to the surface,
- *hardness estimation* of surface material (only rough approach),
- *numerical calculation* on the 1. areal ratio (%) of outcrops, 2. average horizontal diameter of outcrops, 3. average distance of outcrops from each other, 4. height of outcrops, 5. number of crater outcrops, 6. areal density of non-crater outcrops,
- number and types of *remote observations*, to get insight into how much and what type of background information is available to characterize the site,
- *subsurface settings*: general information on subsurface layers, type of their origin, presence of resemble subsurface structures in the surrounding terrain for geological context,
- the *astrobiology potential* was characterized by the signs of former liquid water (based on published results) and information on the preservation potential.

Beyond the above mentioned parameters, at the four sites the surface area was also classified to define the following terrain types:

- *dangerous terrains*: that rover should not cross (dune cover, and steep or very rough terrain),
- *interesting but uncrossable outcrops*: terrains that the rover is not able to cross but are still interesting for close-up analysis from several meters away or in-situ analysis from cm distance (such as steep walls that the rover is not able to climb but can analyze),
- *possible locations of subsurface sampling:* terrains where the rover is able to go, and based on the extrapolation the interesting layer is in shallow depth or directly exposed on the surface. In the latter case, ideal samples are still not expected right at the surface, as solar UV, atmospheric oxidants, and galactic cosmic rays plus energetic solar particles penetrate to a certain depth and change the original materials inside the sediments. The depth where samples relevant to earlier conditions could be present is taken to be 2 meters in the case of ExoMars Rover (EXM).

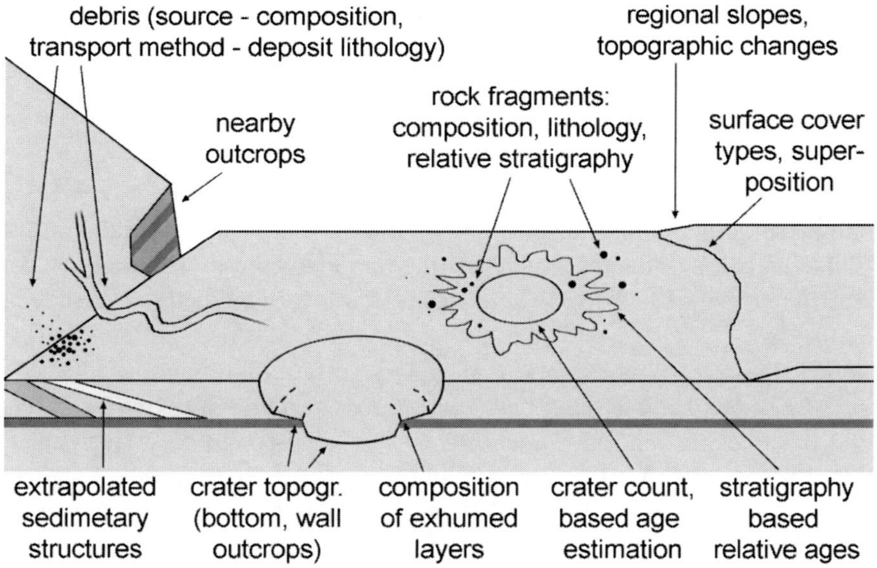

Figure 56. Visual context showing how various surface data could be used to extrapolate shallow subsurface features on Mars.

The four example MSL landing site terrains were analyzed by GIS and graphical software (JMARS, Surfer, ENVI, Photoshop) and characterized according to the above mentioned issues, analyzing surface features (Figure 56). The morphometric measurements were made on HiRISE images, and HiRISE based DTMs and MOLA data.

Steep terrain sections were identified in many cases, especially at small scales by their appearance and shadows. The errors in the distance estimations are around 10%, all the measured values were used only for statistical purposes.

The aim here is a rough characterization of the terrain and the identification of the general size and distribution of outcrops. In this chapter the EXM mission was used estimating its drilling ability as a reference, which has been conceived to acquire soil samples down to maximum 2 m depth. This drill equipment is able to penetrate the soil and acquire 1 cm diameter, 3 cm length core samples.

The drill is designed to be compatible with different soils up to a compressive strength of 150 MPa. First the overview of our current knowledge is presented, than in the discussion section the four possible MSL landing sites are analyzed based on the findings outlined below. In the conclusion the points are listed where our current knowledge and methodology is good enough to characterize a landing site for subsurface analysis, along with those points that need improvement, and/or almost entirely missing yet.

Overview of Current Knowledge

Several factors were identified to characterize the shallow subsurface conditions and to target the drilling activity and also interpret the results of a lander on Mars. The factors are grouped into three categories: 1. accessibility of the subsurface, 2. information on the subsurface and 3. astrobiology potential of the subsurface. These factors may be used as a check list during the planning of a mission and they also help to compare different candidate sites, as well as point to the important but lacking information.

ACCESSIBILITY OF THE SUBSURFACE

The subsurface accessibility depends on: 1. the drilling capacity (Manthri 2007) of the mission, 2. the thickness of the layer covering the important subsurface target, 3. the number, area and horizontal size of locations where subsurface layer is directly accessible, and 4. the hardness of the rocks.

Our knowledge in drilling of various planetary surfaces is limited (Zacny et al. 2009). Only three *successful* drilling have been performed beyond Earth so far, excluding the sampling of the Phoenix lander. (Phoenix was able to produce only fragments and their collection was the task of another piece of equipment). Another but failed attempt was the Beagle-2 mission (Pillinger and Sims 1998). Viking Lander-2 during the sample acquisition excavated to depth 0.14 m, and the average cohesion for crusty to cloddy material there ranged between 0 and 3.2 kPa (Moore and Jakosky 1989). Drill activity beyond the Earth were realizes during the Luna (Coish and Taylor 1997, Norman et al. 1978), Apollo (Nishiizumi et al. 1984, Taylor et al. 1997), and Venera (Surkov et al. 1982) missions. Most information on drilling activity is from the Moon, where the top layer of the regolith was powder-like, while in the case of Mars we often have a duricrust with sulphates cemented grains (Scott and Saaj 2009). This duricrust is fragile and will not create difficulties in drilling activity. For the so-called 'drift material' on Mars the average cohesion at the Viking Lander-2 site was around 1.6±1.2 kPa (Moore and Jakosky 1989), while the penetration of the footpad into blocky material was consistent with a lunar regolith simulant with bulk density around 2300 kg/m^3 (Moore et al. 1977).

On Earth uniaxial compressive strength values are often used to characterize the possibility of drilling certain rock types. This is the strength value at which the target rock fails in compression (Sign and Goel 1999) range from <7 MPa to >186 MPs for soft sediments and hard rock respectively (William and Gary 2005).

Theoretical assumptions and laboratory experiments of analog samples on Earth help to estimate rock strength on Mars. Basalts usually have >130 MPa strength, and show large variation of this value [19], which may produce difficulties during the drilling process (Grady 1999). Unfortunately, there is very little data on the surface strength of Martian rocks. Among the rock types in Table 11, basalts, various sandstones, and sulphates are the most relevant candidates on Mars. In a mission aimed at sampling astrobiologically interesting targets one has to focus on water related and sedimentary rock types.

The hardness of a sedimentary rock may change substantially by weathering, subsurface metamorphism, and cementation/metasomatism by subsurface fluids. *Weathering* might decrease the rock strength, due to increasing the ratio of pore voids, and change the microstructure (Turul 2004). On Earth, substantially weathered and strong bedrock basalts showed a range of compressive strength between 0.25 and 100 MPa. The Martian surface is weakly weathered in general and assuming it has a strength similar to the strongest basalts on Earth, its uniaxial compressive strength value could be around 100-200 MPa. Some basalts, most notably those at the Viking-1 lander landing site were highly vesicular, which decreases the compressive strength (Gates 2008). On Earth, based on laboratory and field tests, a correlation exists between the vesicularity of the basalt and its compressive strength, where the increase of vesicularity from 0.1 to 0.4 decreases the compressive strength to roughly half of the original value. As a result Martian vesicular rocks might have substantially lower strength than non-porous rocks.

Regarding *metamorphism*, compaction starts at 1-4 km depth, which changes the fine-scale structure of the rocks but does not provide high enough pressure for substantial mineral alterations, although the bottom part of Interior Layered Deposits might lose H_2O from their mineral structure. On Earth metamorphism may harden the lithified sediments, but the process usually requires burial under several km thick rock layers. On Earth regional metamorphism occurs at depths between 5 and 20 km (Turner and Verhoogen 1951). On Mars the lithostatic pressure from an overburdening rock body is smaller due to the surface gravity is only 0.4 times that of Earth. One could find metamorphosed rocks on the Martian surface only after exhumation from depths in excess of 10-15 km. This value requires the erosion and transportation of substantial thick rock units to exhume such previously deeply buried sediments, although impact craters may also expose them. As there is no sign of global plate recycling on Mars (Chicarro et al. 1985) (although very early spreading-like process might have been present (Connerney et al. 2005)), great level of metamorphism is not expected. Only in the case of polar caps, where the ice layer is up to 3-4 km thick, might the ice disappear from there due to climate changes in the past (Toon et al. 1980, Mustard et al. 2001) exhuming previously buried units.

The other important hardening process is *metasomatism/cementation* by geofluids, when hydrothermal solutions introduce new minerals and harden the sediments as they precipitate in solid phase. A few observations are available from Mars that suggest such processes might have happened on the planet. Silica rich material was observed by the rover Spirit in Gusev crater (Squyres et al. 2008) demonstrating earlier hydrothermal systems, although that material had not cemented the surface regolith. Observations of post-volcanic hydrothermal alteration also point to migrating subsurface fluids that could produce hardening, but there are few observations and estimations of the magnitude of volcanic hydrothermal alteration (Skok, et al. 2010, Scott et al. 1998). There are fault lines – expressed as long protruding linear features on the sediments of Mars – formed by differential erosion. The fault filling rocks were not eroded probably because subsurface fluids hardened them after precipitation, cementing the minerals.

INFORMATION ON THE SUBSURFACE

For planning the research work beforehand landing and during exploration, it is important to have "enough" observations (imaging, spectral, radar etc,) to determine the important locations and their accessibility, as well as to interpret the in-situ results and put them into context. Here the main focus is not on the raw data (images, spectra, topographic data), but on the geological information provided by their analysis. Important factors here are the followings: 1. do outcrops extending into deep strata (including craters) exist in the region, and if so, how deep are they (here the number and exhumed depth of these craters are important), 2. what kind of subsurface geological setting could there be, and 3. are the proposed subsurface formations known from other locations on the planet: Could they fit into the context of regional or global geological history and planetary evolution?

Astrobiological Considerations

In this section we address the factors that are important regarding the astrobiological potential of the target terrain, i.e. its habitability for extreme organisms resemble to those are present on the Earth. As future missions will likely focus on the astrobiological issues, this topic is classified into a separate group. Here the past and present habitability potential and the preservation possibilities are analyzed.

The *past habitability* can only be roughly estimated based on our present knowledge. Using the observable signatures, the past temperature, water activity (available water), and pH could be roughly estimated based on morphology, mineralogy and chemistry. The morphological issues may point to the past presence of liquid water with the observation of channels, sediments, deltas, terraces, and in some cases volumetric estimations could also be made. Various mineral and chemical fingerprints may also point to the presence, the pH and partially to the duration of these wet environments. The duration could be also estimated by model computations using thermodynamic modelling of the freezing process, (Newsom et al. 1996) but here great uncertainty exists.

Regarding the *present habitability*, the available H_2O could be estimated in the shallow subsurface region. There are many sites on Mars where shallow subsurface ice is present in great quantities, like ice cores of glaciers, the low latitude '"wet" areas discovered by neutron spectrometry, and the seasonally changing absorbed shallow H_2O (Kuzmin et al. 2009). Modelling can also help to estimate the habitability in the presence or recent past. Trapped H_2O could be present even under the climate of Mars today if hygroscopic salts are there and trap atmospheric humidity (Davila et al. 2010). Such materials could be detected by remote observations from orbit, even in the shallow subsurface at locations where they are exhumed (crater ejecta, inner crater wall, steep outcrops), as in the case of the chloride bearing areas in the southern hemisphere (Osterloo et al. 2008).

Ancient organic materials, isotopic markers, and biogenic morphologic features could be searched for in the subsurface during the mission (Westall 2008, Westall 2010). Regarding the successful identification of biosignatures, beyond the capability of detectors, the following factors are important to analyze the targets beforehand: 1. *accumulation plus concentration* of biogenic materials (sediment transportation and deposition) and 2. *preservation potential* (chemical decomposition, mechanical deformation).

Analyzing the accumulation of a specific fraction of surface materials, fluvial deposits and lacustrine sediments are important as they may increase the concentration of various biomarkers. Important issue in targeting is that sedimentary deposits are often located at the topographically lowest point, where the duration of the wet condition is probably the longest. In case of freezing salty solutions, brine bodies with the highest concentration of salts that decrease the melting point are also present at the lowest topographic level. The fine grains of sediments here help the preservation by providing large surfaces for absorption (Figure 55); and on Earth small grains are often surrounded by an organic cover (Westall 2005).

Regarding the preservation potential the following factors should be accounted for:

- Clays are suggested to have good preservation potentials and were found at Eberswalde, Holden and Jezero craters (Ehlmann 2008), although the method to extract organic materials from clay rich environment could be difficult (Martins et al. 2009). At the same time, amino-acids and amines appear to be preserved for geologically long periods of time on Earth inside sulphate sediments (Aubrey et al. 2006), which are abundant on Mars.
- The burial of organic sediments influences the preservation of biogenic materials by:
  - *Shielding from various radiations*. The electromagnetic radiation is only a small concern, although solar UV decomposes organic material, a thin layer of a protecting mineral, with thickness in the order of mm (Córdoba-Jabonero et al. 2005, Cockell and Raven 2004) could shield against them (Abrevaya et al. 2011). High energy particles from solar particle events and galactic cosmic rays are larger problem, as they penetrate roughly a metre below the surface (Dartnell et al. 2006).
  - *Shielding from aggressive surface chemistry* is also important. During the second half of the planetary evolution of Mars (in the Amazonian period from about 2.9-3.3 billion years till today) the surface conditions were usually similar to the current ones (except for short climatic changes). It is reasonable to suppose that aggressive oxidants have been present on the surface during this period. Surface oxidants may penetrate deeper by the climate change related migration of H_2O and by atmospheric diffusion.
  - *Large overburden weight* may cause decomposition and metamorphism. Substantial overburden pressure from the weight of rocks may decompose organic material, metamorphose rocks and minerals wiping out morphologic biosignatures.
- There is an ideal burial depth at which the decomposition is still small by the pressure and the temperature rise (from overburden layers plus geothermal gradient), but deep enough to shield against radiations. This depth depends both on the radiation shielding and permeability (free path for atmospheric diffusion) of the material.

Age estimation is also important to identify in which Martian era a given terrain formed, or how long ago got it exhumed. Dating also helps to correlate certain surface units with other terrains on the planet, putting the analyzed region into context (Hargitai 2008). Although impact crater density based age estimation has improved considerably in recent years, there are still many uncertainties, especially with small areas and small craters, influence by secondary craters. Improvement probably could be done soon as the growing number of HiRISE (High Resolution Imaging Science Experiment onboard Mars Reconnaissance Orbiter probe) and CTX (Context Camera onboard Mars Reconnaissance Orbiter probe) images provide the necessary data for this.

The above factors influence the results of searching for biosignatures; because of the smaller internal activity, lower temperature and drier environment, organic materials could have been preserved on Mars better than on Earth over a long time scale (Kanavarioti and Mancinelli 1989), and as a result the information on surface age is helpful in the planning of in-situ activity.

DISCUSSION

To get a better insight into the specific issues of landing site selection aimed at the analysis of subsurface, the four candidate landing sites of MSL (Golombek et al. 2003) (test areas hereafter: Eberswalde, Holden, Gale crater and Mawrth Vallis region) were characterized. Although this work uses the parameters and aims of a mission like the EXM, experiences from the analysis of MSL landing sites are useful as these sites are scientifically important and fulfil the engineering constraint of a safe landing. The aim here is to identify the important information available today and useful for planning subsurface analysis. To survey these terrains HiRISE images and topographic data were analyzed. Although these images cover only small parts of the target regions, they may give characteristics examples.

At the first test area, the subsurface target in *Eberswalde crater* (23.9S 326.5E) was the light toned layer in the distributary lobes of a delta (Pondrelli et al. 2008, Lewis and Aharonson 2006, Schieber 2008), as it is likely sediment from the ancient lake. This layer shows a polygonal crack pattern, possibly shrinkage cracks in clay-rich strata or due to dehydration of water bearing minerals. Subsurface access type examples in Eberswalde crater are visible in Figure 57. Regarding the trafficability of the terrain, 32% of the area was covered by sand, often in isolated patches, but it would not be a problem to drive around them. About 3% of the analyzed area is covered by exposed and often elongated outcrops in kilometres apart. The height difference at the outcrops may reach 100 m, and as a result nice large outcrop walls could be surveyed. Characteristic outcrop type examples in Eberswalde crater are visible in Figure 57. in subsets c1)-c4). Here c1) shows large extent and undulating surfaced outcrops, c2) large and smoother, c3) elongated outcrops, and c4) elongated plus curved outcrop examples. In general this area has relatively isolated outcrops with diameter between 28 and 1376 m, with the average value of 192 m. and distance from each other were between 5 and 2772 m (average 182 m). Similar analysis was realized for the other three landing sites, with classification of surface units, determination of the size, shape and average distance for each others of the outcrops that could be identified from above. The result of the analysis is visible in the Table 10 of result section where these landing sites could be compared to each other.

RESULTS: COMPARISON OF LANDING SITES

Comparing candidate sites, beside morphology, the spatial density and depth of craters also gives information on outcrops, just like surface pattern (mechanically eroded, chemically weathered, desiccated etc.) might point to the preservation capacity of the target rock. The comparison may point to weak issues, which help to focus the future research on improving landing site selection. The rationality of this analysis is that the most promising outcrops could be identified before the mission, as at the four landing sites show characteristic differences in various parameters.

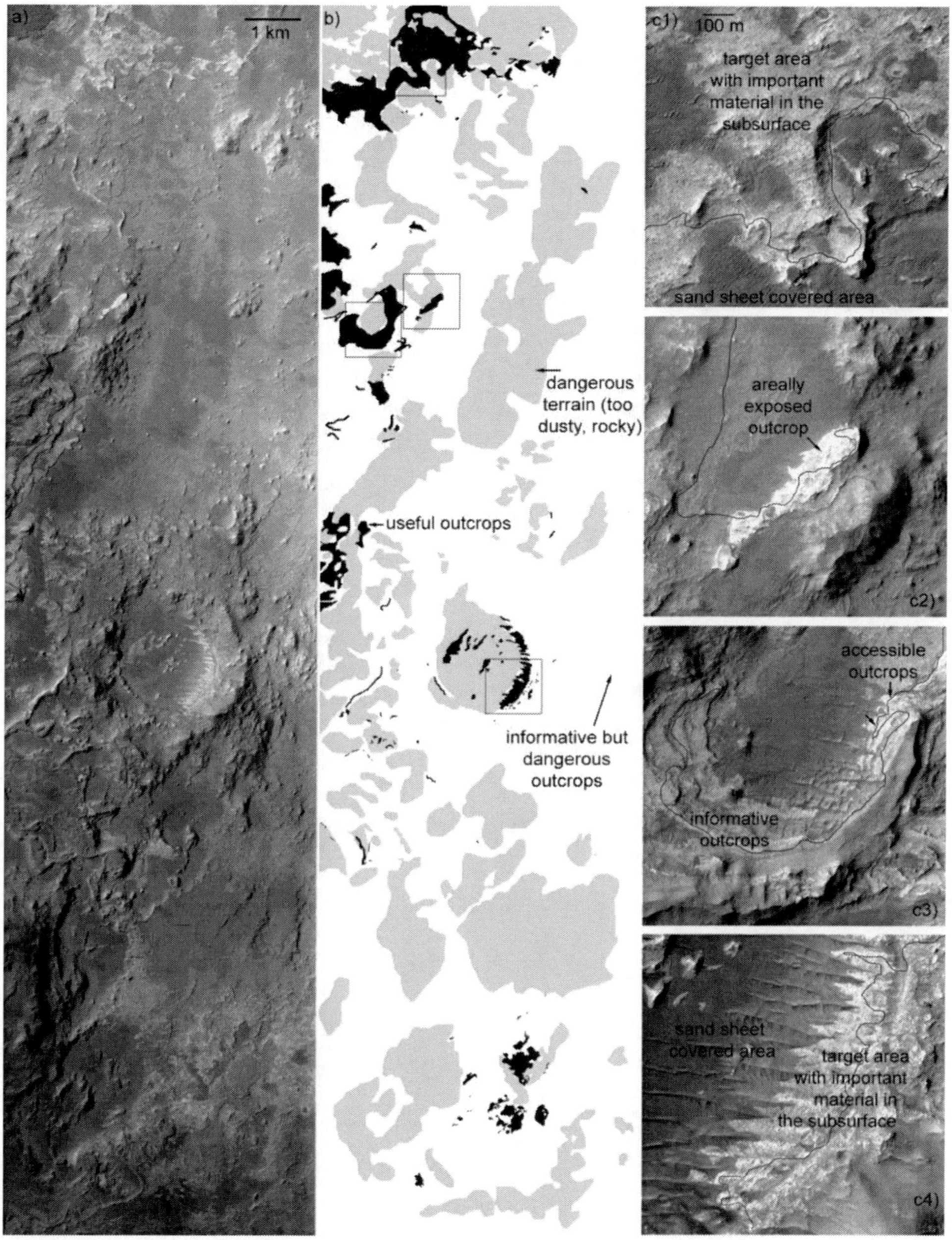

Figure 57. The test area in Eberswalde crater: a) HiRISE image ESP_019757_1560), b) surface unit types classified according to accessibility and scientific interest, c1)-c4) magnified insets sohoe characteristic features with example rover traverses.

The graphical comparison of the outcrop parameters at the four landing sites is shown in Figure 58. To separate the overlapping data points, and to make their identification easier, the same types of parameters are indicated in two separate diagrams: Eberswalde and Holden craters on the left (a, c subsets) and Gale plus Mawrth regions on the right (b, d subsets). In the top of the figure the long versus the short axis of the outcrops is presented that show how much the outcrops are elongated, in addition to the data about their average diameter. At the bottom the average area and distances of the outcrops from each other are indicated.

Analyzing Figure 58 subsets a) and b), it can be seen that at Eberswalde and Holden (hereafter fluvial-type) the outcrops are elongated and sized between 28 and 1938 m, while at Gale and Mawrth area they are roughly isometric and scattered in a larger range between 653 and 11415 m in diameter. Current rover technology (MERs, MSL) with trafficability distance may be able to cross the largest outcrops. The characteristic distance between different outcrops may be useful information during the planning phase. Outcrop area (that is related to the possibility of finding the best location of sampling at one outcrop) is larger in the case of Gale and Mawrth than the two fluvial type locations.

Regarding the distance of neighbouring outcrops there is no great difference, although at the fluvial sites the average difference is somewhat smaller (182 m and 128 m) while in Gale an Mawrth region it is somewhat larger (106 m and 384 m).

For all four sites, the distances between individual outcrops ranged between 5 and 2772 m. This means a rover with the capability mentioned above would be able to visit several of them.

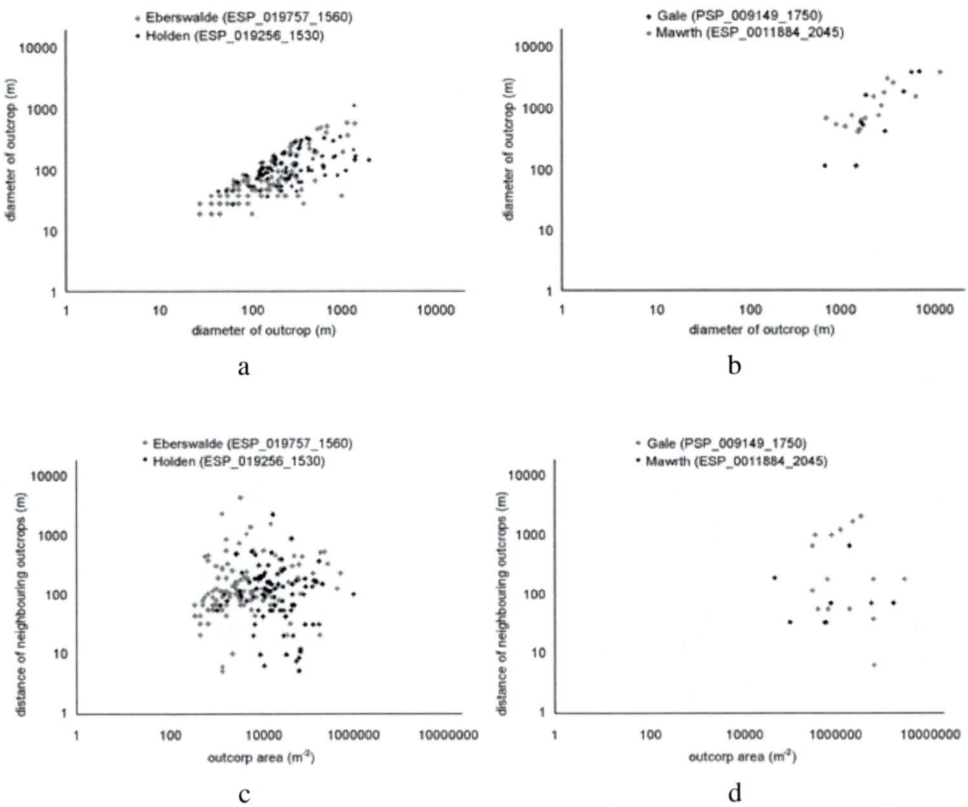

Figure 58. Size distribution of outcrops at the analyzed landing sites. Subsets a) and b) show the large versus small diameter, while subsets c) and d) show the average diameter versus the distance of neighbouring outcrops. Eberswalde and Holden are visible at left, Gale and Mawrth are visible at right.

The separation of the four landing sites into two groups (one for Eberswalde plus Holden, and the other for Gale plus Mawrth) is reasonable as the size and distance of outcrops there resemble to each other inside each group, but differs from that of the other group. The above-mentioned parameters may influence the planning of the field work and the sampling. It is useful to estimate: how much information could be gained from the analysis of one outcrop, how much time is reasonable to spend at a given outcrop, how much is required to reach the next close outcrop, and how many of them could be visited during the mission. If there seems to be only a few useful locations, then the rover may spend more time at one site.

Alternately, if there are a large number of interesting outcrops (like in the high sedimentary mound of Gale crater), the visit of more locations and a shorter time for analysis at each might be more reasonable. Based on the above analysis, in Eberswale and Holden craters only a small part of the analyzed terrain show realistic possibilities of sampling the shallow subsurface unit.

Fusing the theoretical background and the determined parameters at the four landing sites, it can be defined where are the available data sufficient or weak to reasonably evaluate the subsurface access and possible results.

Conclusion

In the conclusion, first the general findings are listed emphasizing those topics that require more focused research to help targeting a subsurface sampling mission.

Secondly the parameters calculated for the four candidate MSL landing sites is presented to highlight the type of differences between landing sites focusing on subsurface sampling. Finally, the benefits of this method are summarized, with notes about the possible future direction of in-situ research on Mars.

General Conclusion

- To target the in-situ work and identify the ideal locations from remote images the best information is available for: 1. the dune coverage of the terrain, 2. the cap layer distribution, 3. the identification of useful outcrops, and 4. the general description of the subsurface settings.
- Where the knowledge is limited but may be improved substantially with the further analysis of available data are: 1. to put the analyzed features into better global context, 2. improve the crater density based age estimations above all the role of small craters (as the outcrops are small) and exhumed surfaces, and 3. there are several methods to define the former presence of liquid water, but they are not sophisticated enough to estimate the duration of the wet period (see for details Kereszturi 2012b).
- Our knowledge is the most limited and only analog data is available for the hardness of rock units and the preservation potential of biogenic features on Mars. Deeply buried and later exhumed materials are not expected on Mars because of the lack of plate tectonism, and drilling probably will not encounter rocks substantially hardened by high pressures and temperatures. Although few observations suggest mineral crystallization and deposition from geofluids might happen on Mars that made the rocks harder.

To fill the missing points in the above mentioned list the following works would be useful and help in the targeting of future subsurface sampling missions: 1. analyze the rock strength change under Martian style weathering in laboratory, by analyzing Martian meteorites, and by designing analog tests and Mars simulation chamber experiments (de Vera et al. 2008, Ehrenfreund et al. 2003), 2. improve the age estimation method and the role of secondary cratering comparing terrains where secondary fields are present and absent, 3. analyze target terrains with more focus on the regional context, 4. improve the estimation of astrobiology potential, above all with new laboratory tests analyzing the preservation potential in different environments and also with more focused inspection on the volumetric and temporal issues of wet environments.

The size, distance and areal density of these outcrops show characteristic differences between the analyze sites. In the case of Eberswalde and Holden crater, 3 and 9% of the analyze surface looked like as accessible outcrops, which were size between ~100-400 m (and often show an elongated shape), while the characteristic distance between outcrops was around 100-200 m. The other two sites: Gale crater and the Mawrth Vallis regions showed larger areal extent (46 and 35%) and fewer outcrops in total number.

These outcrops were more isometric in shape and have average diameters of ~3000 m. These suggest different mission strategy should be applied to different terrain types, and rovers with different capabilities are ideal for the work at certain location.

In earlier missions landing sites were chosen after the design of the mission was nearly finished. Difference exist between the sites regarding the exploration strategy (Kereszturi 2011b), and it might be taken more properly into account for missions aimed at subsurface sampling beforehand, as the four sites suggest the subsurface access have substantial differences in spatial parameters (size, distance of outcrops, dust cover etc.). During the planning phase of a mission it would be useful to outline the required exploration method, to differentiate between or rank the landing sites when the aim is the subsurface access, and also to determine the weakest points where analog or laboratory data would help in the design of the research strategy and the in-situ fine targeting.

The difference between the method outlined here (that contains only elements but not a sophisticated landing site selection process) and other earlier landing site selections are the followings: 1. this method helps to collect information necessary to determine the rationality and comparison between possible landing sites with the aim of subsurface sampling, 2. it gives examples of how the location, size, and areal distribution of outcrops could be parameterized and compared to each other, and 3. it points out information that is weak and thus more research is necessary.

The advantage of this method is that it tries to take into account the information on the subsurface, the disadvantage relative to earlier landing site selections is this method is only just now taking shape and will require more work to finalize.

Even if all the factors to characterize a landing site listed here were well known (which is far from reality), there are still many uncertainties regarding the rationality of one certain landing site for subsurface analysis. But the overview and comparison is important and may help to select different sites and to help the planning process for missions like ExoMars Rover (Vago et al. 2006, Kereszturi 2011a). Landing accuracy will increase in the future, and accurate analysis of proposed landing sites like that shown here will make more sense.

ACKNOWLEDGMENTS

This work was supported by the OTKA PD 105970 project of the Hungarian Academy of Sciences. The author thanks the help of Kelsi Singer (Department of Earth and Planetary Sciences, Washington University in St. Louis, US) and David Hyder (Astrophysics Research Institute - Liverpool John Moores University, UK) in the final compilation of the paper. This chapter is partly based on the experience and discussion at the European Landing site selection for future exploration missions organized under Europlanet JRA1 project at Lorentz Centre and ESTEC (Zegers et al 2011).

REFERENCES

Abrevaya, X. C., Paulino-Lima, I. G., Galante, D., Rodrigues, F., Mauas, P. J. D., Cortón, E., Santos Lage, C. A. (2011) Comparative Survival Analysis of Deinococcus radiodurans and the Haloarchaea Natrialba magadii and Haloferax volcanii Exposed to Vacuum Ultraviolet Irradiation. *Astrobiology* 11, 1034-1040.

Anderson, F. Scott, Haldemann, A. F. C., Bridges, N. T., Golombek, M. P., Parker, T. J., Neumann, G. (2003). Analysis of MOLA data for the Mars Exploration Rover landing sites. *Journal of Geophysical Research* 108 (E12) ROV 25-1, CiteID 8084.

Arvidson, R., Adams, D., Bonfiglio, G., Christensen, P., Cull, S., Golombek, M., Guinn, J., Guinness, E., Heet, T., Kirk, R., Knudson, A., Malin, M., Mellon, M., McEwen, A., Mushkin, A., Parker, T., Seelos, F., Seelos, K., Smith, P., Spencer, D., Stein, T., Tamppari, L. (2008). Mars Exploration Program 2007 Phoenix landing site selection and characteristics. *Journal of Geophysical Research* 113(E6), CiteID E00A03.

Aubrey, A., Cleaves, H. J., Chalmers, J. H., Skelley, A. M., Mathies, R. A., Grunthaner, F. J., Ehrenfreund, P., Bada, J. L. (2006). Sulfate minerals and organic compounds on Mars, *Geology* 34, 357.

Bandfield, J. L. (2007). High-resolution subsurface water-ice distributions on Mars, *Nature* 447, 64-67.

Bishop, J. L., Dobrea, E. Z., McKeown, N. K., Parente, M., Ehlmann, B. L., Michalski, J. R., Milliken, R. E., Poulet, F., Swayze, G. A., Mustard, J. F., Murchie, S. L., Bibring, J. P. (2008). Phyllosilicate Diversity and Past Aqueous Activity Revealed at Mawrth Vallis, Mars, *Science* 321(5890), 830.

Bridges, J. C., Seabrook, A. M., Rothery, D. A., Kim, J. R., Pillinger, C. T., Sims, M. R., Golombek, M. P., Duxbury, T., Head, J. W., Haldemann, A. F. C., Mitchell, K. L., Muller, J.-P., Lewis, S. R., Moncrieff, C., Wright, I. P., Grady, M. M., Morley, J. G. (2003). Selection of the landing site in Isidis Planitia of Mars probe Beagle 2. *Journal of Geophysical Research* 108 (E1), CiteID 5001.

Chicarro, A. F., Schultz, P. H., Masson, P. (1985). Global and regional ridge patterns on Mars, *Icarus* 63, 153-174.

Cockell, C. S., Raven, J. A. (2004). Zones of photosynthetic potential on Mars and the early Earth, *Icarus* 169, 300-310.

Coish, R. A., Taylor, L. A. (1978). Mineralogy and petrology of basaltic fragments from the Luna 24 drill core. In: *Mare Crisium: The view from Luna 24*, New York, Pergamon Press, Inc. 403-417.

Connerney, J. E. P., Acuña, M. H., Ness, N. F., Kletetschka, G., Mitchell, D. L., Lin, R. P., Reme, H. (2005). From the Cover: Tectonic implications of Mars crustal magnetism, *Proceedings of the National Academy of Science* 102(42), 14970-14975.

Córdoba-Jabonero, C., Zorzano, M. P., Selsis, F., Patel, M. R., Cockell, C. S. (2005). Radiative habitable zones in Martian polar environments. *Icarus* 165, 253-276.

Dartnell, L. R., Desorgher, L., Ward, J. M., Coates, A. (2006). J. Modelling the surface and subsurface Martian radiation environment: Implications for astrobiology. *Geophysical Research Letters* 34, L02207.

Davila, A. F., Duport, L. G., Melchiorri, R., Jänchen, J., Valea, S., de los Rios, A., Fairén, A.G., Möhlmann, D., McKay, C.P., Ascaso, C., Wierzchos, J. (2010). Hygroscopic Salts and the Potential for Life on Mars, *Astrobiology* 10, 617-628.

De Vera, J. P., Rettberg, P., Ott, S. (2008). Life at the limits: capacities of isolated and cultured lichen symbionts to resist extreme environmental stresses. *Orig. of Life and Evol. of Biosph.* 38, 457-468.

Edgett, K. S., Malin, M. C. (2001). Rock stratigraphy in Gale crater, Mars, *32nd LPSC* abstract 1005.

Edmonds, D. A., Shaw, J. B., Mohrig, D. (2011). Topset-dominated deltas: A new model for river delta stratigraphy. *Geology* 39(12), 1175-1178.

Ehlmann, B. L., Mustard, J. F., Fassett, C. I., Schon, S. C., Head, J. W., Des Marais, D. J., Grant, J. A., Murchie, S. L. (2008). Clay minerals in delta deposits and organic preservation potential on Mars, *Nature Geoscience* 1, 355-358.

Ehrenfreund, P., ten Kate, I. L., Ruiterkamp, R., Botta, O., Lehmann, B., Boudin, N., Foing, B. H. (2003). Mars Simulation Chamber 2 - goals, instrumentation and methods, *EGS-AGU Joint Assembly* abstract #4298.

ElShafie, A., Chevrier, V. F., Ulrich, R., Roe, L. (2010). Penetration testing for the Optical Probe for Regolith Analysis (OPRA), *Advances in Space Research* 46, 327–336.

Gates, W. C. B. (2008). Estimating the Unconfined Compressive Strength of Vesicular Basalts via Bulk Specific Gravity of the Rock Sample. *42nd US Rock Mechanics Symposium* abstract ARMA 08-131.

Gibson, E. K., Pillinger, C. T., Wright, I. P., Morse, A., Stewart, J., Morgan, G., Praine, I., Leigh, D., Sims, M. R., and Pullan, D. (2003). Beagle 2: Seeking the signatures of life on Mars. In: *Sixth International Conference on Mars*, abstract 3005.

Golombek, M., Cook, R. A., Moore, H. J., Parker, T. J. (1997). Selection of the Mars Pathfinder landing site. *Journal of Geophysical Research* 102, 3967-3988.

Golombek, M., Grant, J., Parker, T. J., Kass, D. M., Crisp, J. A., Squyres, S. W., Haldemann, A: F. C., Adler, M., Lee, W. J., Bridges, N. T., Arvidson, R. E., Carr, M. H., Kirk, R. L., Knocke, P. C., Roncoli, R. B., Weitz, C. M., Schofield, J. T., Zurek, R. W., Christensen, P. R., Fergason, R. L., Anderson, F. S., Rice, J. W. (2003). Selection of the Mars Exploration Rover landing sites, *Journal of Geophysical Research* 108(E12), CiteID 8072.

Golombek, M., Grant, J., Vasavada, A. R., Grotzinger, J., Watkins, M., Kipp, D., Noe Dobrea E., Griffes, J., Parker T. (2011). Final four landing sites for the Mars Science Laboratory. *42nd LPSC* abstract 1520.

Golombek, M. P., Moore, H. J., Haldemann, A. F. C., Parker, T. J., Schofield, J. T. (1999). Assessment of Mars Pathfinder landing site predictions. *Journal of Geophysical Research* 104 (E4), 8585-8594.

Gose, W. A., Morris, R. V. (1977). Depositional history of the Apollo 16 deep drill core. In: *8th Lunar Science Conference, Proceedings* 3 (A78-41551 18-91) New York, Pergamon Press, Inc., 2909-2928.

Grady D. E. (1999). Spall Properties of Solenhofen Limestone and Dresser Basalt, Shock Compression of Condensed Matter 1255-1258.

Grant, J. A., Golombek, M. P., Parker, T. J., Crisp, J. A., Squyres, S. W., Weitz, C. M. (2004). Selecting landing sites for the 2003 Mars Exploration Rovers. *Planetary and Space Science* 52, 11-21.

Grant, J. A. and Wilson, S. A. (2011a). Late alluvial fan formation in southern Margaritifer Terra, Mars, *42nd LPSC* abstract 2048.

Grant, J. A. and Wilson, S. A. (2011b). Late alluvial fan formation in southern Margaritifer Terra, Mars. *Geophysical Research Letters* 38, CiteID L08201.

Hargitai, H. (2008). Topographic map of Mars, Lambert projection.

Hauber, E., Gwinner, K., Stesky, R., Fueten, F., Michael G., Reiss, D., Zegers, T., Hoffmann, H., Jaumann, R., van Gasselt, S., Neukum, G., and the HRSC Co-Investigator Team (2005). Interior layered deposits in Valles Marineris, Mars: insights from 3D-data obtained by the High Resolution Stereo Camera (HRSC), *37th LPSC* abstract 1760.

Jerolmack, D. J., Mohrig, D., Zuber, M. T., Byrne, S. (2004). A minimum time for the formation of Holden Northeast fan, Mars. *Geophysical Research Letters* 31(21), CiteID L21701.

Kanavarioti, A., Mancinelli, R. L. (1989). Could organic matter have been preserved on Mars for 3.5 billion years? *Icarus* 84, 196-202.

Kereszturi, A. (2011a). Extrapolation of shallow subsurface structures form orbital data – case study for ExoMars Rover mission, *42th LPSC* abstract 1351.

Kereszturi, A. (2011b). Geologic field work on Mars: distance and time issues during surface exploration. *Acta Astronautica* 68, 1686-1701

Kereszturi, A. (2012a) Crater wall outcrop analysis for targeting subsurface sampling on Mars, submitted to *Planetary and Space Science*.

Kereszturi, A. (2012b). Review of wet environment types on Mars with focus on duration and volumetric issues, *Astrobiology* accepted for publication.

Kuzmin, R. O., Zabalueva, E. V., Christensen, R. P. (2009). Estimation and mapping of wintertime increase in water ice content of the Martian surface soil based on seasonal Thermal Emission Spectrometer thermal inertia variations. *Journal of Geophysical Research* 114(E4), CiteID E04011.

Le Deit, L., Hauber, E., Fueten, F., Pondrelli, M., Zegers, T., van Gasselt, S., Massé, M., Verpoorter, C., and the ISSI ILD team (2011). Geological comparison of the Gale crater mound to other equatorial layered deposits (LEDs) on Mars. *42nd LPSC* abstract 1857.

Lewis, K. W., Aharonson, O. (2006). Stratigraphic analysis of the distributary fan in Eberswalde crater using stereo imagery, *Journal of Geophysical Research* 111(E6), CiteID E06001.

Manthri, S. (2007). Experimental investigation of PCD compact core drill performance on basalt simulating sustainable dry drilling on Mars (Thesis) The Graduate School, University of Kentucky.

Martins, Z., Sephton, M. A., Foing, B. H., Ehrenfreund, P. (2011). Extraction of amino acids from soils close to the Mars Desert Research Station (MDRS), Utah. *Astrobiology* 10, 231 - 238.

Moore, H. J. and Jakosky, B. M. (1989). Viking landing sites, remote-sensing observations, and physical properties of Martian surface materials. *Icarus* 81, p. 164-184.

Moore, H., Hutton, R., Scott, R., Spitzer, C., and Shorthill, R. (1977). Surface materials of the Viking landing sites. *Journal of Geophysical Research* 82(B28).

Mustard, J. F., Cooper, C. D., Rifkin, M. K. (2001). Evidence for recent climate change on Mars from the identification of youthful near-surface ground ice, *Nature* 412, 411-414.

Navarathinam, N., Lee, R., Borschiov, K., Quine, B. (2011). Northern light drill for Mars: Performance evaluation. *Acta Astronautica* 68, 1234–1241.

Newsom, H. E., Brittelle, G. E., Hibbitts, C. A., Crossey, L. J., Kudo, A. M. (1996). Impact crater lakes on Mars, *Journal of Geophysical Research* 101, 14951-14956.

Nishiizumi, K., Arnold, J. R., Elmore, D., Ma, X. Z. (1984). Be-10 and Cl-36 depth profiles in an Apollo 15 drill core. *Earth and Planetary Science Letters* 70, 157-163.

Noe Dobrea, E., Mustard, J., Bibring, J-P., Bishop, J., Carter, J., Ehlmann, B., Loizeau, D., Mangold, N., McKeown, N., Michalski, J., Parente, M., Poulet, F., Wray J. (2010). Geology and planetary context of the MSL Landing Site at Mawrth Vallis. *4th MSL Landing Site Workshop*

Norman, M., Coish, R. A., Taylor, L. A. (1978). Glasses in the Luna 24 core and petrogenesis of ferrobasalts. In*: Mare Crisium: The view from Luna 24,* New York, Pergamon Press, 281-289.

Osterloo, M. M., Hamilton, V. E., Bandfield, J. L., Glotch, T. D., Baldridge, A. M., Christensen, P. R., Tornabene, L. L., Anderson, F. S. (2008). Chloride-Bearing Materials in the Southern Highlands of Mars. *Science* 319, 1651.

Pelkey, S. M., Jakosky, B. M., Christensen, P. R. (2004). Surficial properties in Gale Crater, Mars, from Mars Odyssey THEMIS data, *Icarus* 167(2), 244-270.

Pillinger, C. T., Sims, M. R. (1998). Beagle 2: A Lander for Mars, *Meteoritics and Planetary Science* 33, A123-A133.

Pondrelli, M., Baliva, A., Di Lorenzo, S., Marinangeli, L., Rossi, A. P. (2005). Complex evolution of paleolacustrine systems on Mars: An example from the Holden crater, *Journal of Geophysical Research* 110(E4), CiteID E04016.

Pondrelli, M., Rossi, A. P., Marinangeli, L., Hauber, E., Gwinner, K., Baliva, A., di Lorenzo, S. (2008). Evolution and depositional environments of the Eberswalde fan delta, Mars, *Icarus* 197, 429-451.

Rossi, A. P., Pondrelli, M., van Gasselt, S., Zegers, T., Hauber, E., Neukum, G. (2008). Gale crater bulge: a candidate multi-stage large spring mound, 39th *LPSC* abstract 1611.

Schieber, J. (2008). Features of geologic of interest within the proposed Eberswalde crater Mars Science Lab (MSL) landing ellipse. 39th *LPSC* abstract 2117.

Scott, G., Saaj, C. (2009). Measuring and Simulating the Effect of Variations in Soil Properties on Microrover Trafficability, *AIAA Space 2009 Conference and Exposition* Pasadena.

Scott, E. D., Wilson, L. (1998). Evidence for fluvial features produced by sill emplacement on the upper flanks of Ascraeus Mons volcano, Mars. *29th LPSC* abstract 1355.

Sign, B., Goel, R. K. (1999). Rock mass classification, A practical approach in civil engineering, Elsevier, Oxford.

Skok, J. R., Mustard, J. F., Ehlmann, B. L., Milliken, R. E., Murchie, S. L. (2010). Silica deposits in the Nili Patera caldera on the Syrtis Major volcanic complex on Mars. *Nature Geoscience* 3, 838–841.

Squyres, S. W., Arvidson, R. E., Ruff, S., Gellert, R., Morris, R. V., Ming, D. W., Crumpler, L., Farmer, J. D., Des Marais, J. D., Yen, A., McLennan, S. M., Calvin, W., Bell, J. F., Clark, B. C., Wang, A., McCoy, T. J., Schmidt, M. E., de Souza P. A. (2008). Detection of Silica-Rich Deposits on Mars. *Science* 320, 1063.

Surkov, Y. A., Moskaleva, L. P., Shcheglov, O. P., Kharyukova, V. P., Manvelyan, O. S., Smirnov, G. G. (1982). Element composition of Venus rocks - Preliminary results from Venera 13 and Venera 14. *Soviet Astronomy Letters* 8, 237-240.

Taylor, G. J., Keil, K., Warner, R. D. (1977). Petrology of Apollo 17 deep drill core. I - Depositional history based on modal analyses of 70009, 70008, and 70007. In: *8th Lunar Science Conference, Proceedings* 3., New York, Pergamon Press, Inc. 3195-3222.

Thomson, B. J., Bridges, N. T., Milliken, R., Hook, S.. Baldridge, A., Crowley, J., Marion, G. (2008). Signs of aqueous activity in Gale Crater, Mars as viewed by Mars Reconnaissance Orbiter, *American Geophysical Union, Fall Meeting* abstract #P41B-1365.

Toon, O. B., Pollack, J. B., Ward, W., Burns, J. A., Bilski, K. (1980). The astronomical theory of climatic change on Mars, *Icarus* 44, 552-44,607.

Turner, F. J., Verhoogen, J. (1951). Igneous and Metamorphic Petrology, McGrow Hill Book Company, New York, Toronto, London.

Turul, A. (2004). The effect of weathering on pore geometry and compressive strength of selected rock types from Turkey. *Engineering Geology* 5(3-4), 215-227.

Vago, J., Gardini, B., Kminek, G., Baglioni, P., Gianfiglio, G., Santovincenzo, A., Silvia, B. M., van Winnendael, M. (2006). ExoMars - searching for life on the Red Planet, *ESA Bulletin* (ISSN 0376-4265) 126, 16-23.

Westall, F. (2008). Morphological biosignatures in terrestrial and extraterrestrial materials, *Space Science Reviews* 135, 95-114.

Westall, F. (2005). Life on the early Earth: A sedimentary view, *Science* 434, 366-367.

Westall, F. and Cavalazzi, B. (2010). *Biosignatures in rocks, Encyclopedia of Geobiology* (Eds.) V. Thiel, J. Reitner, Springer, Berlin, in press.

Westall, F., Brack, A., Hofmann, B., Horneck, G., Kurat, G., Maxwell, J., Ori, G. G., Pillinger C., Raulin, F., Thomas, N., Fitton, B., Clancy, P., Prieur, D., Vassaux, D. (2000). An ESA study for the search for life on Mars. *Planetary and Space Science* 48, 181-202.

William, C. L., Gary, J. P. (2005). *Standard handbook of petroleum and natural gas engineering,* Elsevier, 2nd edition 194-255.

Wood, C. A., Head, J. W., Cintala, M. J. (1977). Crater degradation on Mercury and the Moon: Clues to surface evolution, *8th LPSC* p. 3503-3520.

Wray, J. J., Ehlmann, B. L., Squyres, S. W., Mustard, J. F., Kirk, R. L. (2008). Compositional stratigraphy of clay-bearing layered deposits at Mawrth Vallis, Mars, *Geophysical Research Letters* 35, CiteID L12202.

Zacny, K., Bar-Cohen, Y., Davis, K, Coste, P., Paulsen, G., Sherrit,S., George, J., Derkowski, B., Gorevan, S., Boucher, D., Guerrero, J., Kubota, T., Thomson, B. J., Stanley, S., Thomas, P., Lan, N., McKay, C., Onstot, T.C., Stoker, C., Glass, B., Wakabayashi, S., Whyte, L., Visentin, G., Re, E., Richter, L., Badescu, M., Bao, X., Fincher, R., Hoshino,

T., Magnani, P., Menon, C. (2009). Extraterrestrial Drilling and Excavation. In: Y. Bar-Cohen, K. Zancy, *Drilling in extreme environments*. Wiley-VCH Verlag GmbH, p. 347–557.

Zegers, T. et al. (2011). Summary outcome and recommendations workshop on landing sites for exploration missions leiden/noordwijk, Europlanet special document

In: Mars
Editor: Alberto G. Fairén

ISBN: 978-1-62618-102-1

Chapter 9

OVERVIEW OF DESIGN ANALYSIS OF A SPACE MISSION FOR SAMPLE RETURNING FROM MARS

***Giuseppe Pezzella**[*1]* and *Antonio Viviani*[2]**
[1]Centro Italiano Ricerche Aerospaziali, Caserta, Italy
[2]Dipartimento di Ingegneria Aerospaziale e Meccanica,
Seconda Università degli Studi di Napoli, Naples, Italy

ABSTRACT

Sample return missions aim for returning extraterrestrial material sampled in the outer space (e.g., beyond the Moon), for example from comets or planets as Mars to Earth. Indeed, an important step forward for Space Exploration activities, and for a more accurate knowledge of the Earth and other planets, is to develop the capability to send vehicles into space which select, collect and finally return samples to Earth where to perform their analysis. In this framework this chapter performs an overview on design method approach which can be used to address the design of a generic sample return capsule (SRC). Generally speaking, the design process demands of satisfying some set of requirements, usually in some optimal manner. To attain this optimal solution, or at least to satisfy as many as possible of the imposed requirements, usually requires tradeoffs between individual elements or systems. To mediate these tradeoffs requires an engineering familiarity and literacy, if not outright talent, with all of the systems and engineering disciplines involved. Spacecraft represent particularly broad challenges, in that a wide range of disciplines is involved – communications, power, thermal control, propulsion and so on. Arguably, planetary probes are even broader, in that all the usual spacecraft disciplines are involved, plus several aspects related to delivery to and operation in planetary environments, such as aerothermodynamics, soil mechanics and so on. It is crucial that the requirements be articulated in a manner that adequately captures the intent of the 'customer'. Once mission requirements and constraints are established, the SRC design analysis starts taking into account for design results, available both from literature and previous missions, throughout similarity and scaling processes. This phase is most important considering that it represents the initial condition from which the SRC design evolves and matures. It takes the form of an assessment study, followed by a

* Centro Italiano Ricerche Aerospaziali (g.pezzella@cira.it)

phase of study, development, launch and operation. This process may require several iterations to obtain an optimization of SRC design.

DESIGN PROCESS FOR SAMPLE RETURN VEHICLES

The analytical, computational and experimental methods described in this chapter can be used to address the design of a generic sample return vehicle (SRV). Generally speaking, the design process demands different stages which may require several iterations to obtain an optimization of SRV design. For instance, a typical scheme that synthesizes the design process of a SRV is shown in Figure 59.

Indeed, once mission requirements and constraints are established, the SRV design analysis starts taking into account for design results, available both from literature and previous missions, throughout similarity and scaling processes. This phase is most important considering that it represents the initial condition from which the SRV design evolves and matures. For example, the design starts basing on initial assumptions as SRV configuration, mass, stability considerations during entry, descent and loading environment. Hence, the ballistic parameter is estimated from size, weight and drag coefficient early evaluated. This value of ballistic coefficient is then used to calculate atmospheric trajectories and perform parametric analysis of SRV re-entry flight. In particular, Figure 59 also shows that each design level must take care of mission constraints. As a consequence, the design process evolves in a loop until mission constraints are fulfilled and then the design results are available.

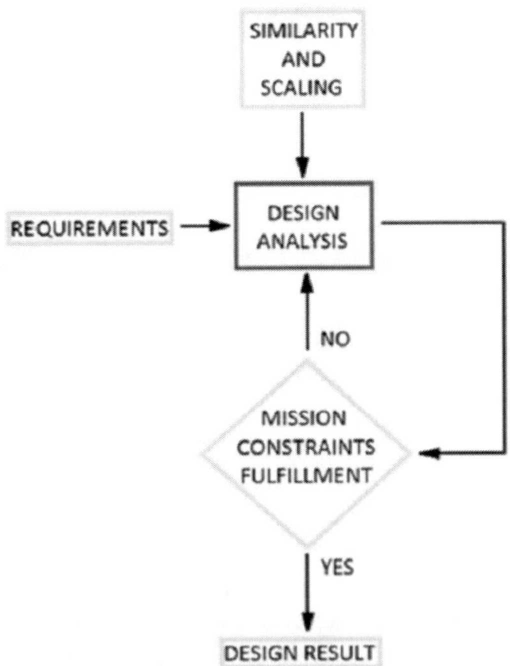

Figure 59. Typical design process scheme.

REQUIREMENTS AND CONSTRAINTS

In the framework of the design of any space mission both requirements and constraints must be considered. Example of constraints for SRV are the curve limits of the re-entry corridor such as re-entry velocity fixed to about 12 km/s for an altitude of 200 Km, maximum deceleration below 100 g, maximum heat flux below 15 MW/m^2 and total heat load below 300-350 MJ/m^2 [1]. Hence, once the heat flux has been estimated, one can calculate the stagnation pressure, thus evaluating approximate TPS layout and thickness.

Typical requirements limits are:

- Landing
- Maximal carrying capacity with minimal total volume
- Stability during entry, descent and landing
- Maximum heat flux below ... MW/m^2
- Maximum heat load ... MJ/m^2
- Maximum g-load below ...g
- Impact velocity ~ ... m/s
- Maximum TPS ratio: ...% _ Corresponding to a maximum energy load of ... MJ/m^2

On the other hand, typical constraints are:

- The landing on Earth shall be performed with a fully passive entry (e.g., no parachute) for the SRV.
- Initial entry velocity into Earth is (given by a mission analysis and it is equal to) 12.3 km/s for an altitude of 120 km.

Moreover, one of the main considerations in entry vehicle design is if a parachute system should be included.

The advantages of parachute recovery are:

a) entry vehicle tracking at high altitude is easier;
b) the probe or canister has a lower ground impact velocity;
c) recovery operations are easier (the probe is easier to find).

Disadvantages may include:

a) slight increase in mass;
b) increased mechanical complexity with the separation of the probe from the aft-cover and parachute;
c) slight decrease in static stability margin.

A small parachute may be able to decelerate the entry vehicle adequately. Even though impact-area drift may increase significantly, ground radar should be able to track the probe and parachute to ensure efficient recovery operations. In this case, the landing on Earth shall be performed with a fully passive ballistic entry (no parachute).

SIMILARITY AND SCALING

Generally speaking, a preliminary design of any space mission is obtained comparing the requirements of this mission with the requirements of previous ones. Therefore, it is useful to individuate missions which have similar characteristics to our and use them to obtain a work line.

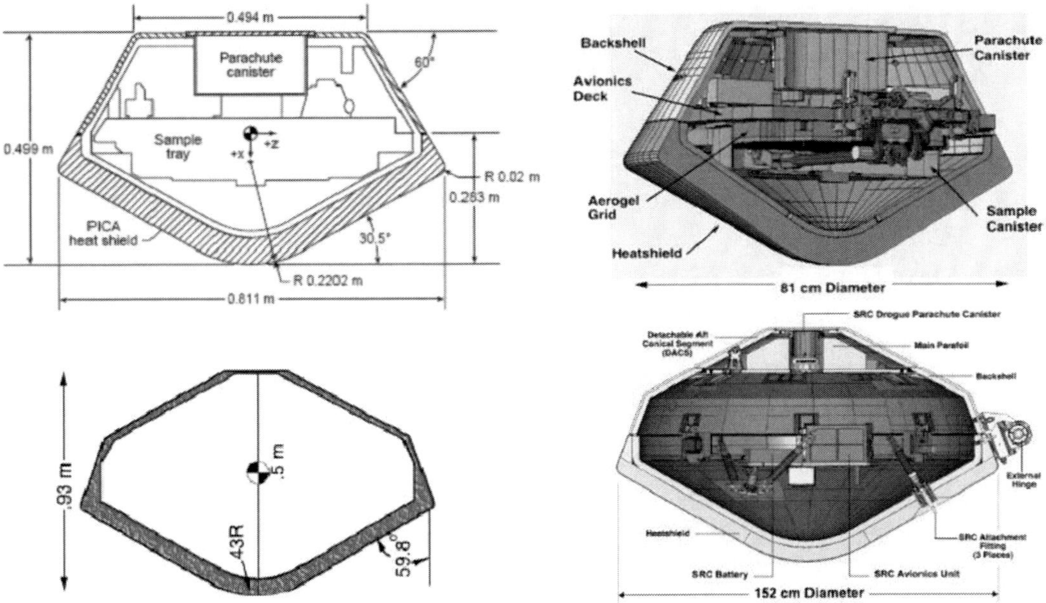

Figure 60. Stardust (up) and Genesis (down) sample return capsules[2].

Table 12. Overview of some fundamental SRV parameters [2]

	Stardust	Genesis	Hayabusa	Marco Polo
R_{nose} m/cone angle °	0.23/60	0.43/60	0.202/45	0.41/45
V_{entry} km/s	12.6	11	11.3	11.8
γ_{entry} °	-8.2	-8	-13.8	-12
M kg	45.8	210	16.3	76
Ballistic coef kg/m²	60	80	113.5	70
Area m²	0.56	1.78	0.128	0.95
$\dot{Q}_{max}$ MW/m²	13	7	15	11
$\int_0^{t(end)} \dot{Q}dt$ MJ/m²	360	166	320	210
TPS material / mass frac.	PICA-15/ 22%	C-C/ 18%	C phenol/ 43%	PICA?/ 38%

Obviously, some data (capsule size, re-entry velocity, etc..) will be different to ours, so we can adapt these to our mission through a scaling process.

In the Table 12 some examples of the entry capsules which can be taken as a guideline are summarized [1],[2]. Example of SRC configurations are shown in Figure 60 and Figure 61.

Figure 60 shows an overview of both internal and external layout of Stardust and Genesis sample return capsules [2]. An overview of main mission parameters such as trajectory, vehicle geometry, aerothermal environment, and TPS for both Stardust and Genesis SRV can be recognized in Tables 13 and 14, respectively [3]. It is worth to note that the SRV layout is an extremely important design issue. For instance, it determines the position of the vehicle centre-of-mass (CoM). In fact, if a non-spherical entry shell is used, the centre-of-mass of the configuration must lie adequately below the aeroshell's centre-of-pressure (CoP), which in turn moves the spacecraft and its dense components (batteries, etc.) closer to the leading face of the entry shell. If the offset between the CoM and CoP is made too small, then the vehicle may be unstable to disturbances and make large pitching movements, exposing non-shielded parts to the hotter airflow. As an example, a craft's transition from supersonic to subsonic speed causes changes in the wake flow which in turn can be coupled to the spacecraft, destabilizing it. Some entry vehicles, such as that of the Genesis sample-return mission, are designed to deploy small drogue parachutes at supersonic speeds to provide extra stability through the transonic region [4].

Table 13. Overview of the Stardust main parameters [3]

Trajectory		Geometry		Aero/thermal		TPS	
Entry angle	−8.2° ±0.08° @ 125 km	*Shape*	Blunt-nosed 60° half-angle cone	*Trim L/D (specify trim α)*	0	*Material designation*	PICA-15
Inertial entry velocity	12.8 km/s @ 125 km	*Nose radius*	0.23 m initial	*Ballistic coeff (Ablated).*	60.0 kg/m² 60.4 kg/m²	*Thickness*	5.82 cm
Relative entry velocity	12.6 km/s @ 125 km	*Base area (Ablated)*	0.52 m² 0.50 m²	*Stagnation heating rate*	(non-ablating) 1200 W/cm²	*Ablating? Ejected?*	Yes No
Velocity at peak heat	11.1 km/s	*Vehicle mass*	45.8 kg	*Integrated heat load*	36,000 J/cm²	*Resin mat. Matrix mat.*	Phenolic Carbon fiber
Control method	Ballistic	*TPS mass fraction, inc. insul.*	22%	*Radiative heat flux*	130 W/cm²	*Resin dens. Matrix density*	109 kg/m³ 160 kg/m³
Center of Gravity, X_{cg}/D	.35	*Payload mass*		*PH stag. pressure*	0.275 atm	*Total material density*	250 kg/m³ approx.

Figures 61 to 65 highlight examples of SRV design study results. Figure 61 shows a typical SRV configuration under investigation in Europe so far [2]. Such a configuration relies on a sphero-conical aeroshell with a 45 deg half cone angle, a 1.1 m diameter front shield and a smaller back-cover.

Table 14. Overview of the Genesis main parameters [3]

Entry angle	-8°	*Shape*	59.81° blunt cone	*Trim L/D (specify trim α)*	0	*Material designation*	Fore: Carbon-carbon Aft: SLA-561V
Inertial entry velocity	11.0 km/s	*Nose radius*	.43 m	*Ballistic coeff.*	80 kg/m^3	*Thickness*	6 cm
Relative entry velocity	10.8 km/s	*Base area*	1.78 m^2	*Stagnation heating rate*	700 W/cm^2	*Ablating? Ejected?*	Partially No
Velocity at peak heat	9.2 km/s	*Vehicle mass*	210 kg	*Integrated heat load*	16,600 J/cm^2	*Resin mat. Matrix mat.*	
Control method	Spin-stabilized aero-ballistic	*TPS mass fraction, inc. insul.*	~18%	*Radiative heat flux*	30 W/cm^2	*Resin dens. Matrix density*	
Center of Gravity, X_{CG}/D	.33	*Payload mass*		*PH stag. pressure*		*Total material density*	

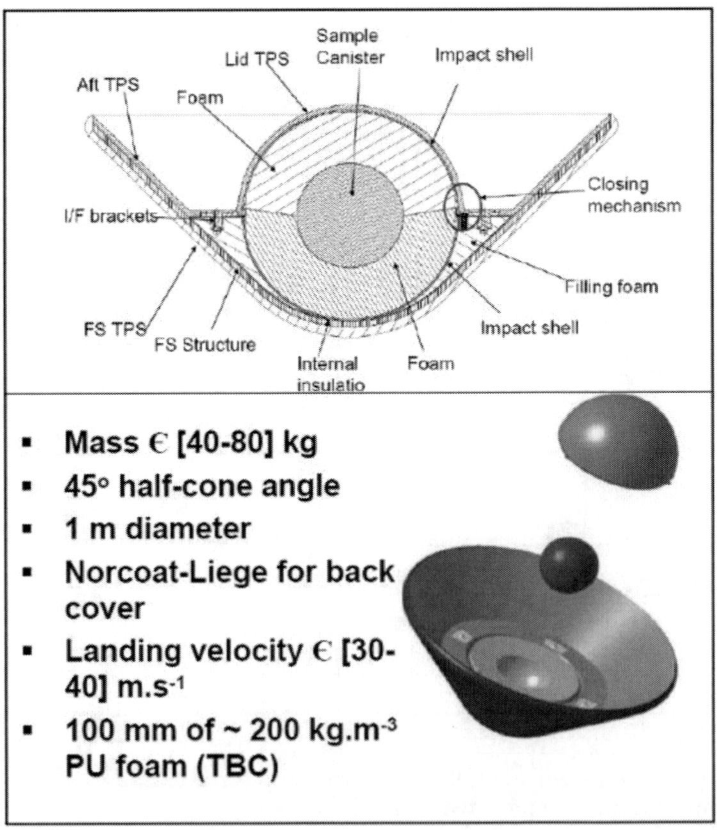

Figure 61. Example of SRV study results [2].

Table 15. Overview of the Hayabusa main parameters [3]

Trajectory		Geometry		Aero/thermal		TPS	
Entry angle	-13.8°	*Shape*	45° sphere-cone	*Trim L/D (specify trim α)*	0	*Material designation*	Carbon-phenolic
Inertial entry velocity	11.7 km/s	*Nose radius*	.202 m	*Ballistic coeff.*	113.5 kg/m³	*Thickness*	3.0 cm
Relative entry velocity	11.3 km/s	*Base area*	.128 m²	*Stagnation heating rate*	1500 W/cm²	*Ablating? Ejected?*	Yes Yes
Velocity at peak heat	10.2 km/s	*Vehicle mass*	16.27 kg	*Integrated heat load*	32,000 J/cm²	*Resin mat. Matrix mat.*	
Control method	none	*TPS mass fraction, inc. insul.*	43%	*Radiative heat flux*	300 W/cm²	*Resin dens. Matrix density*	
Center of Gravity, X_{CG}/D	0.28	*Payload mass*	1.04 kg	*PH stag. pressure*	0.61 atm	*Total material density*	1400 kg/m³

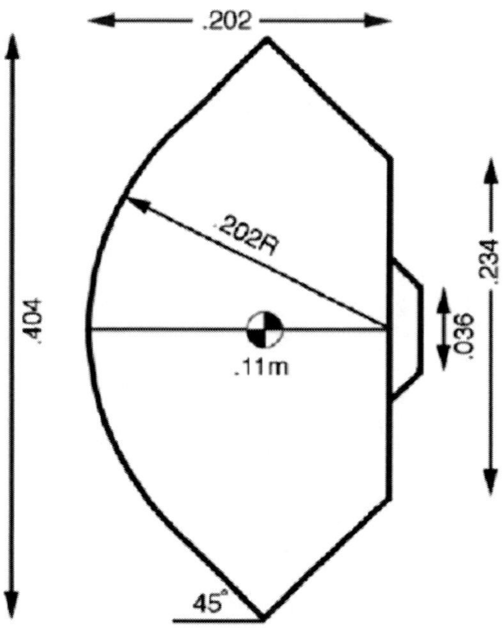

Figure 62. HAYABUSA sample return capsule[3].

In Figure 62 is reported the configuration of the Hayabusa sample return capsule, whose main mission parameters are summarized in Table 15[3].

As far as Flight Mechanics is concerned, Figure 63 shows a typical re-entry corridor of a SRV reported in the flight path angle (FPA) ballistic coefficient map. Of course the green-surface represents the flyable conditions during descent. This re-entry envelope refers to an atmospheric velocity of 11.8 km/s, at -12 deg FPA, landing load on the sample of 800 g, peak heat flux of 11.3 MW/m^2, heat load of 209 MJ/m^2, and entry duration of 484 s [2].

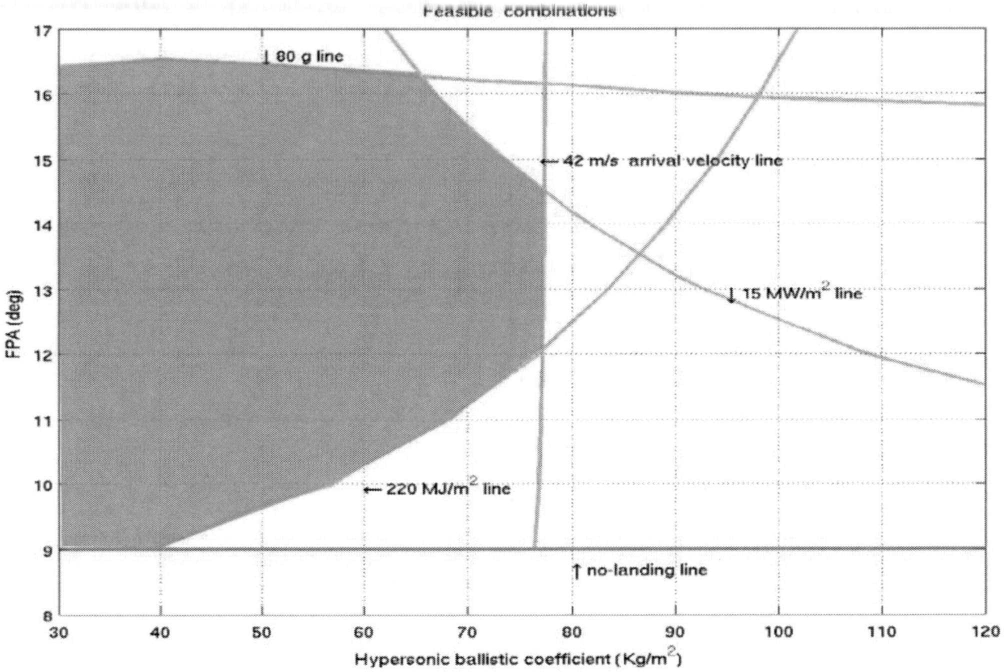

Figure 63. Example of SRV study results: re-entry corridor[2].

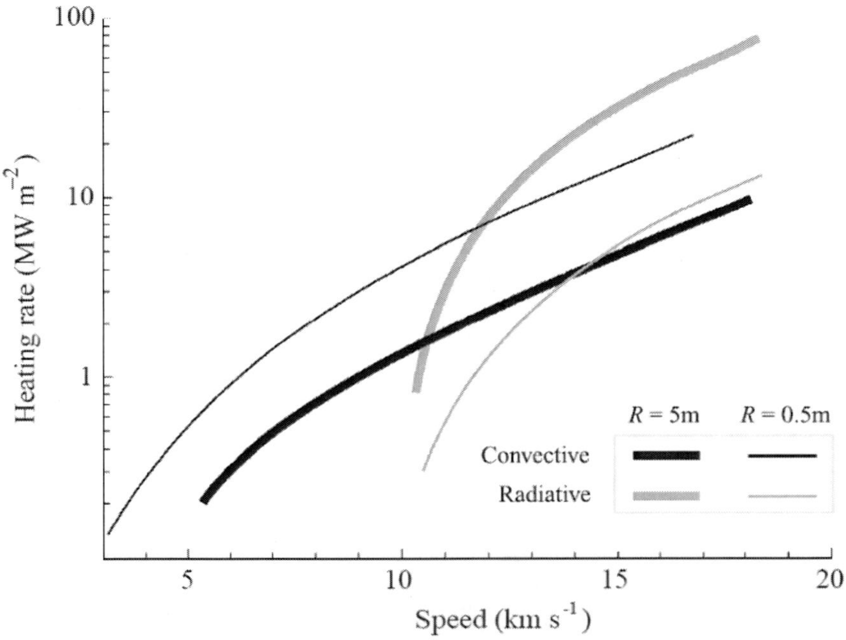

Figure 64. Radiative and convective heating rates limits for Earth entry[4].

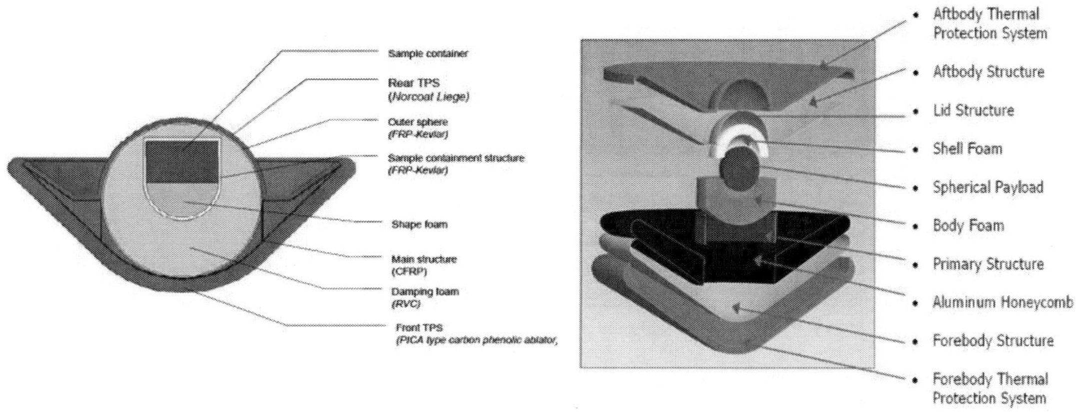

Figure 65. Example of SRV study results: TPS layout[2].

The lower boundary of the re-entry corridor in the hypersonic phase of descent is the heat flux limit (both convective and radiative). Indeed, at high entry speeds the temperature rise in the shock wave around the spacecraft may be sufficiently intense for radiant heating from the hot gas to be equivalent to the convective heating rate. For Earth entry this occurs at speeds above 10 km/s for bluff objects, as is shown in Figure 64[4].

In particular, Figure 64 shows radiative and convective heating rates compared for two spheres of different radii entering the Earth's atmosphere. Note that the same equivalence in the heating processes occurs at higher speeds for objects with smaller radii, but for such spacecraft the temperatures in the shock would be far higher, potentially compromising the temperature limits of the TPS.

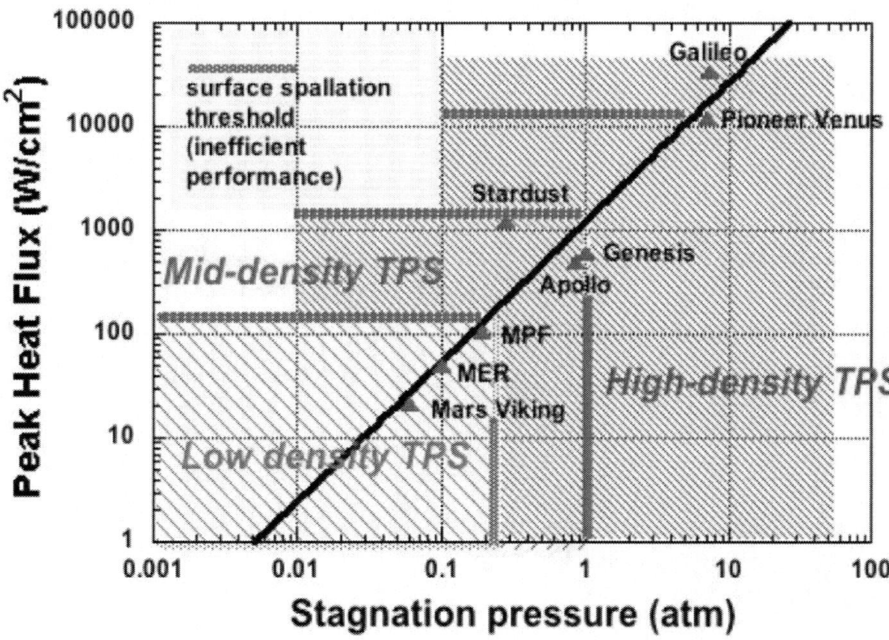

Figure 66. Mission environments for ablative TPS applications[5].

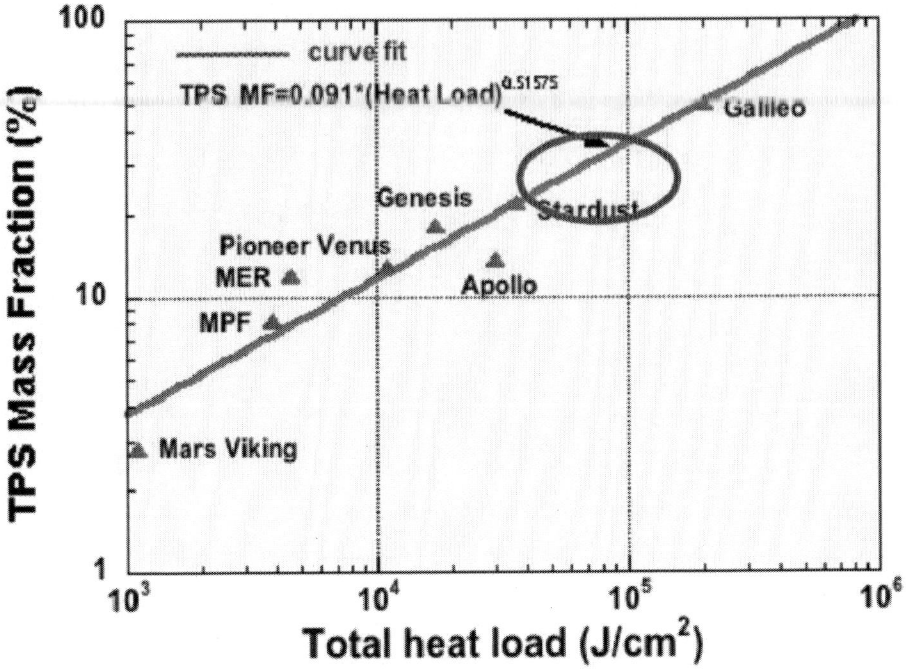

Figure 67. TPS mass fraction for ablative heat shield missions[5].

Table 16. Properties of some TPS materials [6]

Material	Density, kg/m^3	Thermal Conductivity,[a] W/m-K	Emissivity,[a] ε	Melt Temp., K
AVCOAT 5026 (virgin)	529	0.242	0.67	1920
AVCOAT 5026 (char)	264	—	0.49	—
Carbon-Phenolic (Acusil I)	480	0.112	0.95	2000
Reinforced Carbon-Carbon (RCC)	1580	5.05 (normal) 7.88 (parallel)	0.54 to 0.9	2030[b]
Teflon	2190	0.251	—	—
Shuttle Tile (LI-900)	144	0.047	0.88	1755[b]
Beryllium	1840	170	—	810[b]
Oak	610	0.146	—	—

* Squire, 2006 (TPSX database).

[a] Thermal conductivity and emissivity are functions of temperature.

[b] Single-use temperature limit.

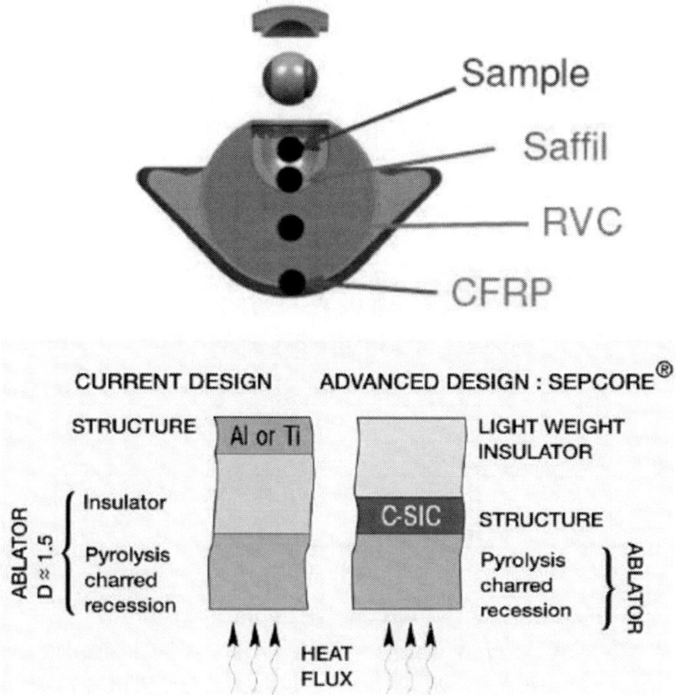

Figure 68. Internal and External (lower side) layout of Marco Polo TPS[2].

Table 17. Examples of mass budget for SRV [2]

Element	**Mass (Kg)**
Heat shield	11.19
Aftbody TPS	1.96
Primary Structure	4.71
Secondary Structures	2.96
Canisters (2)	7.20
Receptacles	1.36
Energy Absorbing Mat.	2.52
Beacons (2)	0.60
Mechanisms	1.90
Sensors and cables	0.20
Miscellaneous	1.14
Launch Total	28.53
Entry Total	35.73
Launch Total (+25%)	**35.66**
Entry Total (+25%)	**42.86**

Table 18. Examples of mass budget for SRV [2]

Subsystem	Basic, kg	Equipment margin		Nominal, kg	System margin		Total, kg
		%	kg		%	kg	
Thermal Protection	2.1	20.0	0.4	2.5			
Stabilizer	1.2	20.0	0.2	1.4			
Parachute System	2.6	20.0	0.5	3.1			
Shock Absorber	0.4	20.0	0.1	0.5			
Control Unit	1.7	20.0	0.3	2.0			
Recovery Equipment	1.1	20.0	0.2	1.3			
Power Supply System	0.5	20.0	0.1	0.6			
Samples Container	2.5	20.0	0.5	3.0			
Cabling	0.5	20.0	0.1	0.6			
Structure	1.3	20.0	0.3	1.6			
Total mass	13.9		2.8	**16.7**	20.0	3.3	20.0

A possible internal and external TPS layout is shown in Figure 65.

In this example, the TPS layout refers to PICA-like (260 kg.m^{-3}) front TPS, Norcoat-Liege rear TPS, 45 deg half-cone angle, 1.1 m base diameter, 200 mm Reticulated Vitreous Carbon (RVC) foam, and 76 kg capsule[2].

Figure 66, together with Figure 67, summarize the thermal protection systems for several space missions. As we can see, there are two regression curves that can be used for the TPS preliminary design [5].

In fact, once stagnation pressure and peak heat flux are known, Figure 66 suggests what kind of TPS must be used; while Figure 67 gives an idea of TPS mass fraction once the total heat load is known.

Examples of SRV thermal design adopted for the Marco Polo mission is reported in Figure 68. As shown, the TPS internal layout of the Marco Polo SRV is made of Saffil, RVC, and CFRP (Carbon Fiber Reinforced Polymer); while the TPS external layout is reported on the lower side of Figure 68[2].Some TPS materials are summarized in Table 16, where their properties are also reported [6].

As far as capsule mass budget is concerned, in Tables 17 and 18 are summarized examples of the impact of each capsule subsystem on the overall vehicle mass. In particular, Table 17refers to Mars Sample Return Study; while Table 18 to the Mercury Sample Return mission study [2].

DESIGN ANALYSIS

Within each design loop, several Sciences are involved. From Structural Mechanism, Electronics to Fluid Dynamics and etc. Indeed, the SRV is a result of a Multidisciplinary Design Optimization (MDO) [7].

As a consequence, the design analysis evolves through “trade-off” analysis. For example, requirements and constraints impact on design in terms of trade-off analysis that accounts for:

- Heat flux
- Heat input
- Drag
- Stability
- Landing accuracy
- Landing conditions
- Accommodation into launcher fairing
- Mass

that in turns depends on mission parameters as:

- Entry angle
- Entry velocity and etc.

Anyway, a detailed description of the flowfield past a re-entry vehicle is mandatory to address the impact of several design issues [8],[9]. For instance, Computational Fluid Dynamics (CFD) analysis is necessary for the determination of convective and radiative heat fluxes, and these fluxes have a direct impact on the selection of the material for the TPS and on the determination of the TPS thickness distribution as well [8].Furthermore, the detailed knowledge of flowfield past the re-entering vehicle allows designer to evaluate SRV aerodynamic performances as well as static and dynamic stability in order to improved Flight Mechanics analysis and so on [10].

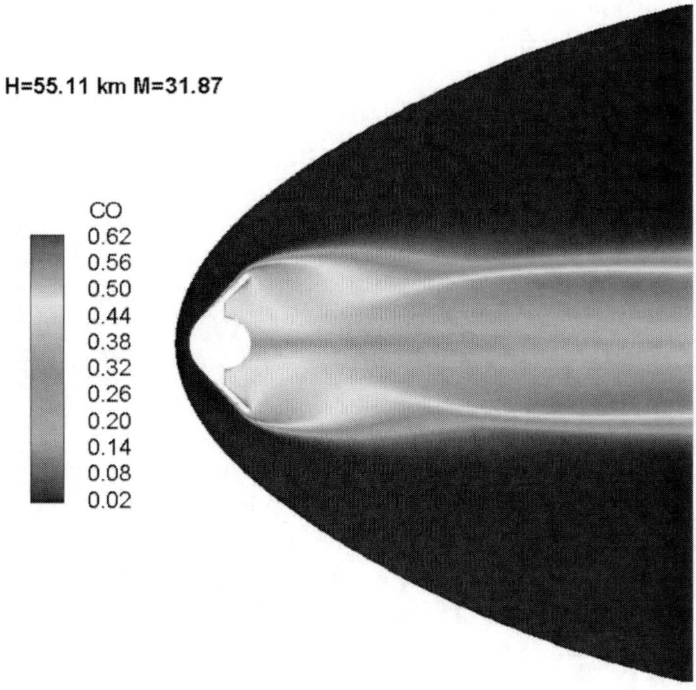

Figure 69. CO mass fraction distribution at 55.11 km altitude and M_{∞}=31.87.

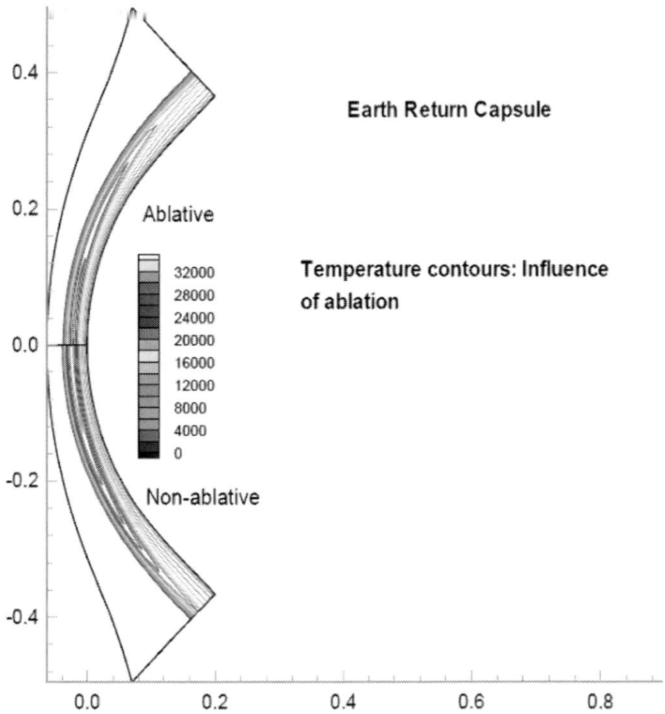

Figure 70. Static temperature contours (K). Influence of ablation[12].

Example of design analysis, can be found in Figure 69 where is shown the flowfield of CO species, blowing from the heat shield made of carbon-phenolic material, past a sphere-cone aeroshell (close to that shown in Figure 61) flying at 55.11 km altitude and M_∞=31.87. Note that CO is one of the main ablation products (for carbon based materials), and it lowers significantly the SRV wall enthalpy [11].

Figure 69 also highlights that TPS ablation demands particular care within flowfield CFD analysis because of it may influence SRV aerothermal performance considering that: heat shield roughness promotes turbulence transition; TPS ablation determines convective and radiative aeroheating blockage because of ablation reduces the surface gradients of temperature and that of various species mass fractions, causing a decrease of convective and diffusive heat-fluxes; heat shield ablation influences the flowfield past the SRV by means of introduction of new species in the boundary layer, that in turn may increase flowfield ionization [12].

Further example of design analysis is shown in Figure 70 where the influence of ablation on flowfield and shock stand-off distance for a high-speed Earth entry is recognized. In particular, the top figure represents non-equilibrium predictions with ablation. The bottom figure shows non-equilibrium without ablation [12]. This comparison puts in evidence the influence of ablation on results of non-equilibrium simulations. As shown, the temperature predicted in the ablative case is lower than in the non-ablative one [12]. Moreover, the shock wave is closer from the capsule in the ablative simulation. Therefore, ablation has a large influence on the non-equilibrium results. As a consequence, ablation process must be carefully taken into account for the design of the SRV thermal protection shield.

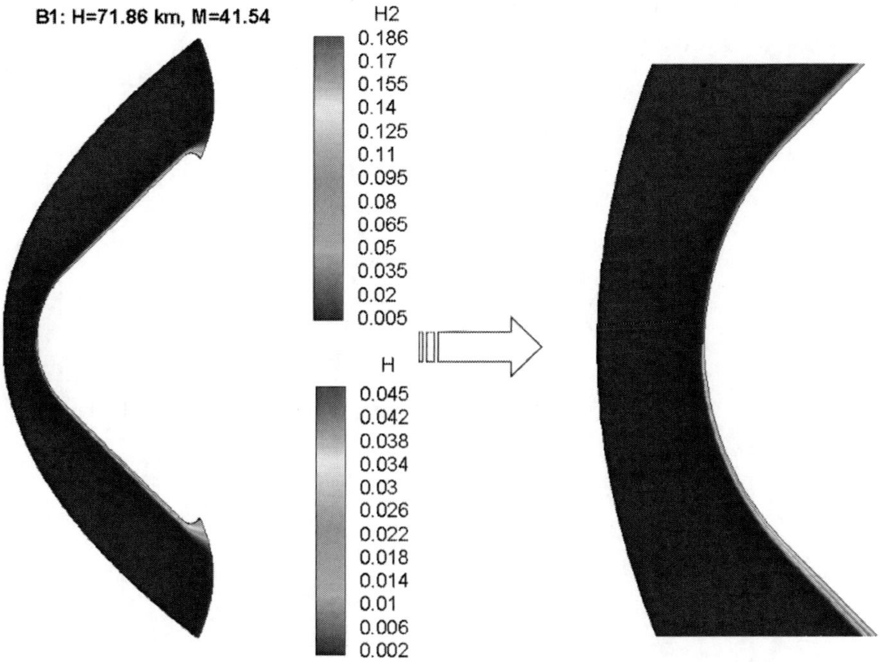

Figure 71. H_2 and H mass fractions at H_∞=71.86 km and M_∞=41.54.

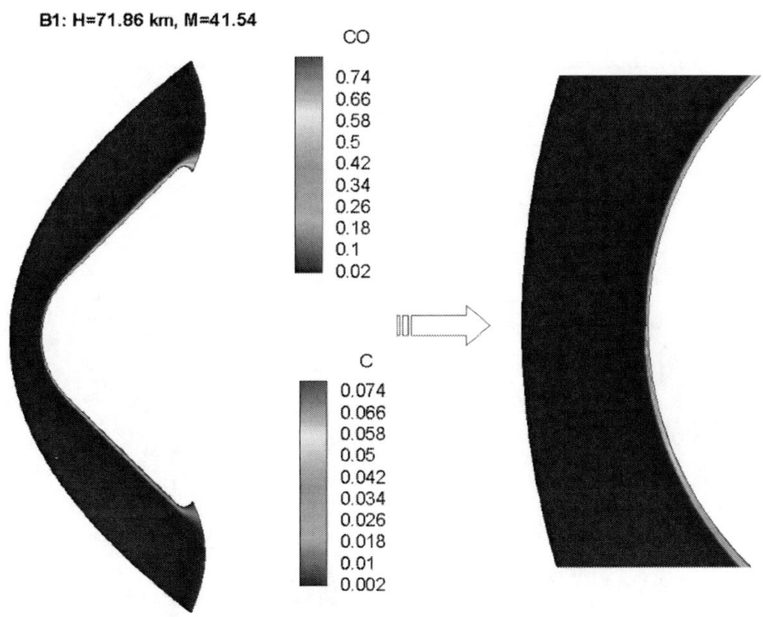

Figure 72. CO and C mass fractions at H_∞=71.86 km and M_∞=41.54.

Generally speaking, the ablation products in the boundary layer mainly consist of H_2, C_2H, C_2H_2, CO, C, and H [11].

For example, the mass fraction fields of H_2, H CO and C that take place in the shock layer in front of a 1.1 m diameter spherically-blunted 45 deg half-angle fore body (i.e., close to that shown in Figure 65) flying at 71.86 km altitude and M_∞=41.54 are reported in Figure 71 and Figure 72, respectively.

The species produced by the blowing in the shock-layer are plotted, along with the stagnation streamline in Figure 73.

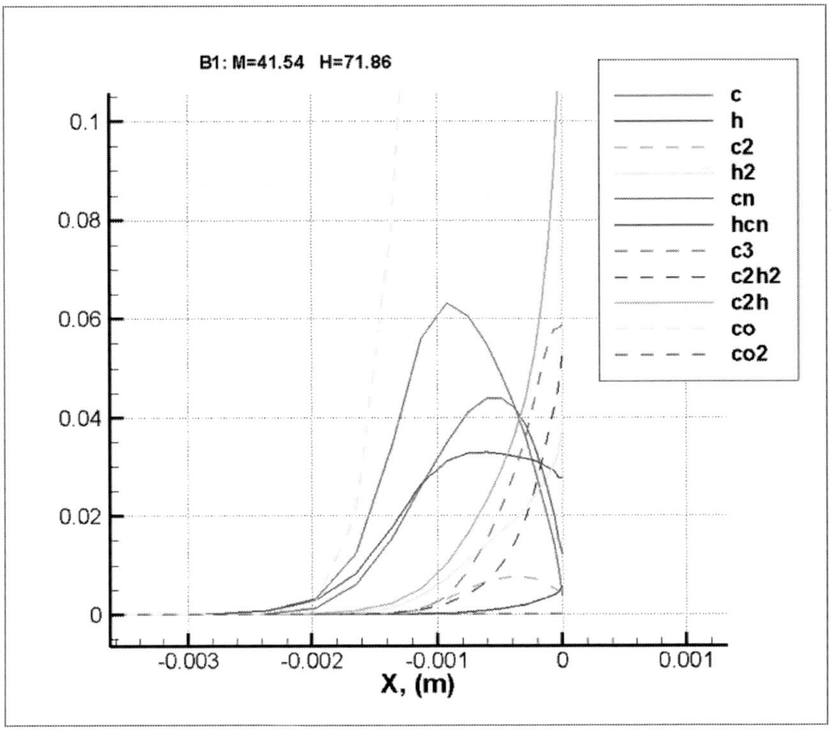

Figure 73. Ablation products on the stagnation streamline at H_∞=71.86 km and M_∞=41.54.

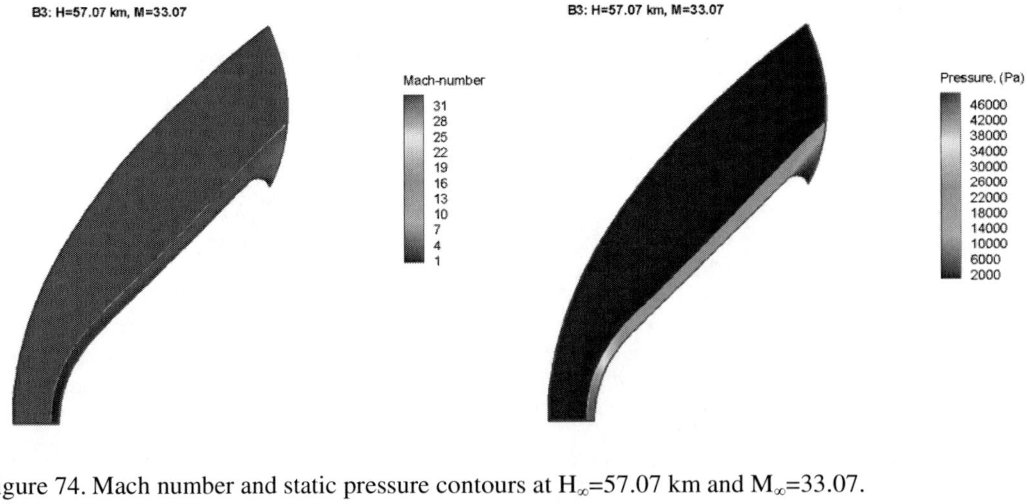

Figure 74. Mach number and static pressure contours at H_∞=57.07 km and M_∞=33.07.

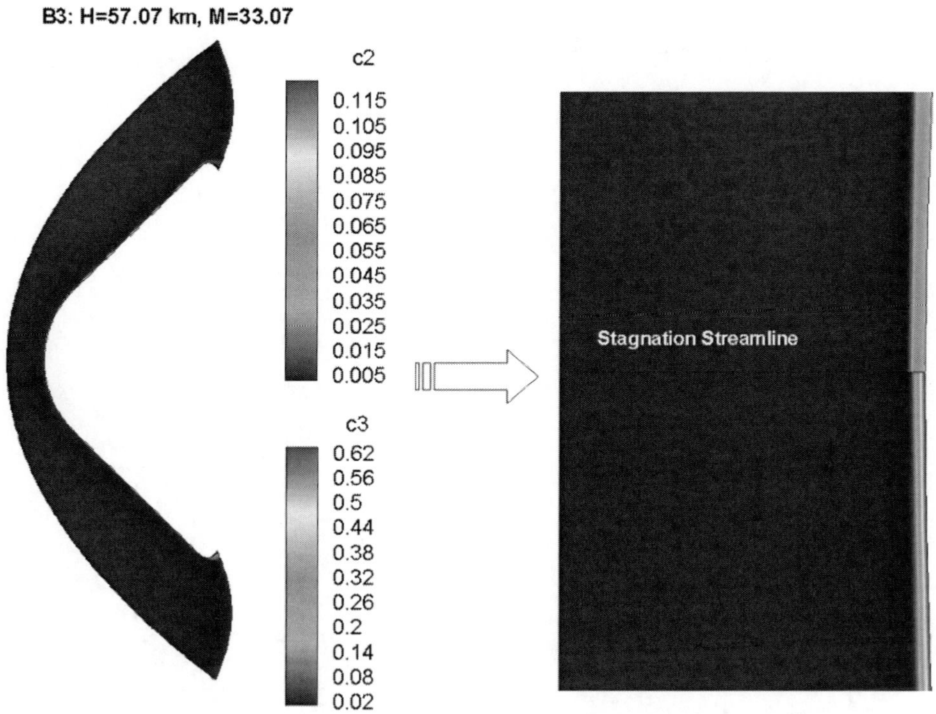

Figure 75. C_2 and C_3 mass fractions at H_∞=57.07 km and M_∞=33.07.

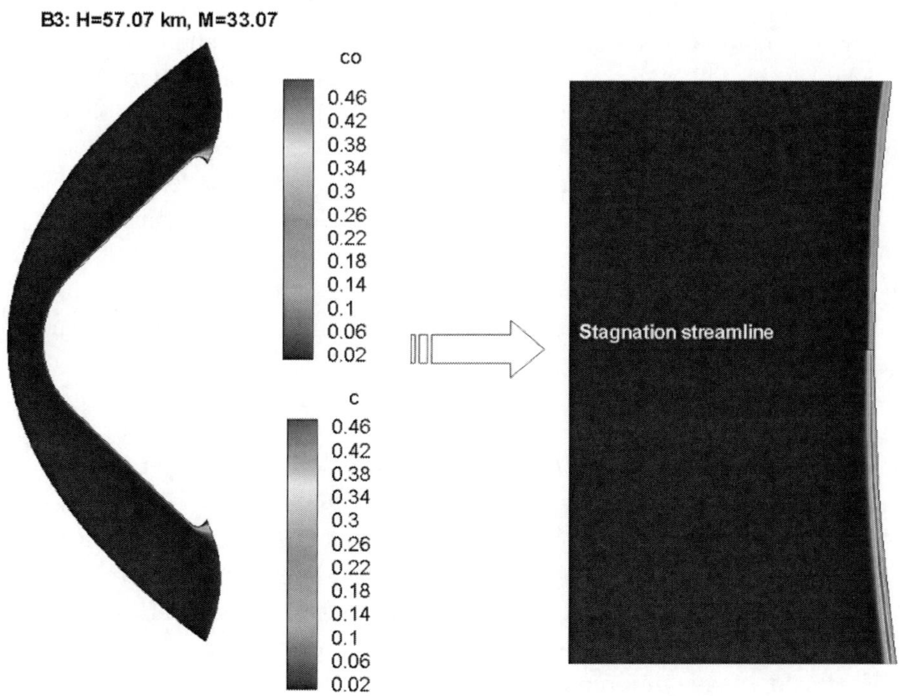

Figure 76. CO and C mass fractions at H_∞=57.07 km and M_∞=33.07.

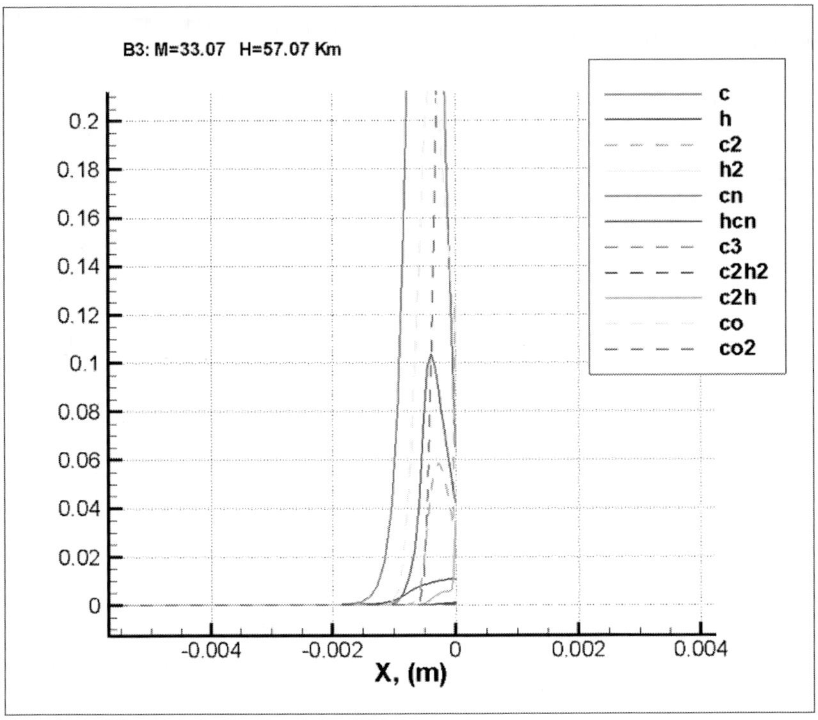

Figure 77. Ablation products on the stagnation streamline at H_∞=57.07 km and M_∞=33.07.

As shown, the boundary layer in the case of TPM species blowing is significantly different with respect to the non-ablating case [11]. In fact, as the wall is approached the level of all of the air species fall as the presence of the blown gases becomes more dominant.

Moreover, it is evident from the profiles that some of the ablation products are undergoing chemical change and it is interesting to note that although most of the ablation products are restricted to the boundary layer there is a significant level of some of these products, specifically C, at positions slightly beyond the edge of the boundary layer.

As further example, flowfield features regarding free stream conditions equal to H_∞=57.07 km and M_∞=33.07 are recognized from Figure 74 to Figure 76. In Figure 74 static pressure and Mach number fields are shown; while Figure 75 and Figure 76 highlight C_2, C_3,C, CO mass fractions, respectively.

The species produced by the blowing along with the stagnation streamline are plotted in Figure 77.

The knowledge of the ablation products distribution within the shock layer is extremely important since species such as C_2, C_3 and also CO have strong radiative properties. In particular, C_2 and C_3 possess absorption properties while CO is a strong radiator [13].

As shown, the level of C_3 falls rapidly with increasing distance from the body, as it dissociates to form C_2 and C, and also indirectly leads to the formation of CN and HCN. Note that the formation of small amounts of CN induces a more severe heat flux distribution because of a major non-equilibrium radiation contribution (i.e., CN molecule is a highly emissive product). Moreover, the rise in the level of C in the boundary layer is mainly attributable to the dissociation of CO. The relatively slow rate at which the level of H falls with increasing distance from the wall suggests that either H_2 or alternatively some of the hydrocarbons present are dissociating [11],[12].

Another important issue in designing SRV capsule is the coupling of plasma radiative effects and the CFD computation. For example, Figure 78 presents, on the left, the front shield shock layer temperature field and, on the right, temperature profiles along with the stagnation line of the Huygens probe with and without coupling effects. As shown, a reduction of the stand-off distance can be observed as well as a faster decrease of the temperature for the coupled computations due to the removal from the flowfield by radiation phenomenon [13].

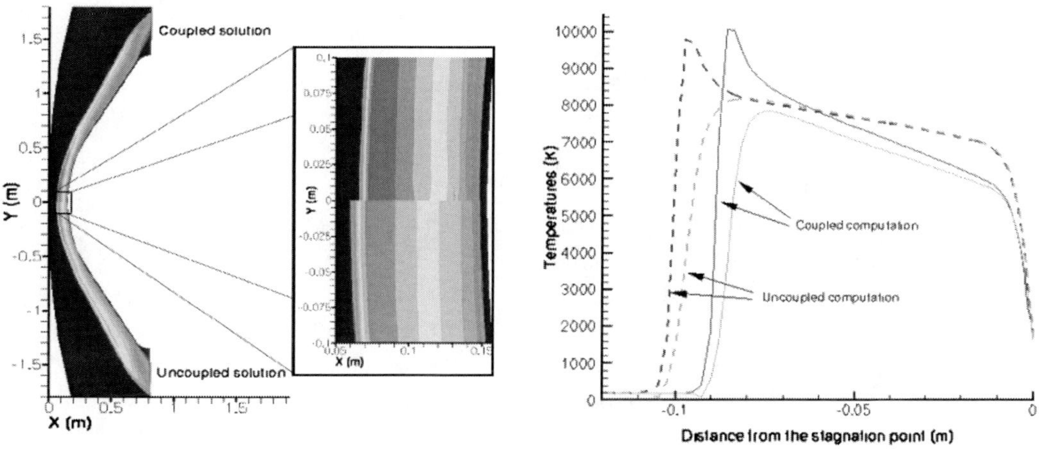

Figure 78. Temperature contours: influence of ablation[13].

Conclusion

In this chapter, a partial overview of the analysis approach typically followed within sample return vehicle design has been presented. A more systematic survey was out of the scope of the paper, but would be required for the preparation of ambitious sample return projects. This work gives some indications concerning main design topics to address in designing such a kind of vehicle for planetary exploration.

References

[1] Design Overview of an Asteroid Sample Return Capsule. *The Institute of Space and Astronautical Science. Report SP No. 17* March 2003.

[2] Re-entry technologies for Sample Return capsules. ESA/ESTEC, Aerothermodynamic Section. *Alpbach*, 25/7/08

[3] Planetary Mission Entry Vehicles Quick Reference Guide, Version 3.0 *NASA*/SP-2006-3401

[4] J., Garry, J. R. C., Lorenz, R. D., Kerzhanovich, V. V., *"Planetary Landers and Entry Probes"*. Cambridge University Press. 2007.

[5] Laub, B., Venkatapathy, E., *"Thermal Protection System Technology and facilities needs for demanding future planetary mission"* International Workshop on Planetary Probe Atmospheric Entry and Descent Trajectory Analysis and Science, Lisbon, Portugal, 6-9 October 2003

[6] Notes on Earth Atmospheric Entry for Mars Sample Return Missions. *NASA*/TP–2006-213486. 2006

[7] Bertin, J. Hypersonic Aerothermodynamics. *AIAA Educational Series,* Washington. 1994

[8] *Predictions of Radiative and Convective Fluxes at Stagnation Point for an Earth High-Speed Re-Entry*. SP667-2009

[9] *Hypersonic and High Temperature Gas Dynamics*. McGraw-Hill Book Company, New York 1989

[10] Mitcheltree, R., "A passive Earth-Entry Capsule for Mars Sample Return. *51st International Astronautics Federation Congress, Rio de Janeiro*. 1998. IAF-00-Q.3.04.

[11] Modelling Gas/Surface Interaction Processes of Ablating Wall Boundaries Associated with Planetary Entry. *ESASP*.367..311R. 1995

[12] Reynier, Ph., Marieu, V., Marraffa, L., *"Aerothermodynamics developments for planetary missions"* Stuttgart, November 17 - 19, 1999. SSAE 99

[13] Mazoué, F., Graciano, J., Tournebize, L., Marraffa, L., "Updated Analysis of the Radiative Heat Flux During The Entry of The Huygens Probe in the Titan Atmosphere" *Proceedings of the Fifth European Symposium on Aerothermodynamics for Space Vehicles, 8-11 November 2005*, Cologne, Germany. (ESA SP-563, February 2005).

In: Mars
Editor: Alberto G. Fairén
ISBN: 978-1-62618-102-1

Chapter 10

AERODYNAMICS AND AEROTHERMODYNAMICS ASSESSMENT OF A MARS SAMPLE RETURN MISSION

Antonio Viviani[1] **and Giuseppe Pezzella**[2]*
[1]Dipartimento di Ingegneria Aerospaziale e Meccanica,
Seconda Università degli Studi di Napoli, Naples, Italy
[2]Centro Italiano Ricerche Aerospaziali, Caserta, Italy

ABSTRACT

An important step forward for Space Exploration activities and for a more accurate knowledge of the Earth, and universe is to develop the capability to send vehicles into space which select, collect and finally return samples from Mars to Earth where to perform their analysis.

To return these samples to Earth, however, very high-speed re-entry trajectories must be performed. Therefore, the design of an Earth entry sample return vehicle (SRV) requires strong technological bases and relies on a good understanding of the loading environment encountered during the super-orbital re-entry.

A high speed Earth entry vehicle has the following characteristics:

- Entry velocity higher or equal to 11.7 km/s (compared to 7.5 km/s for the US Space Shuttle);
- Very high heat fluxes (more than 10 MW/m^2) and heat loads (in the range of 500MJ/m^2), where the radiative part is rather important.

In this framework a capsule configuration has been selected and a re-entry trajectory has been computed to address the SRV design.

* Dipartimento di Ingegneria Aerospaziale e Meccanica, Seconda Università degli Studi di Napoli (antonio.viviani@unina2.it)

Capsule Geometry

The selected shape is a sphero-conical aeroshell with a 45 deg half cone angle, a 1.1 m diameter front shield and a smaller back-cover that is shown in Figure 79. The layout and design of the capsule assure a safe return of the sample canister, relying on a fully passive concept [1].

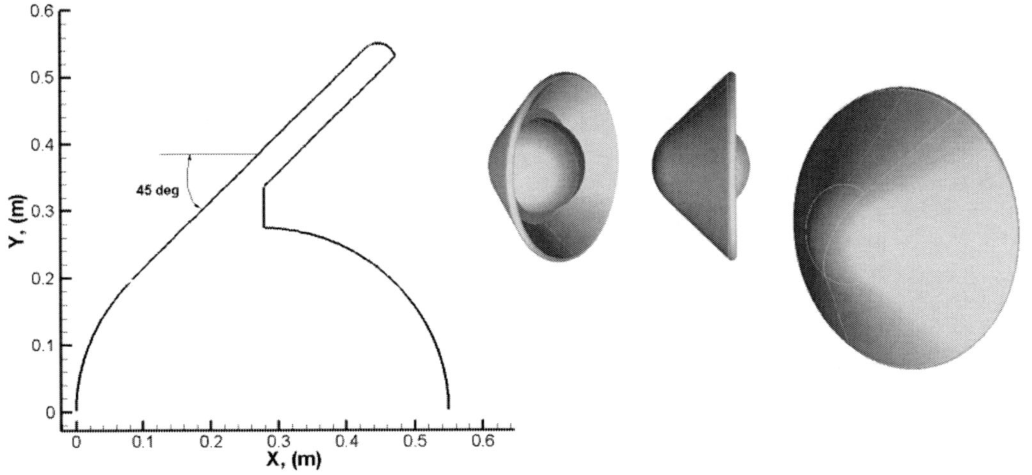

Figure 79. Capsule geometry with quotes.

The selection of such a kind of aeroshell was the result of a trade-off design among hypersonic drag (heating), subsonic drag (impact velocity) and subsonic stability (available crush stroke) [1]. The SRV nose radius is equal to 0.275 m; while at the shoulder the radius of curvature is equal about 0.0275 m.

The capsule vehicle, including margins, is estimated to weigh about 50 kg. The CoG is estimated at 26.5% (CoG/SRV diameter) from the nose relatively to the SRV diameter. This value is almost compliant with the stability requirements of such a shape.

Reference Re-Entry Flight Scenario

The flight design scenario of the SRV concept is shown in Figure 80. It refers to an Earth entry. In particular, this sizing descent trajectory is named max heat flux re-entry flight because of it is a steep descent (i.e., -12.5 deg of flight path angle) and, therefore, is characterized by the maximum heat flux peak.

Such a trajectory allows predicting the flowfield past the proposed vehicle. Knowledge of the vehicle flowfield at several points along with the trajectory enables computation of vehicle surface pressures, aerodynamic forces, and heat transfer characteristics. Mach number and Reynolds number histories are also necessary to characterize the vehicle boundary layer (laminar, transitional or turbulent); provide initial conditions for CFD solutions; and establish appropriate requirements for ground testing. The aerothermodynamic environment for the capsule is also provided. In fact, the evolutions of heat fluxes (convective and radiative part) are presented in Figure 81.

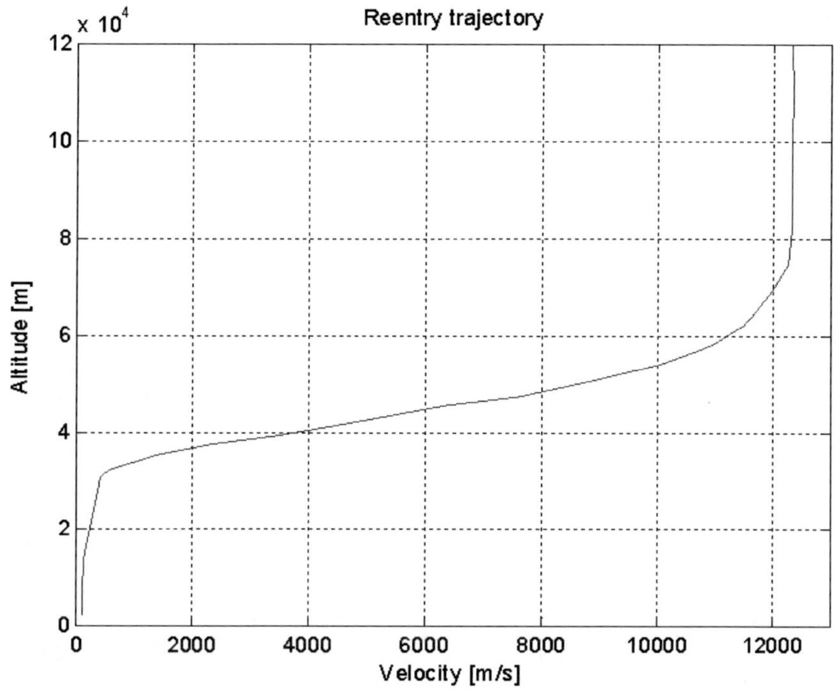

Figure 80. Altitude evolution vs. velocity .

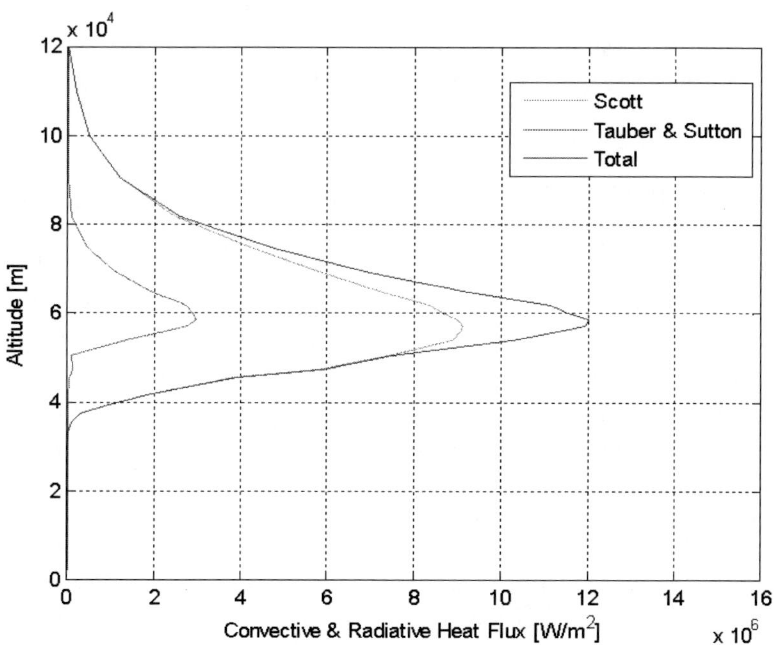

Figure 81. Heat Flux vs. altitude.

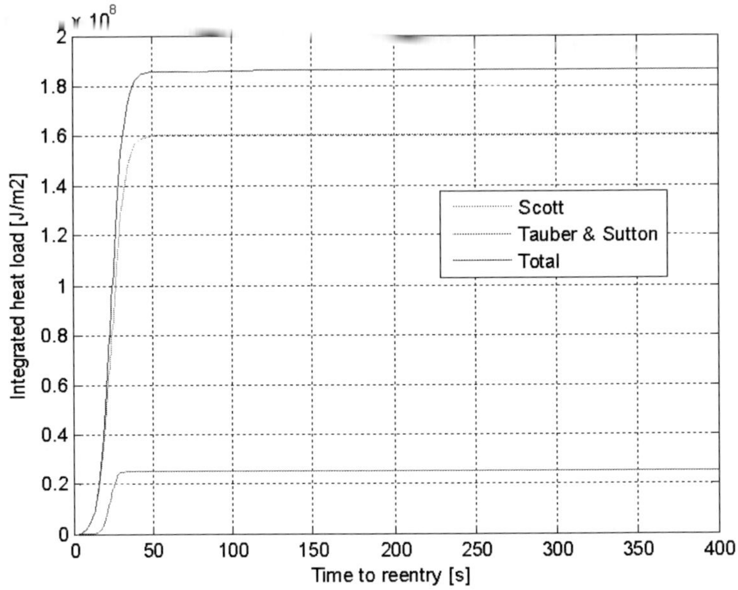

Figure 82. Time histories of the integrated heat load for both convective and radiative heat flux profiles.

In the preliminary design the convective and radiative heat fluxes were estimated by using analytical engineering correlations such as Scott relationship for convective heat flux and Tauber-Sutton one for the radiative heat flux estimation [2][3]. To the heat flux profile corresponds the integrated heat load of Figure 82.

All the estimations provided in Figure 81 and Figure 82 are extremely important for designing the capsule heat shield. Indeed, the aeroheating environment dictates, in fact, the type and size of the thermal protection system (TPS) to be used. Peak heat rate generally determines the range of possible thermal protection material (TPM) while the integrated heat load determines the thickness and hence the mass of the heat shield.

Design Approach and Used Tools

A summary review of the aerodynamic (AEDB) and aerothermodynamic (ATDB) characteristics of the SRV concept is performed. These evaluations are aimed to address the aerothermal load the vehicle has to withstand during re-entry at super-orbital speed. For instance, the SRV concept has a number of extreme loading flight conditions for which analyses are required. They must return from outer space, fly trimmed throughout hypersonic and supersonic regimes until landing is gained. In this work the spacecraft aerodynamic coefficients have been provided for different flight conditions along with the sizing re-entry trajectory, according to the trajectory-based design approach. The heat flux distributions, both convective and radiative one, are also provided for each trajectory design point. It is worth to note that heat flux distribution herein provided are extremely relevant for the SRV design considering that high speed planetary entries are often characterized by strong radiation and ablation effects. This is particularly the case for sample return missions. Therefore, in order to improve the state-of-the-art for such projects, computations are performed accounting for both radiation and ablation effects on SRV aerothermal performances. In fact, one of the most challenging topic of SRV design is to evaluate the blocking effect induced by the blowing of pyrolysis gases on the heat–flux and on the shock layer.

In particular, the appraisal of the vehicle AEDB was performed by means of both engineering-based tools and CFD computations; while the SRV ATDB was accomplished by means of Navier-Stokes (NS) computations in thermo-chemical non-equilibrium conditions. Then results of those CFD simulations are provided to PARADE code for the estimation of the radiative heat fluxes the capsule has to withstand during the super-orbital re-entry flight [4]. The range between Mach 3 and Mach 41.54 was analyzed. In the present analysis only continuum regime (supersonic and hypersonic speed ranges) with the air modeled as a mixture of several gases (including also those coming from heat shield ablation) has been studied. In the following paragraphs the tools used for the analysis are described.

Aerodynamic and Aerothermodynamic Analyses Tools

The aerodynamic analyses were performed by means of engineering level methods and more reliable CFD computations.

Engineering-Based Tool

Engineering based aerodynamic analyses were extensively performed by using a 3D Panel Method code developed by CIRA (SIM–Surface Impact Method) in the frame of its research activities on preliminary design of re-entry vehicles [5]. This tool at high supersonic and hypersonic speeds is able to accomplish the aerodynamic and aerothermodynamic analyses of a complex re-entry vehicle configuration by using simplified approaches as local surface inclination methods and approximate one-dimensional boundary-layer methods, thus avoiding the time consuming and complex grid generation and computation processes of CFD. The vehicle surface is approximated by a system of planar panels; the lowest level of geometry used in the analysis is a quadrilateral element. The pressure acting on each panel is evaluated by user-specified compression-expansion and approximate boundary-layer methods. The methods to be used in calculating the pressure in impact and shadow regions may be specified independently and can be selected by the user; several methods are available (see Table 19).

In order to predict viscous contribution to aerodynamic forces and moments the shear force is determined on each vehicle panel. The skin friction is estimated based on the assumption of a laminar or turbulent flat plate as:

$$\Delta C_{D,friction} = c_f \frac{S_{wet}}{S_{ref}}$$

where c_f is the skin friction coefficient, S_{wet} is the panel wetted area and S_{ref} is the reference vehicle's surfaces. Reference temperature and reference enthalpy methods are available for both laminar and turbulent flows. The viscous calculation is performed along streamlines, and the results are then interpolated to the panel centroids.

As far as the base drag is concerned, the following simple formula was used [5]:

$$\Delta C_{D_{base}} = -c_{P_{base}} \frac{S_{base}}{S_{ref}}$$

where

$$c_{P_{base}} = -\frac{1}{M_\infty^2}$$

in hypersonic flow conditions and S_{base} is the vehicle's base surface.

Table 19. Available methods for inviscid analysis [5]

Impact flow	Shadow flow
Modified Newtonian	Newtonian
Modified Newtonian/ Prandtl-Meyer	Modified Newtonian/ Prandtl-Meyer
Tangent wedge	Prandtl-Meyer empirical
Tangent wedge empirical	OSU blunt body empirical
Tangent cone empirical	Van Dyke unified method
OSU blunt body	High Mach no. base pressure
Van Dyke unified method	Shock expansion method
Blunt body skin friction	Input pressure coefficient
Shock expansion method	Free molecular flow
Free molecular flow	
Input pressure coefficient	
Hankey flat surface empirical	
Delta wing empirical	
Dahlem-Buck empirical	
Blast wave	
Modified tangent code	

CFD-Based Tool

Numerical tool used to carry out the CFD analyses is the CIRA code H3NS [6]. It solves the flowfield governing equations, including chemical and vibrational non-equilibrium, with a finite volume approach. The fluid is treated as a mixture of perfect gases and the energy exchange between vibrational and translational modes (T_V) is modelled with the classical Landau-Teller non-equilibrium equation, with average relaxation times taken from the Millikan-White theory modified by Park. For what concerns transport coefficients, the viscosity of the single species is evaluated by a fit of collision integrals calculated by Yun and Mason, the thermal conductivity is calculated by means of the Eucken law; the viscosity and thermal conductivity of the gas mixture are then calculated by using the semi-empirical Wilke formulas. The diffusion of the multi-component gas is computed through a sum rule of the binary diffusivities of each couple of species (from the tabulated collision integrals of Yun and Mason). Transport coefficients, in the hypothesis of an ideal gas, are derived from Sutherland law, suitably modified to take into account low temperature conditions. With respect to the numerical formulation, conservation equations, in integral form, are discretized with a finite volume, cell centred, technique. Eulerian fluxes are computed with a flux difference splitting upwind scheme. Second order formulation is obtained by means of an ENO-like reconstruction of cell interface values. Viscous flux is computed with a classical centred scheme i.e. computing the gradients of flow variables at cell interfaces by means of Gauss theorem. Integration in time is performed by employing an explicit multistage Runge-Kutta algorithm coupled with an implicit evaluation of the species and vibration energies source terms. Also a parallel version of the code is currently available.

Flow Radiation Assessment Tool

At very high reentry speeds (~12 km/s) temperature in the shock layer reaches values higher than 10000 K. At those energy levels radiation contributes significantly to the capsule aerothermal loading.

The flow radiative heat flux at the SRV wall has been computed through the code PARADE [4]. PARADE has been developed in collaboration between *Fluid Gravity Engineering Ltd* (UK) and the *Institute of Raumfahrtsysteme* from Stuttgart under an ESA/ESTEC contract in 1996 for air species first. The code is used to compute flowfield emission and absorption, between the shock layer and the surface of the probe. The spectral emission and absorption are determined as function of transition level (from upper level to lower level) and emitting population of this level. The population can be derived from the Quasi-Steady-State (QSS) method or by a Boltzmann method in order to take into account the non-equilibrium or equilibrium regime, respectively. The radiative computations have been performed with the Boltzmann assumption for the determination of the population of the excited molecular states.

Indeed, due to the very high temperatures reached in the shock layer caused by the strong bow shock in front of the capsule, the gas not only dissociates and ionizes but can also emits radiated energy which travels across the entire flowfield interacting with the gas itself. At the entry velocity foreseen for the capsule, this contribution, that is generally very low with respect to other "energies" in the flowfield, cannot be neglecting because it can cause an additional source of heat load at the wall to be taken into account [7]. From a general point of view, the flowfield and the radiative field are coupled: the radiated energy travels across the flowfield, interacts with the fluid dynamic field which changes its configuration and causes the change of the radiated energy, and so on. From a mathematical point of view, this interaction is represented by a source term in the classical Navier-Stoke equation for the energy conservation which represents the quantity of radiated energy travelling across the flowfield.

In the present case, however, we consider a simplified, non-coupled, approach: it assumes that the energy emitted mainly just behind the shock region, travels towards the wall, being in part absorbed by the gas in the shock layer, without changing its structure and arrives at the wall increasing the total heat flux. Different results presented in literature provided that this uncoupled approach is conservative if applied to the radiative heat flux estimation at the wall [7]. In order to compute the contribution to the wall due to the radiative processes in the field, it is necessary therefore to know the radiative properties in the gas which have been computed through PARADE starting from the results of the fluid dynamic computations (in terms of gas composition and temperature). The Radiative Heat Transfer equation has been then solved using the one dimensional tangent slab approximation (radiation properties are assumed to vary only in the direction normal to the wall). The intensity of radiation at a given wavelength λ satisfied the equation of radiative transfer.

$$\frac{\partial I_\lambda}{\partial s} = j_\lambda - k_\lambda I_\lambda$$

where j_λ and k_λ are respectively the emission and absorption coefficient. The calculation of the radiative heat flux has been done considering a spectral region between 100 and 40000 nm using 50000 spectral grid location. Emission and absorption coefficients have computed through PARADE. These coefficients are integrated along straight lines toward the wall according to the above equation to compute I_λ. Radiative heat flux at a given wavelength λ is then obtained through integration of I_λ over the solid angle whereas the total radiative heat flux is obtained through integration over the spectrum of interest.

Loading Conditions Scenario

All the aerothermal analyses provided hereinafter refer to the flight loading condition scenario foreseen for the SRV concept, as summarized in the following paragraphs. Then CFD simulations are performed at several discrete points within the SRV flight scenario (i.e., trajectory-based approach), and the results are used to address vehicle AEDB and ATDB. In particular, eighteen trajectory points (TP) have been considered to perform 18 NS two-dimensional (2D) axy-symmetric non-equilibrium simulations. Those points are shown in Figure 83.

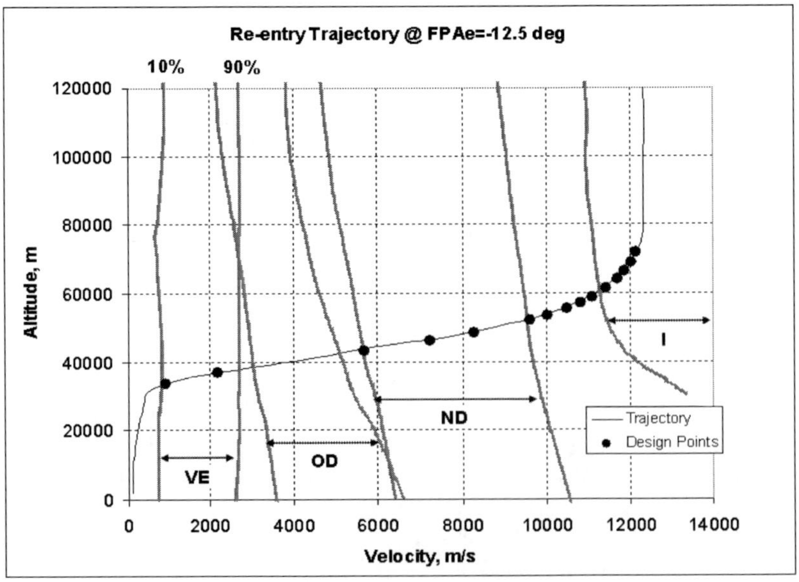

Figure 83. Reference re-entry trajectory with CFD design points.

Further, to take an idea of real gas effects the capsule will experience during descent, Figure 83 also shows the re-entry trajectory superimposed on the fields (from 10% to 90 %) of vibrational excitation, dissociation, and ionization of flow species [8]. Freestream conditions of design points are summarized in Table 20.

Table 20. Freesteam condition of design points

Design Points Id	Altitude [km]	Velocity [m/s]	Pressure [pa]	Density [kg/m³]	Temperature [K]	Mach [-]
R1	71.86	12138	4.14	$6.78x10^{-5}$	212.41	41.54
R7	69.03	12032	6.386	$9.95x10^{-5}$	223.48	40.14
R8	66.25	11880	9.589	$1.42x10^{-4}$	234.37	38.70
R2	63.98	11711	13.19	$1.88x10^{-4}$	243.29	37.45
R9	61.31	11445	18.889	$2.59x10^{-4}$	253.19	35.87
M1/R3	58.73	11099	26.49	$3.57x10^{-4}$	258.26	34.45
R10	57.07	10816	32.835	$4.37x10^{-4}$	261.52	33.07
R4	55.46	10490	40.33	$5.30x10^{-4}$	264.69	32.16
R11	53.53	10022	51.459	$6.67x10^{-4}$	268.49	30.50
M2/R5	52.05	9604	61.84	$7.95x10^{-4}$	270.65	29.12
M3/R6	48.36	8280	97.84	$1.25x10^{-3}$	270.65	25.11
R12	46.04	7230	130.72	$1.70x10^{-3}$	267.03	22.07
M4	43.14	5681	189.53	$2.54x10^{-3}$	259.02	17.61
M5	36.93	2181	437.43	$6.30x10^{-3}$	241.86	7.00
M6	33.62	922	701.11	$1.04x10^{-2}$	232.69	3.02

The gas model to consider is air (13 species) in thermo-chemical non-equilibrium conditions. CFD computations are in turbulent flow conditions for points below 51.9 km altitude. Note that, for the determination of the laminar-to-turbulent (L/T) transition the simplest estimate is made from freestream Reynolds number and here we use a critical value of $5.0x10^5$ based on diameter. This value was chosen for the Atmospheric Reentry Demonstrator (ARD) post-flight data.

$$Re_{\infty,D} = \left(\frac{\rho_\infty u_\infty D}{\mu_\infty}\right) \geq 5.0 \times 10^5$$

So, for the sizing trajectory the transition occurred at 51.9 km altitude.

Atmosphere Modelling

The atmosphere composition considered in the present CFD computations is recognized in Table 21.

Table 21. Complete atmospheric composition for Earth

Species	Mass Fraction
N_2	0.75548
O_2	0.23161
Ar	0.01291
Total	1.0000

Therefore, the specific gas constant for air is $R_{air}=R_0/M=287.182$ J/Kg/K. The model proposed for the air mixture is constituted by 13 species:

o Molecules: N_2, O_2, NO

o Atoms: N, O, Ar

o Ions: N_2+, O_2+, NO+, N+, O+, Ar+ and e-

Thermo-Chemical Models for Air with and Without Ablation

The following chemical kinetics and thermal non-equilibrium models for Earth have been considered

Chemical Kinetics Model

The chemical model proposed for air is based on Park [9]. It is constituted of 13 species and 22 chemical reactions, as recognized in Table 22, where molecules, atoms and e- are a reacting partner (third body) and can be any of the above thirteen species.

Table 22. Chemical kinetics model for air

Reaction Id		$K_f = A\,T^n e^{(-Ta/T)}$			
	Dissociation Reactions	K_f (cm3/mole/s)	n	Ta [k]	Tr (Control Temperature) [K]
R1,1	N_2 + Molecules ↔ N + N + Molecules	$7.00 \cdot 10^{21}$	-1.60	113200	$T^{0.5}T_v^{0.5}$
R1,2	N_2 + Atoms ↔ N + N + Atoms	$3.00 \cdot 10^{22}$	-1.60	113200	$T^{0.5}T_v^{0.5}$
R1,3	$N_2 + e^- \leftrightarrow N + N + e^-$	$3.00 \cdot 10^{24}$	-1.60	113200	$T^{0.5}T_v^{0.5}$
R2,1	O_2 + Molecules ↔ O + O + Molecules	$2.00 \cdot 10^{21}$	-1.50	59500	$T^{0.5}T_v^{0.5}$
R2,2	O_2 + Atoms ↔ O + O + Atoms	$1.00 \cdot 10^{22}$	-1.50	59500	$T^{0.5}T_v^{0.5}$
R3,1	NO + Molecules ↔ O + N + Molecules	$9.64 \cdot 10^{14}$	0	74700	$T^{0.5}T_v^{0.5}$
R3,2	$NO + N_2 \leftrightarrow O + N + N_2$	$1.45 \cdot 10^{15}$	0	74700	$T^{0.5}T_v^{0.5}$
R3,3	NO + Atoms ↔ O + N + Atoms	$9.64 \cdot 10^{14}$	0	74700	$T^{0.5}T_v^{0.5}$
	Exchange Reactions: Neutrals				
R4	$NO + O \leftrightarrow O_2 + N$	$8.40 \cdot 10^{12}$	0	19400	T
R5	$N_2 + O \leftrightarrow NO + N$	$6.40 \cdot 10^{17}$	-1.00	38400	T
	Associative Ionization Reactions				
R6	$N + N \leftrightarrow N_2^+ + e^-$	$2.00 \cdot 10^{13}$	0	67500	T
R7	$O + O \leftrightarrow O_2^+ + e^-$	$1.10 \cdot 10^{13}$	0	80600	T
R8	$N + O \leftrightarrow NO^+ + e^-$	$5.30 \cdot 10^{12}$	0	31900	T
	Electron Impact Ionization Reactions				
R9	$N + e^- \leftrightarrow N^+ + e^- + e^-$	$2.50 \cdot 10^{34}$	-3.82	168200	$T^{0.5}T_e^{0.5}$
R10	$O + e^- \leftrightarrow O^+ + e^- + e^-$	$3.90 \cdot 10^{33}$	-3.78	158500	$T^{0.5}T_e^{0.5}$
R11	$Ar + e^- \leftrightarrow Ar^+ + e^- + e^-$	$2.50 \cdot 10^{34}$	-3.82	181700	$T^{0.5}T_e^{0.5}$
	Exchange Reactions: Ions				
R12	$NO^+ + O_2 \leftrightarrow O_2^+ + NO$	$2.40 \cdot 10^{13}$	0.41	32600	T
R13	$O_2^+ + N_2 \leftrightarrow N_2^+ + O_2$	$9.90 \cdot 10^{12}$	0	40700	T
R14	$O_2^+ + N \leftrightarrow N^+ + O_2$	$8.70 \cdot 10^{13}$	0.14	28600	T
R15	$O_2^+ + O \leftrightarrow O^+ + O_2$	$4.00 \cdot 10^{12}$	-0.09	18000	T

Table 22. (continued)

R16	$N^+ + N_2 \leftrightarrow N_2^+ + N$	$1.00{\cdot}10^{12}$	0.50	12200	T
R17	$O^+ + N_2 \leftrightarrow N_2^+ + O$	$9.10{\cdot}10^{11}$	0.36	22800	T
R18	$O^+ + NO \leftrightarrow N^+ + O_2$	$1.40{\cdot}10^{5}$	1.90	15300	T
R19	$NO^+ + N \leftrightarrow N_2^+ + O$	$7.20{\cdot}10^{13}$	0	35500	T
R20	$NO^+ + N \leftrightarrow O^+ + N_2$	$3.40{\cdot}10^{13}$	-1.08	12800	T
R21	$NO^+ + O \leftrightarrow O_2^+ + N$	$7.20{\cdot}10^{12}$	0.29	48600	T
R22	$NO^+ + O \leftrightarrow N^+ + O_2$	$1.00{\cdot}10^{12}$	0.50	77200	T

Thermal Non-Equilibrium Model

Generally speaking, a three temperature model is suggested: (T, T_v, T_e) [8][10]. T for translational and rotational mode for heavy species; T_v for vibrational modes, same temperature for all molecular species and T_e for translational mode of free electron.

In the present CFD computations, however, only two temperatures model for thermal non-equilibrium model (e.g., T and T_{vib}) has been considered. Vibrational relaxation is modelled using a Landau-Teller formulation, where relaxation times are obtained from Millikan-White with the high temperature correction Park [8].

NUMERICAL RESULTS

CFD design analysis for each TP is presented in this section in terms of flowfield visualisations, surface distributions of pressure, heat flux, temperature and shear stress, stagnation line analysis, and local and global dimensioning parameters for the capsule's design.

To this in mind flight regime assessment was made in order to be sure that all TPs lie within the flow continuum regime to be able to use CFD analysis without slip condition at wall.

Flight Regimes Assessment

Generally speaking, along the atmospheric re-entry trajectory going from upper to lower altitude, three main flow regimes can be recognized [10]. They are the Hypersonic-Supersonic regime, concerning the upper part of descent trajectory, the Transonic regime and the Subsonic regime, regarding the flight within the lower atmospheric layers. Moreover, within the Hypersonic regime two main and heavily overlapping regimes apply. They are the rarefied regime and the continuum regime.

Once SRV vehicle started its descent, the atmospheric density is low enough such that the continuum assumption breaks down and one must begin to consider the general macroscopic mass, force, and energy transfer problem at the vehicle surface. In this regime, two distinctions can be made. They are the re-entry through the upper part of the atmosphere (above 90 to 100 km) and that through the lower part of the Earth atmosphere. In the former the disturbed free molecular regime is completely established. It must be noted that the free molecular regime exists above approximately 130 km. For the latter the slip conditions and temperature jump have to be introduced in the continuum approach to take into account for the rarefaction effects.

When the air density becomes rarefied enough the molecular mean free path (λ) can become as large as the scale of the body itself. This condition is known as Free Molecule Flow (FMF) regime. In this condition, the aerodynamic characteristics of the vehicle are determined by individual, scattered molecular impacts, and must be analyzed on the basis of Kinetic Theory. Therefore, the SRV concept, moving from a very rarefied atmosphere (at entry interface) to a denser atmosphere, shifts from the FMF regime, where individual molecular collisions are important, to the transition one, where slip effects are important, and then to the continuum regime, as represented in Figure 84.

For instance, the similarity parameter that governs these different flow regimes is the Knudsen number, defined as (rarefaction parameter):

$$Kn_{\infty L_{ref}} = \frac{\lambda}{L_{ref}} = 1.25\sqrt{\gamma}\,\frac{M_{\infty}}{Re_{\infty L_{ref}}}$$

where L_{ref} (i.e., 1.1 m – capsule diameter) is the characteristic length of the body and γ is the specific heat ratio. As shown in Figure 84, the maximum Mach number (i.e., 41.54) evaluated by the SRV Flight Mechanics analysis lies in the continuum flow regime. Therefore, all CFD computation will be performed with Navier-Stokes approximation.

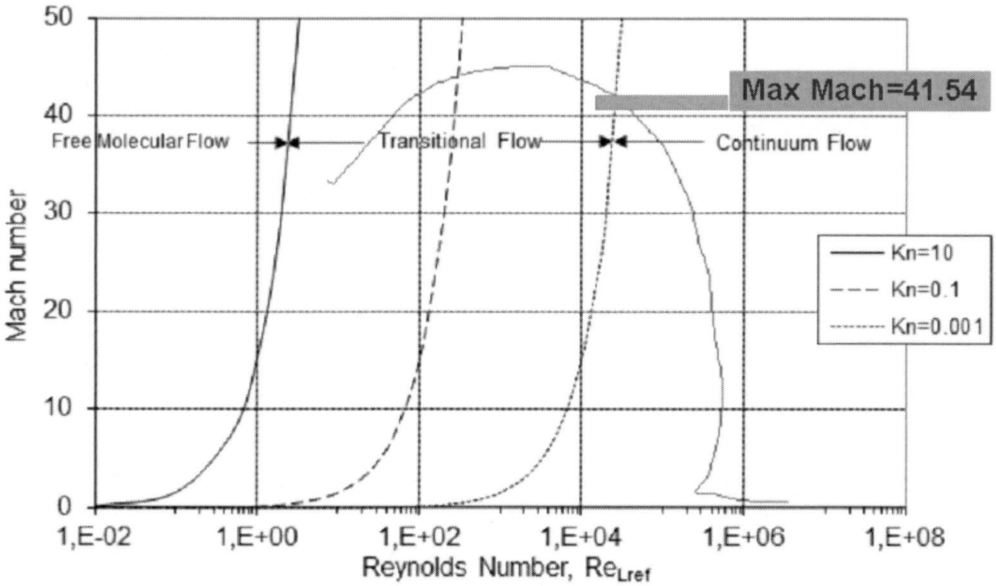

Figure 84. SRV re-entry trajectory in the Ma-Re map with constant Knudsen numbers.

Computational Domains, Wall Boundary Conditions and Convergence Histories

CFD computations have been carried out on multiblock structured grids generated with the commercial tool ICEM-CFD. Two kinds of computational domain, similar to those shown in Figure 85, have been considered.

The effective dimensions of the outflow boundary, axis and outer boundary are modified in each simulation in order to obtain a grid compliant to the flowfield conditions to simulate and are estimated by engineering-based methods. For example, the stand-off distance has been estimated by means of Seiff and Probstein relationship [8].

The strategy adopted was to build the volume mesh be large enough to accommodate the freestream Mach number; while the distribution of grid points in the wall-normal direction are driven by freestream Reynolds number. Hence, the distribution of surface grid points was dictated by the level of resolution desired in various areas of the capsule such as stagnation region and base fillet, according to the computational scopes. For example, wake flows are sensitive to the details of the volume grid construction and it is important that the grid be aligned to anticipated flow features. Furthermore, the grid has sufficient points in the shoulder region to capture the rapid expansion and accurately predict the flow separation point and the angle of the resulting shear layer. There are also enough points in the separated flow region to resolve the vertical structure at the beginning of the wake flowfield. All grids are tailored for the freestream conditions of the TPs summarized in Table 20.

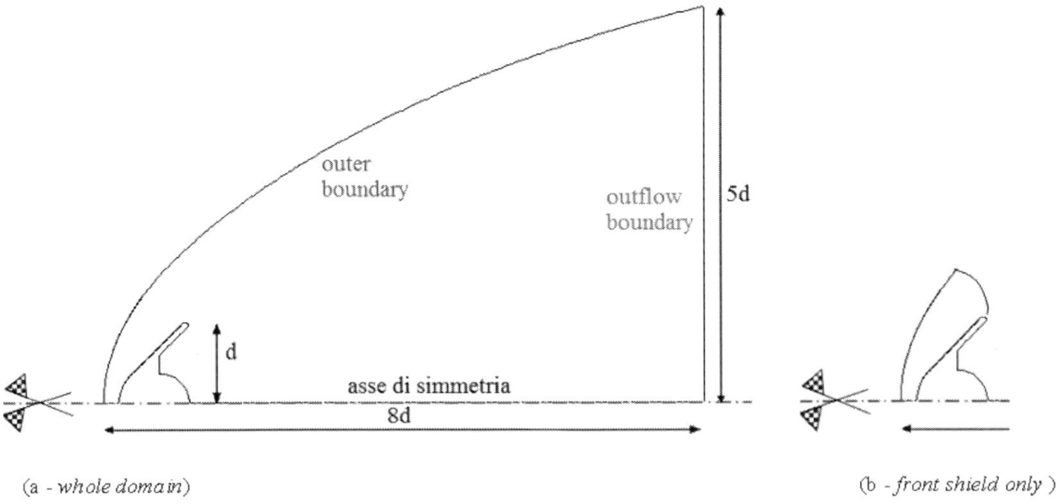

Figure 85. CFD computational domains: both whole domain (a) and front shield only (b).

Close-up views of 2-D axis-symmetric mesh for both whole and front shield domains can be seen in Figure 86 and Figure 87, respectively.

Grid refinement in strong gradient regions of flowfield was made through a solution adaptive approach. As shown, grids for whole computational domain take advantage of the *o-gird* domain decomposition technique around the capsule in order to have meshes block wrapping the SRV and with direction outgoing form the body wall, necessary for the boundary layer stretching.

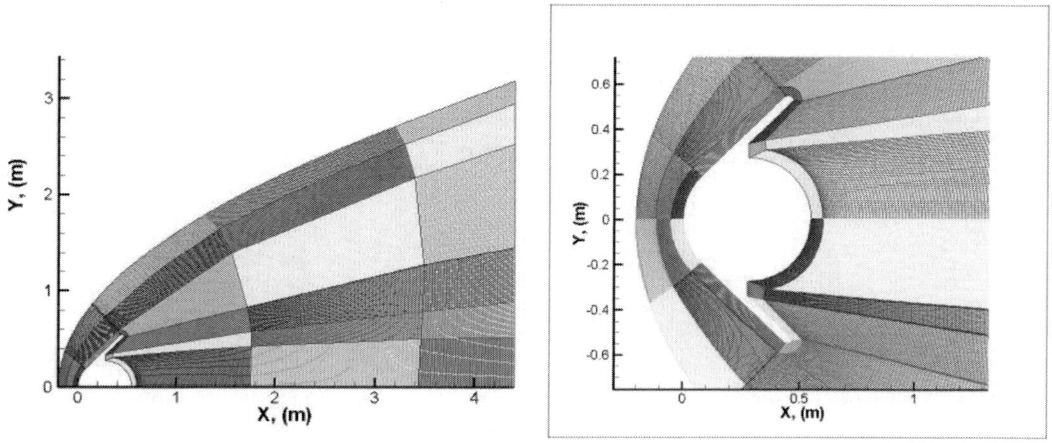

Figure 86. CFD multiblock whole computational domain.

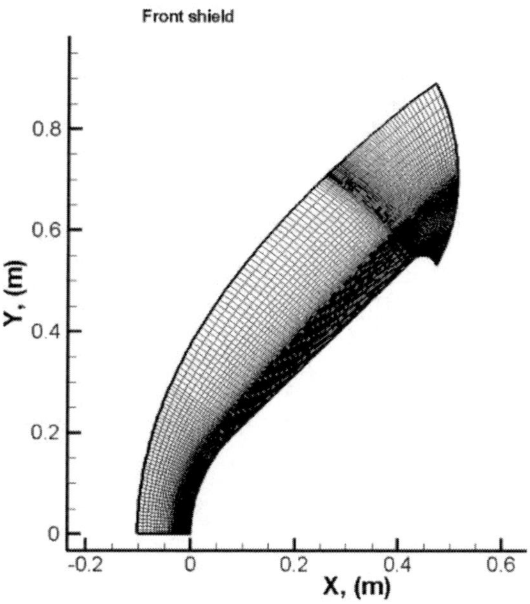

Figure 87. CFD multiblock front shield computational domain.

The grid (front shield case), as that shown in Figure 87, is constituted by 124 x 160 cells (longitudinal x normal to the wall direction) and assures fully spatially converged results. A local refinement has been done in the shock region in front of the capsule to better resolve the steep gradients and reduce possible numerical instabilities.

The generic grid used for whole domain numerical calculations has consisted of about 30 blocks for an overall number of about 72000 cells (half body). In the case of front shield only simulation the generic grid has consisted of 9 blocks for an overall number of about 20000 cells (half body).

In particular, the mesh was initially generated algebraically and then adopted as the solution evolved, aligning the grid with the bow shock and clustering points into the boundary layer. This reduced the spurious oscillation in the stagnation area that are often observed in hypersonic flows, especially for large bluff-body flowfield computations. The minimum spacing at the wall is equal to 10^{-6} m to accurately predict heat transfer at the vehicle surface. When the flow is in turbulent conditions grids use y^+ less than 1; y^+ is defined as:

$$y^+ = \frac{\eta\rho}{\mu}\sqrt{\frac{\tau_w}{\rho}}$$

where the quantities η and τ_w are the surface normal distance and the shear stress, respectively.

All CFD results refer to both converged and grid independent computations. Mesh sensitivity analyses have been carried out on three levels of the structured multi-block grid (L_1, L_2, L_3) and the Richardson extrapolation criteria has been applied [11]. The medium grid (L_2) is obtained by halving the size of cells of the coarse (L_1) grid. Similarly, the finer grid (L_3) is obtained halving the size of cells of the medium grid.

As far as wall boundary conditions are concerned, all simulations are performed with fully catalytic wall assumption i.e. chemical equilibrium at the wall to ensure the same wall conditions for all computations. In particular a wall temperature of T_W=1000K (i.e., isothermal wall) has been also imposed.

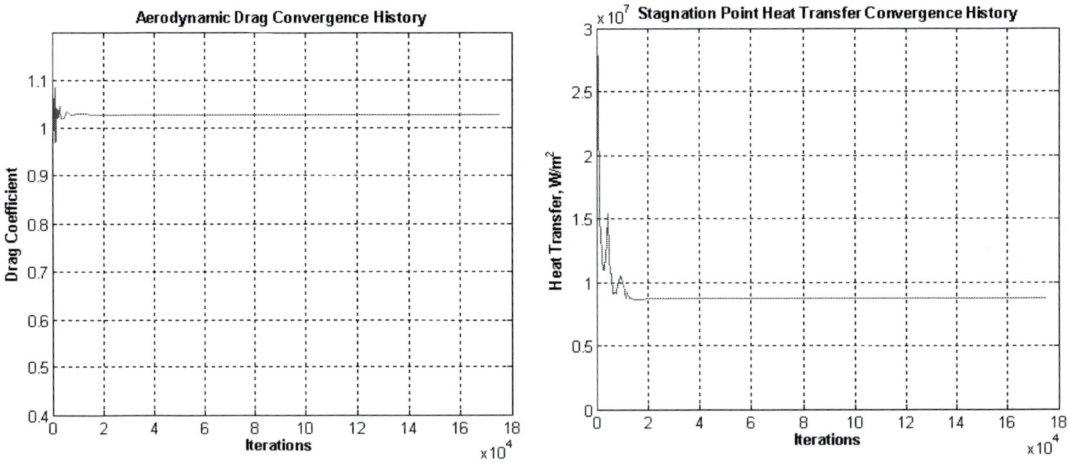

Figure 88. Drag coefficient and stagnation point heat flux vs iterations for M1 simulation.

In order to assess numerical solution convergence, equation residuals and aerodynamic coefficient (i.e., C_D) as well as the stagnation point heat flux have been monitored during iterations. Solution convergence is assumed when equation residuals drop more than three orders of magnitude and both the aerodynamic coefficient and the stagnation point heat flux plots are flat for enough time. For example, Figure 88 shows the aerodynamic and stagnation point heat flux convergence histories for the M1 simulation.

Finally, the convergence histories obtained for all the conditions foreseen in the CFD test matrix are quite similar and, therefore, are left for brevity.

Flowfield Analysis

In the hypersonic portion of the re-entry, the flowfield is dominated by a strong bow shock and is characterized by a thin shock layer. The pressure and Mach number contours for M_{∞}=3.02 (i.e., M6 TP) are shown in Figure 89.

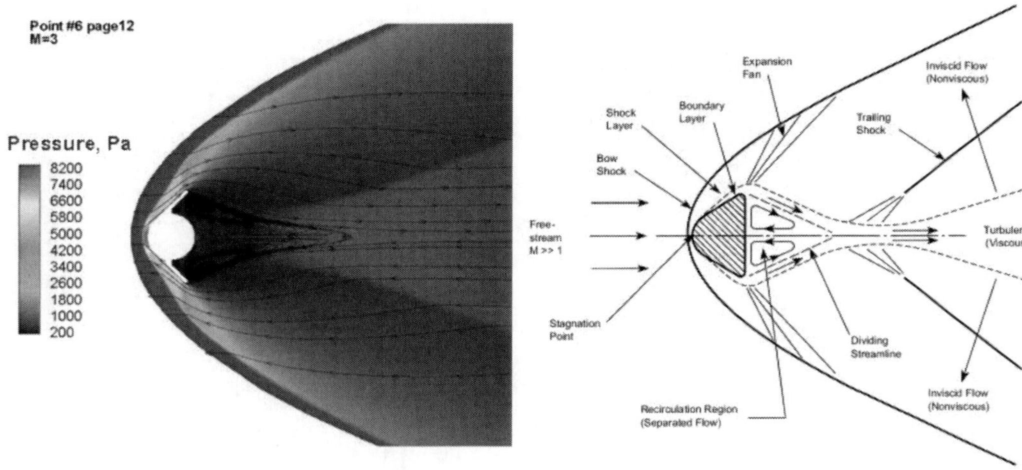

Figure 89. Static Pressure field at M6 TP. Characteristics of hypersonic flow around a blunt body (right side).

Figure 89 also shows, on the right side, characteristics of hypersonic flow around a blunt body as a reference [12]. As one can see, present CFD result is characterized by all the typical hypersonic flowfield features as, for example, shock waves very close to the body surface, thick boundary layers, high temperatures, and aerodynamic coefficients that can be nonlinear functions of angle of attack and etc [8]. In particular, numerical results in figure highlights a strong bow shock in front of the capsule well resolved by the adopted spatial discretization. This strong shock wave causes a large sonic region (see Figure 90), a smooth conical flow along the SRV conical part, and a strong flow expansion at the shoulder. The high angle of cone after the rounded stagnation region (equal to 45 deg) causes the curvature of the shock wave and, therefore, the presence of an entropy layer that affects the results at the wall, as it will be clarified hereinafter. Streamlines are also shown in Figure 89 in order to appreciate the flowfield structure surrounding the vehicle, especially the complexity of the wake flow. A strong base expansion together with a vortex feature can be appreciated as well. In particular, at the rearward facing base of the body the flow separates and creates a region of recirculating flow bounded by dividing streamlines.

Velocity and Mach flowfield contours at M6 TP are summarized in Figure 90.

Figure 91 shows the Mach number and static temperature contours at M5 TP. As one sees, the flow crossing the bow shock suddenly decelerates thus increasing the pressure (see for example Figure 89) and temperature in the shock layer close to the stagnation region.

Figure 91 also displays flow streamlines in order to highlight the complex flow recirculation that takes place in the wake. Here, we can see clearly the shear layer that separates the recirculation zone by the rest of the flowfield, the recompression wave that redirects the shear layer in the direction of the farfield, and the so called wake-core zone (in the terminal zone of the wake), where the flow expands

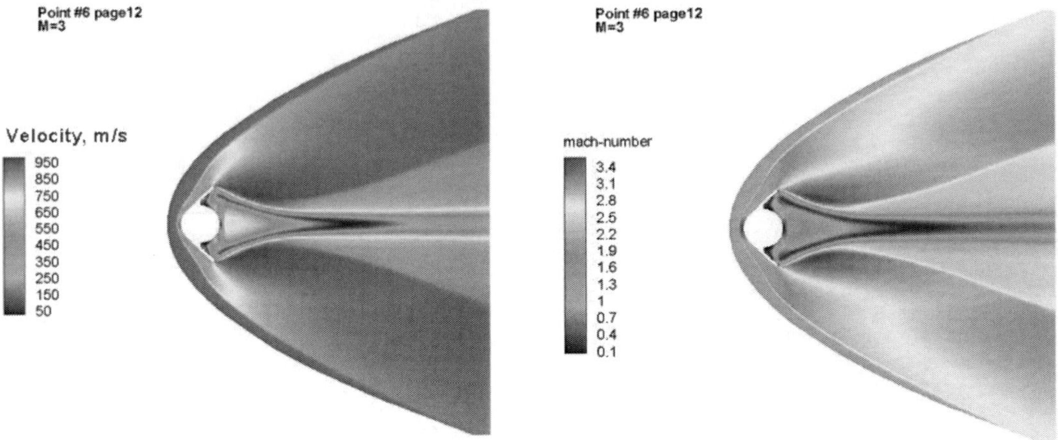

Figure 90. Velocity and Mach flowfield contours at M6 TP.

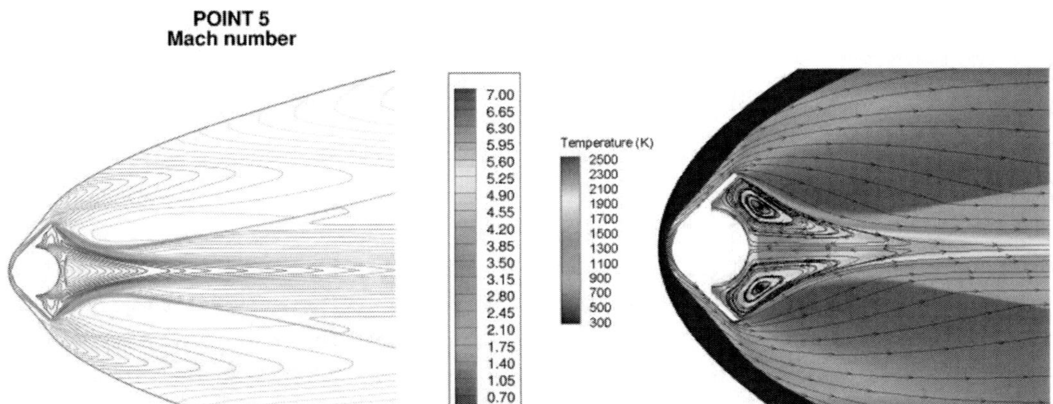

Figure 91. Mach number and static temperature contours at M5 TP.

The Mach number and the static temperature fields past the SRV at M4 TP are shown in Figure 92.

Figure 92 highlights a very strong shock wave in front of re-entering SRV. As a consequence, the air crossing the bow shock is suddenly heated thus reaching a temperature of the order of 8000 K. It also shows that the shear layer, starting from the SRV shoulder, converges on the capsule axis, undergoing normal shock at the neck of this flow, from which develops an oblique trailing shock wave, ultimately forming a viscous core or inner wake. Fluid in the inviscid wake crosses the trailing shock, increasing pressure, temperature, and density thereby, and in continuing downstream this outer wake merges with the inner wake.

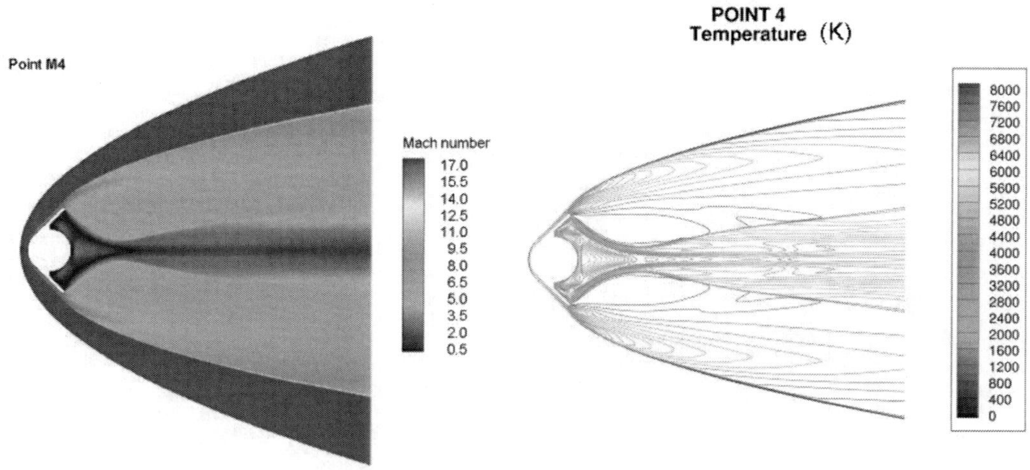

Figure 92. Mach number and static temperature contours at M4 TP.

Illustrated in Figure 92 is also the presence of a rapid expansion as the highly compressed gas flows around the shoulder of the vehicle. This expansion, dominated by inviscid effects, determines the rapidly lowering of the gas translational temperature, density and pressure, while the flow chemical state and temperatures, that characterize the energy in the internal modes, tend to remain frozen and the gas is still dissociated and excited (see Figure 93). This aspect is more important by capsule aeroheating point of view.

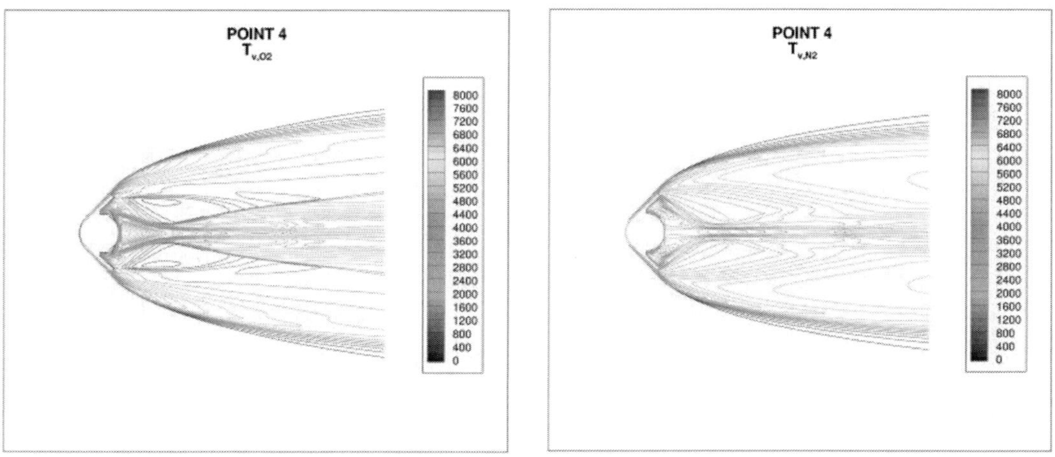

Figure 93. O_2 and N_2 vibrational Temperatures at M4 TP.

In fact, as the gas flows downstream, because of recombination occurs slowly, the vibrational temperature rises still higher with the consequence that the gas can radiate significantly in the after body region.

Figure 94 shows the O_2 and N_2 mass fractions contours past the SRV at M4 TP. The distributions of flowfield species along with the stagnation streamline are reported on the right side of figure. As one can see, at M4 flight conditions Oxygen completely dissociates while Nitrogen starts to dissociate. A comparison among translational and vibrational temperatures is reported in Figure 95 for the point M4.

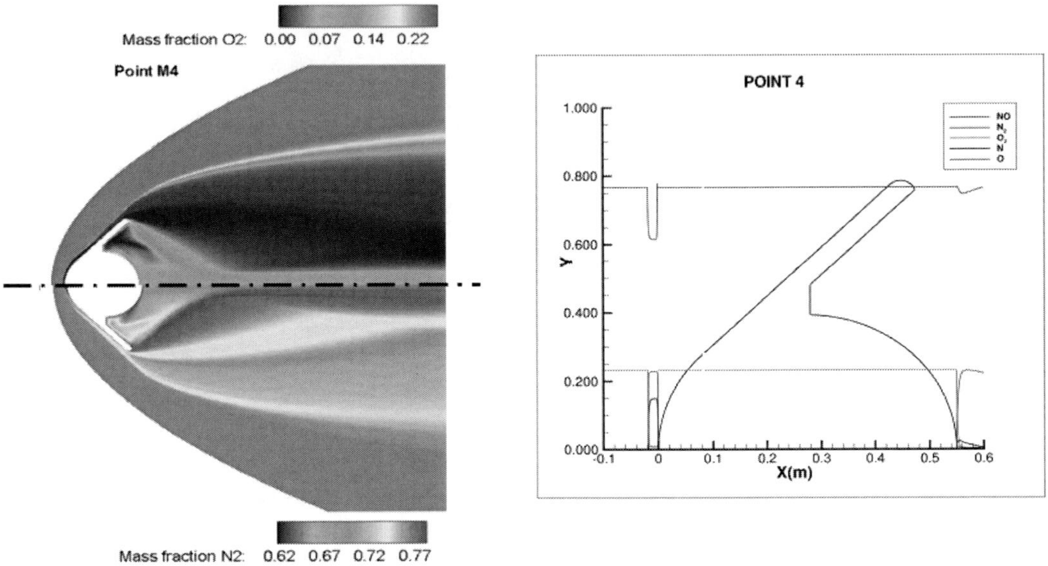

Figure 94. Species mass fractions field at M4 TP.

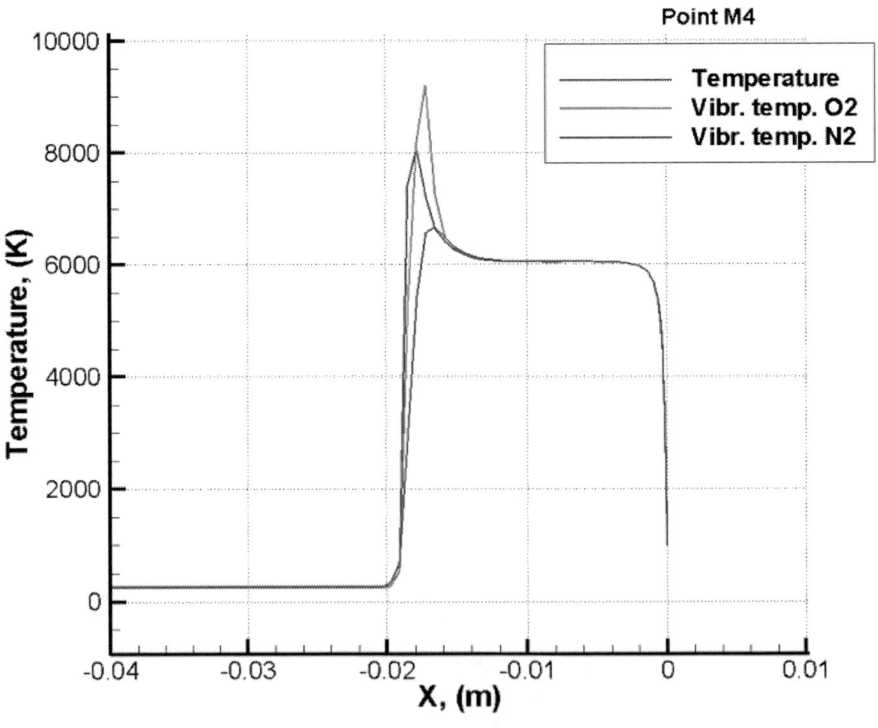

Figure 95. Temperature profiles along with SRV stagnation line at M4 TP.

As shown, the vibrational temperature profiles show that the thermal non-equilibrium region extends only close to the shock region. In fact even if Tv increases much more slowly since it is density dependent, the two temperatures (T,Tv) are nearly equilibrated throughout the shock layer.

The Mach number and temperature contour fields past the SRV at M2 TP is shown in Figure 96.

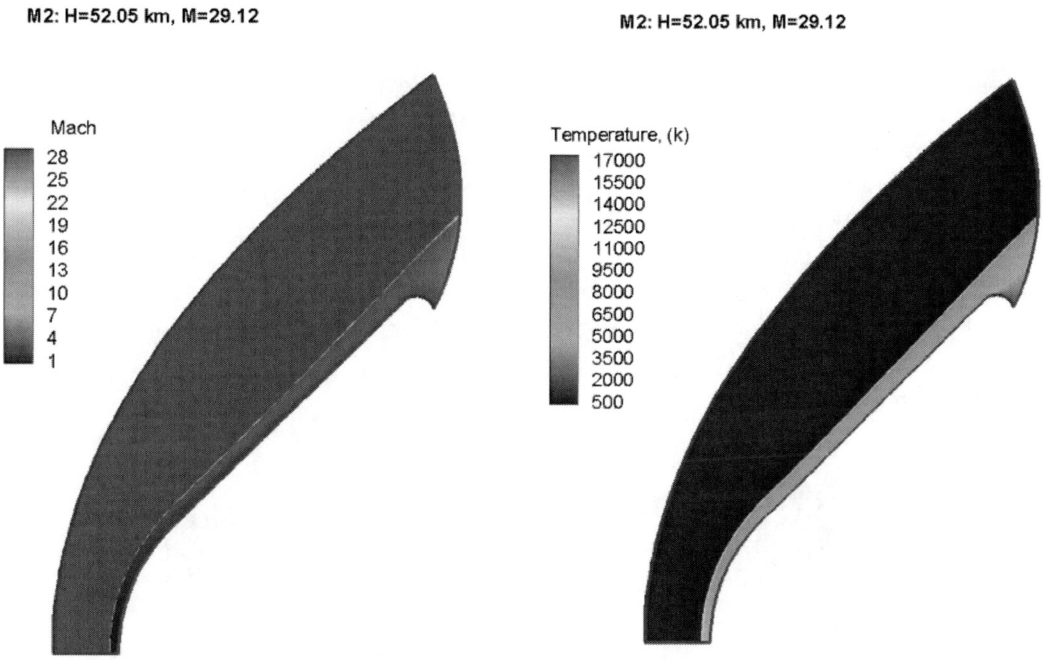

Figure 96. Mach number and Temperature contour fields past the SRV at M3 TP.

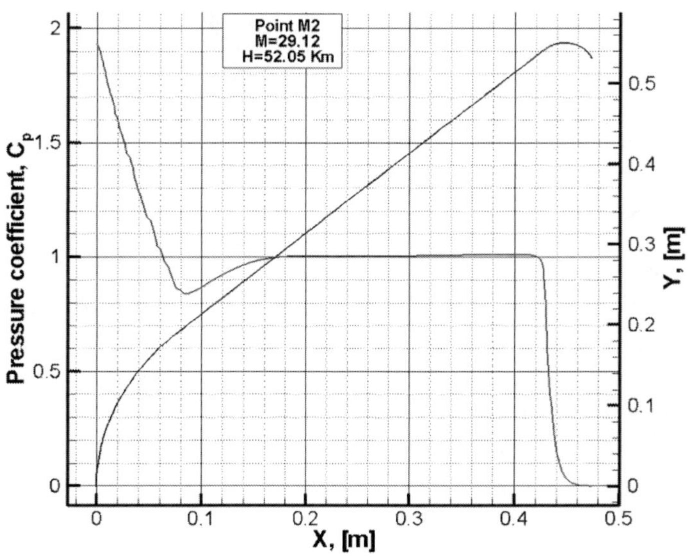

Figure 97. Pressure coefficient at M2 TP.

Note that pressure distributions (and thus aerodynamic forces) are affected by the sonic character of the shock layer. When supersonic, the pressure distributions on the conical flank are flat, which is characteristic of conical flow, when the entire forebody shock layer is subsonic, the elliptic nature of that flow results in higher, more rounded distributions. The pressure coefficient profile along with the SRV centreline is given in Figure 97. As one can see, the effect of local entropy values is clearly shown.

Indeed, the entropy layer, due to the curvature of the bow shock, swallowing causes the decreasing of pressure, at the sphere cone junction, up to reach the quite constant value of pressure at the conical skirt.

Figure 98 shows the static pressure and Mach number fields characterising the shock layer at M1 flight conditions. As shown, a very strong shock wave is generated in front the capsule due to the very high Mach number (i.e., M_∞=34.45). Indeed, as the capsule enters into the Earth's atmosphere directly from the hyperbolic earth-return trajectory, the atmospheric entry speed of SRV is over 12 km/s (see Figure 80). Therefore, the shock layer gas is much more highly heated that the case of a normal earth orbital re-entry. As a consequence, a large contribution to the vehicle aeroheating comes from the radiative heating of the plasma flow within the shock layer. This must be taken into account for the SRV TPS design.

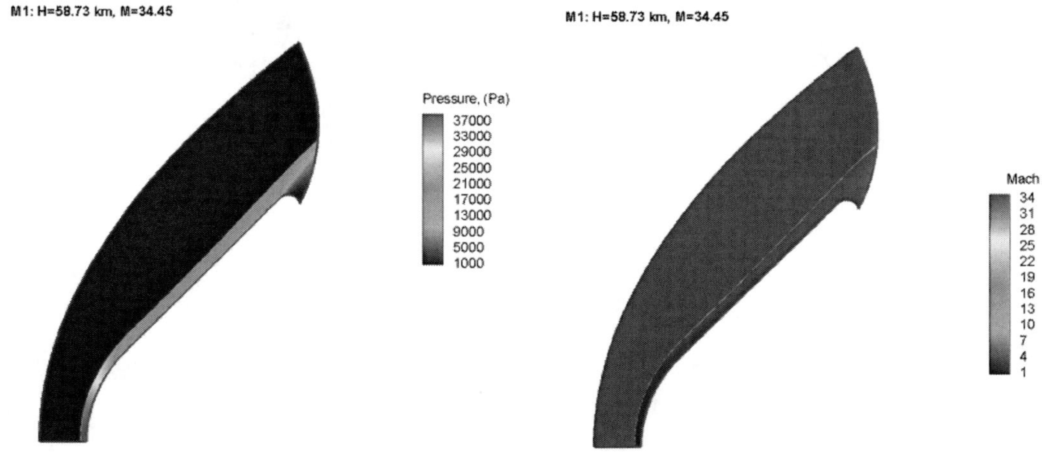

Figure 98. Static Pressure and Mach flowfields past SRV at M1 TP.

The strong bow shock shown in Figure 98 causes molecules of atmosphere to be dissociated and ionized, and consequently the shock layer gas consists of molecules, atoms, ions and electrons. Pressure and temperature distributions along with the capsule's stagnation line are given in Figure 99, in which translational and vibrational temperatures distributions are also highlighted.

As one can see, the temperature behind the bow shock is very high thus causing the complete Oxygen and Nitrogen dissociation inside the shock layer, as Figure 100 and Figure 101 clearly illustrate. In particular, these figures also highlight that a large portion of the shock layer is in thermo-chemical equilibrium (i.e., the temperature profiles are quite flat until the boundary layer is reached) and only near the shock and in the boundary layer there is a departure from this state: the non-equilibrium region is just downstream the bow shock wave, and the size is of the order of the shock wave thickness. Moreover, the temperature trends exhibit a sharp discontinuity at the shock wave with a rapid decreasing behind the bow shock due to finite rate dissociation of molecules. For instance, the translational-rotational temperature, T, reaches the maximum value at about x=14 mm, while the vibrational-electronic temperature, T_v, is still much lower than T. T_v begins to be equilibrated around x=10 mm and continues to be equilibrated until the surface. On the contrary, the thermal non-equilibrium is observed at the region adjacent to the equilibrium one; i.e., from x=14 mm to 10 mm. The equilibrated temperature amounts up to about 11000 K.

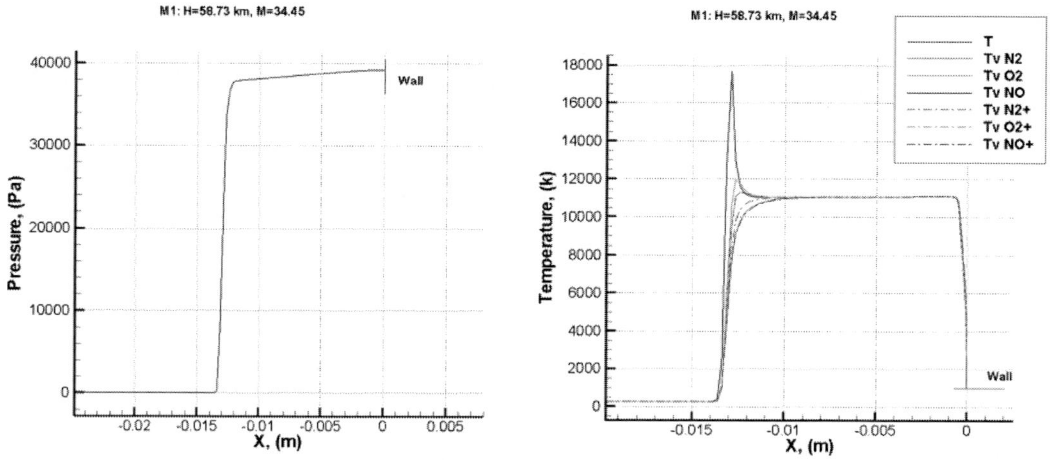

Figure 99. Pressure and Temperatures along with the stagnation line at M1 TP.

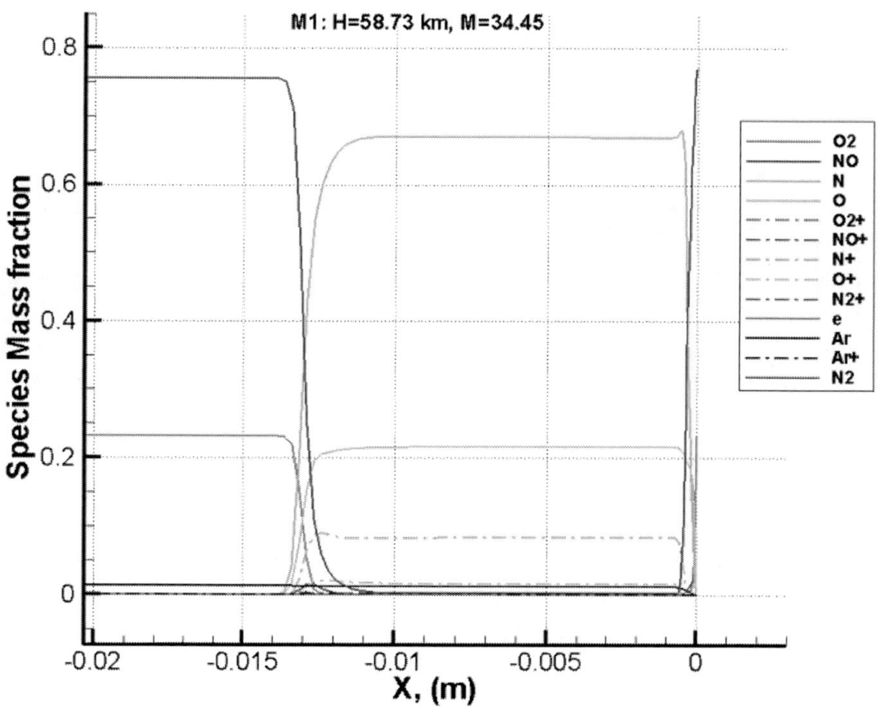

Figure 100. Species mass fraction along with SRV stagnation streamline at M1 TP.

Figure 100 shows the distribution of the mass fraction of the neutral and ionized species along the stagnation line. As shown, N_2 and O_2 molecules dissociate rapidly and, successively, the generation of NO molecule, the atomic nitrogen and oxygen, and the ionization of the molecules such as N_2, O_2 and NO, occur rapidly in the thermal non-equilibrium region.

In general the level of each formed species reaches a state of near chemical equilibrium for a large portion of the shock layer. Then, at the edge of the boundary layer the levels of the species start to change rapidly again as the temperature falls and density rises through the boundary layer: N and O recombine to their molecular forms resulting in an increase in the levels of N_2 and O_2 and a fall in the levels of N and O. The reason for this is that the ionised species recombine with the electrons to form neutral species of N, O, N_2 and O_2, and NO, as shown in Figure 101.

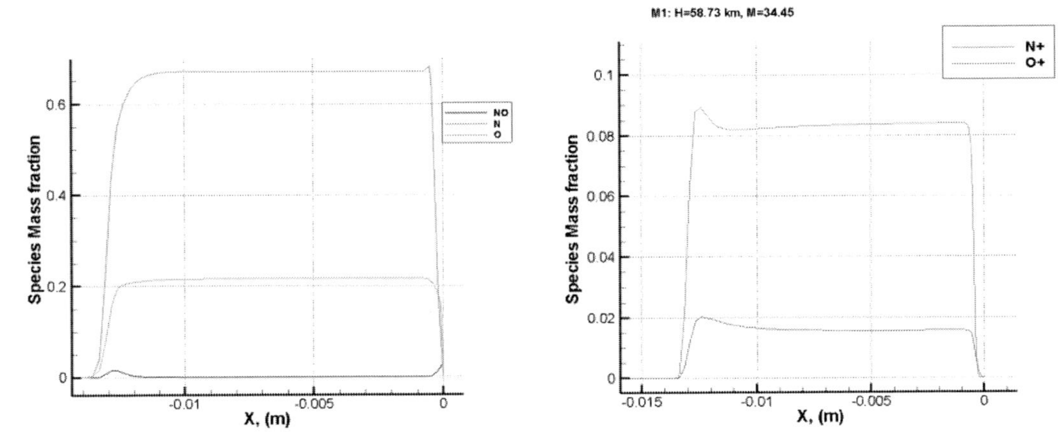

Figure 101. Species mass fractions along with the stagnation line at M1/R3 TP.

Concerning the ionic species, there are substantial levels of O+ and N+ while very low mass fractions of N_2+, O_2+ and NO+. In fact, crossing the shock, O_2 is highly dissociated forming O and O+. N_2 is dissociated to form N and NO by recombination with O but a small fraction of NO is created. A large part of the atomic nitrogen produced by the dissociation of N_2 is ionized in N+. In the Figure 101 O+ and N+ mass fractions are reported. As one can see, the mass fraction of atomic ions such as N+ and O+ amount up to around 0.085 and 0.015, respectively. Note that the ionisation process is very important considering that it has a not negligible impact on radiative heat flux at high velocity flight conditions.

Finally, heat fluxes distribution at wall for the first 3 computations of the CFD matrix, i.e. the variation along the trajectory being constant all the boundary conditions, is shown in Figure 102.

As shown, the trajectory point M2, i.e. the point characterized by the maximum aeroheating, the heat flux is equal to about 8750 kW/m^2 at the stagnation point and 3500 kW/m^2 at the shoulder, where the boundary layer thickness decreases for the effect of the strong expansion.

The skin friction coefficient (C_f) over SRV centreline is summarized in Figure 103 for several trajectory points. As one can see, the peak of C_f is confirmed at the shoulder where boundary layer is extremely thin due to flow rotation around it.

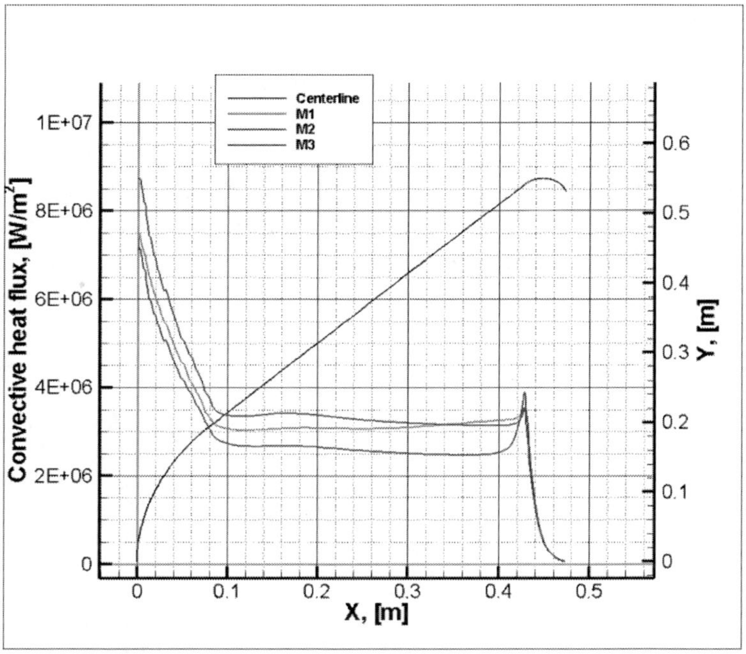

Figure 102. Convective heat flux for M1, M2, and M3 TP.

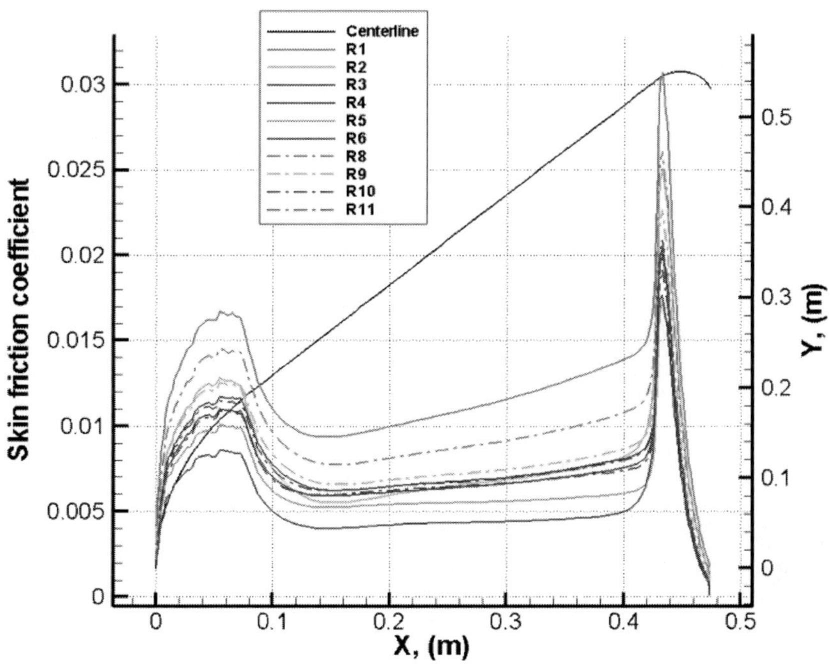

Figure 103. Skin friction coefficient for several R trajectory points.

With the density, the molar fractions and the two temperatures (translational and vibrational) of all the R trajectory points coming from CFD simulations, the radiation heat flux has been computed with PARADE [4].

Results in terms of radiative and convective heat fluxes at the wall are shown in Figure 104. As one can see, the radiative heat flux is greater at the stagnation point, reaching a value of about 6000 kW/m^2 at the R3 TP, and then it decreases along with the SRV front shield. Therefore, this additional contribution of surface heating must be taken into account in designing the SRV thermal protection system.

In particular, after the peak heating at the stagnation point the heat flux decreases along with the surface as the boundary layer develops up to an inflection point that corresponds at the end of the spherical shape of the capsule. Along the conical part it continues to decrease with a different shape and then it increases near the shoulder due to the small radius of curvature and the expansion that causes a reduction of the boundary layer thickness.

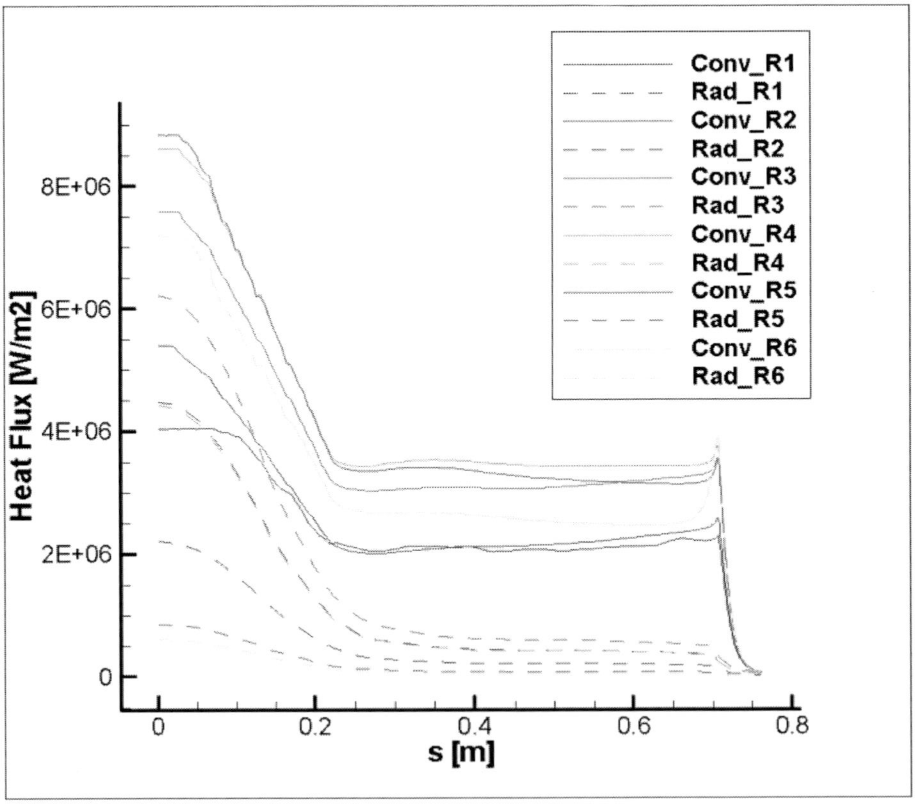

Figure 104. Convective and radiative heat fluxes comparison for R1-R6 TPs.

Radiative heat flux profiles for the remaining R7 to R12 cases are quite the same as those of R1 to R6 reported in Figure 104 and, therefore, are left for brevity. Finally, it is worth to note that no full coupled computations were undertaken since they represent a very prohibitive computational effort.

SRV Aerodynamics

A summary review of SRV aerodynamics is recognized both for engineering-based and CFD-based approaches, respectively.

In the following, the aerodynamic analysis is shown in term of lift (C_L), drag (C_D), and pitching moment (C_M) coefficients (only drag in the CFD analysis) that are calculated according to the following equations:

$$C_L = \frac{L}{\frac{1}{2}\rho_\infty v_\infty^2 S_{ref}} \; ; \; C_D = \frac{D}{\frac{1}{2}\rho_\infty v_\infty^2 S_{ref}} \; ; \; C_M = \frac{M_y}{\frac{1}{2}\rho_\infty v_\infty^2 L_{ref} S_{ref}}$$

The geometric reference parameters that have been chosen in order to make aerodynamic forces non-dimensional coefficients are:

- S_{ref} =0.95 [m^2] (i.e., maximum SRV cross section area);
- L_{ref} =1.1 [m] (i.e., maximum SRV diameter).

Engineering-Based Aerodynamics of SRV

In the present analysis the Surface Impact Methods (SIM) typical of hypersonics, such as Prandtl-Meyer expansion flow and Modified Newtonian theories are widely applied [10]. Analysis has been performed on the whole Mach number profile of re-entry.

Figure 105 shows a typical mesh surface of SRV that has been used for the engineering-level computations.

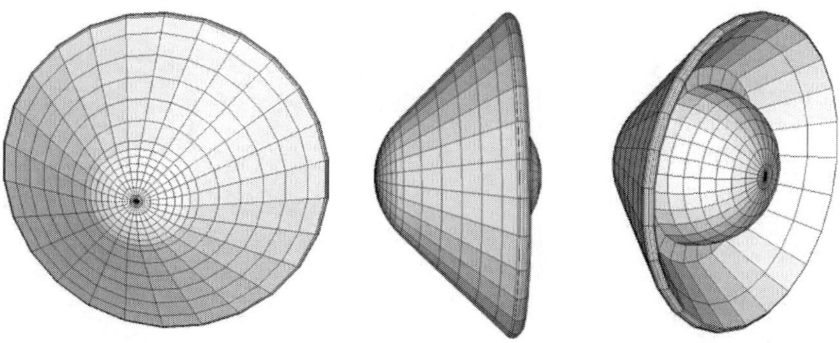

Figure 105. SRV panel mesh.

As an example of engineering level results, the static pressure distributions over the wetted vehicle surface for M_∞=25 and for three different angles of attack: 0, 5 and 10 deg are summarized in Figure 106, respectively.

As shown, when the AoA increases the surface pressure distribution changes thus increasing on the capsule windside, as expected. At the same time flow expansion on the capsule leeside determines locally lower pressure contours.

The lift and drag coefficients for AoA ranging from 0 to 180 deg, in hypersonic conditions, are summarized in Figure 107.

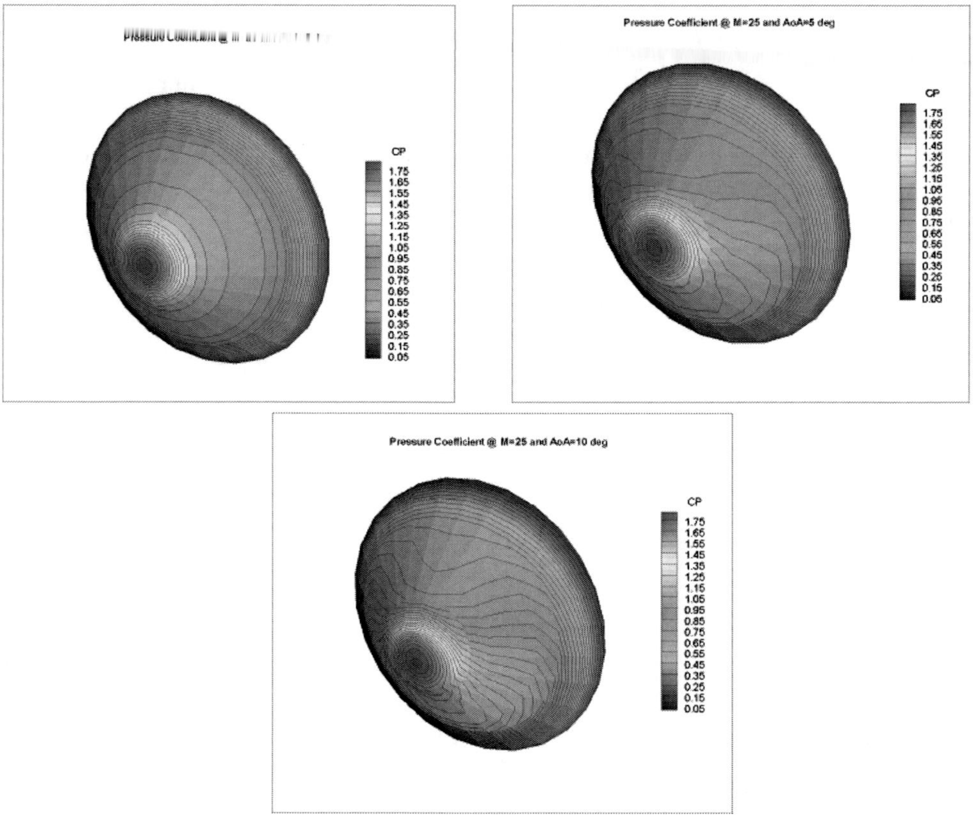

Figure 106. Pressure coefficient at M_{∞}=25 and AoA=0, 5, and 10 deg.

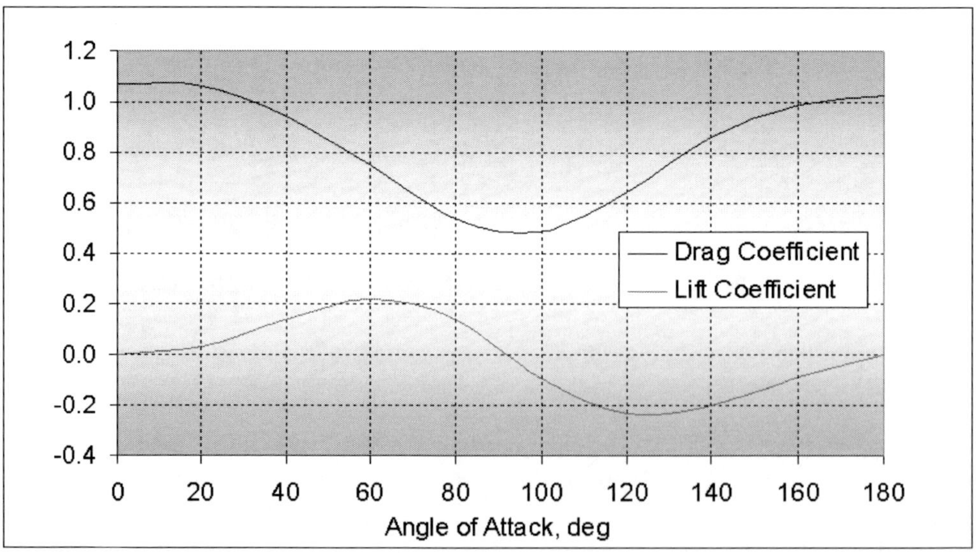

Figure 107. Lift and Drag coefficients of SRV in hypersonic regime.

Results provided for aerodynamic drag agree with analytical results available for blunt cone at hypersonic speed. Indeed, for a blunt cone with a semivertex angle θ_c, nose radius R_N, and base radius R_B, the drag coefficient reads [12]:

$$C_D = 2\sin^2\theta_c + \left(\frac{R_N}{R_B}\right)^2 \cos^4\theta_c$$

C_D is shown as a function of bluntness ratio R_N/R_B and cone-section semivertex angle θ_C in Figure 108.

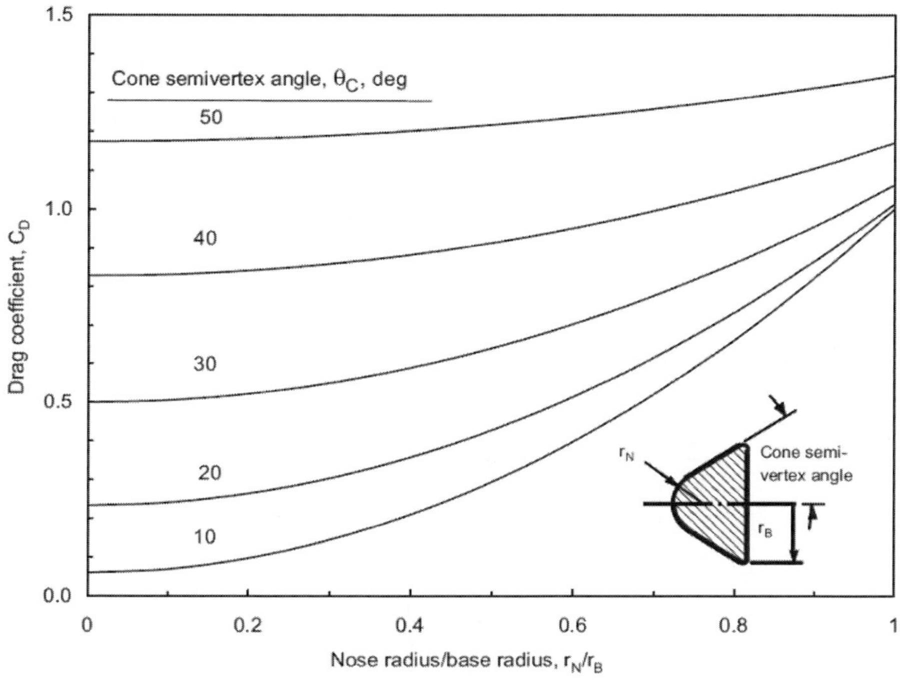

Figure 108. Hypersonic drag coefficient for sphere-cones.

In the present case the nose to base radius ratio is equal to 0.5. Therefore, analytical and engineering-based results compare quite good each other. Moreover, preliminary aerodynamic computations in hypersonic regime were carried out to study also aerodynamic backward static stability of the probe during re-entry trajectory with a center-of-gravity located at 26.5 % of diameter.

Indeed, Figure 109 shows the pitching moment coefficient of SRV in hypersonic regime for AoA varying from 0 to 180 deg. As illustrated in Figure 109, the chosen aeroshape has two trim angles-of-attack where the pitching moment at the center of gravity is equal to zero (Cm=0 at AoA=0 and 180 deg). But, for the 180 deg AoA, the slope of the pitching coefficient (Cm_α) is positive and consequently the aeroshape has only one single static stable position for AoA=0 deg. This ensures that the probe will not perform a backward re-entry in hypersonic regime.

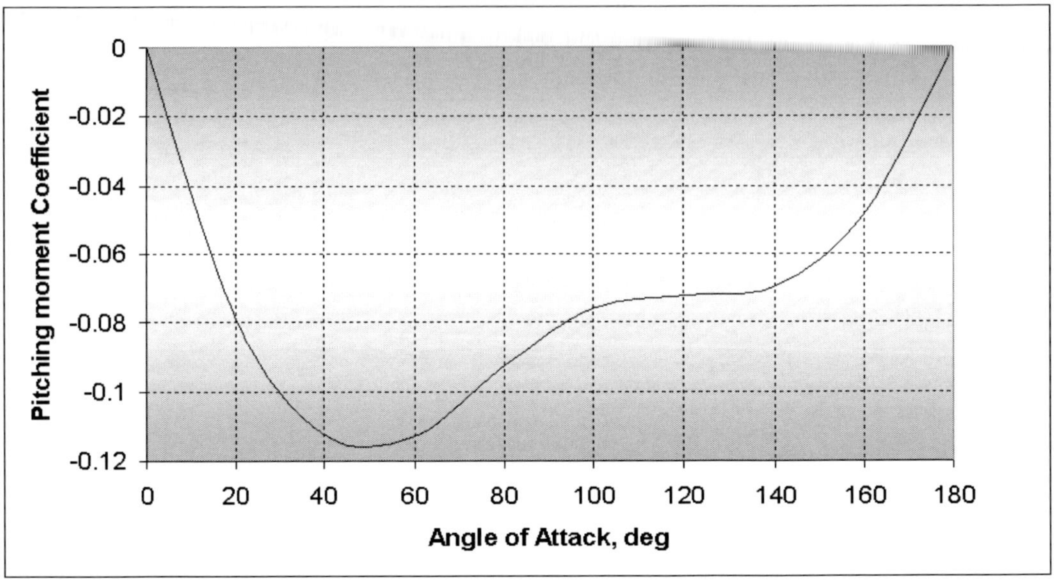

Figure 109. Pitching moment coefficient of SRV in hypersonic regime.

CFD-Based Aerodynamics of SRV

Capsule drag coefficients have been evaluated for all performed CFD computations. In particular, starting from the value of pressure and shear stress at the wall, the drag coefficient has been computed by integrating the contribution of all the elements. The results are summarized in Table 23. As one can see, drag break-down is also reported for those computations performed for whole flowfield domain (i.e., for both SRV front and back shield). The SRV front shield (FS) and back shield (BS) used for drag break-down, are also shown in the table.

Table 23. SRV drag coefficients

Reference parameters:
S_{ref}=0.95 m²
(maximum capsule cross area)
L_{ref}=1.1 m
(capsule diameter)

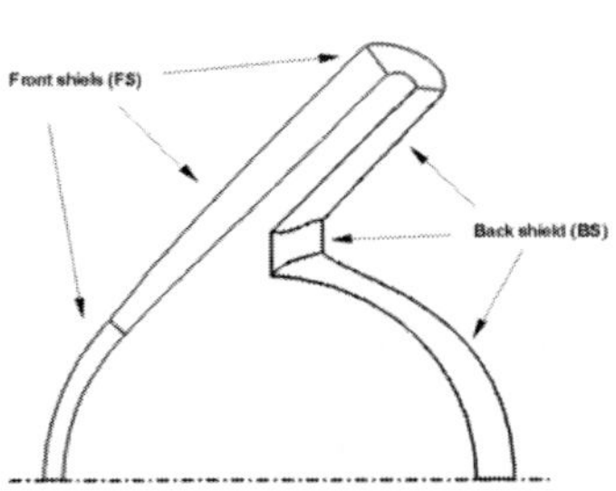

			Drag Coefficient				
			CDf		CDp		CDtot
Points	Mach	Mesh	FS	BC	FS	BS	
M1	34.45	FS	0.0131	na*	1.0196	na	1.0327
M2	29.12	FS	0.0072	na	1.0219	na	1.0291
M3	25.11	FS	0.0055	na	1.0269	na	1.0324
M4	17.61	FS+BS	0.0041	0.0004	1.0276	0.0029	1.0350
M5	7.00	FS+BS	0.0031	0.0008	1.0436	0.0204	1.0679
M6	3.02	FS+BS	0.0024	0.0010	1.1570	0.1111	1.2716
R1	41.54	FS	0.0206	na	1.0092	na	1.0298
R2	37.45	FS	0.0155	na	1.0095	na	1.0250
R3	34.45	FS	0.0131	na	1.0196	na	1.0327
R4	32.16	FS	0.0101	na	1.0172	na	1.0273
R5	29.12	FS	0.0072	na	1.0219	na	1.0291
R6	25.11	FS	0.0055	na	1.0269	na	1.0324
R7	40.14	FS	0.0231	na	1.0133	na	1.0364
R8	38.70	FS	0.0196	na	1.0121	na	1.0317
R9	35.87	FS	0.0138	na	1.0107	na	1.0245
R10	33.07	FS	0.0114	na	1.0170	na	1.0284
R11	30.50	FS	0.0087	na	1.0203	na	1.0290
R12	22.07	FS	0.0070	na	1.0283	na	1.0353

*na= NOT AVAILABLE

As one sees, at very high Mach number conditions, there is a negligible effect of BS on the drag coefficient estimation, as expected. This is due to the fact that the greater contribution to this coefficient is represented by the wave drag due to the pressure acting on the SRV forebody.

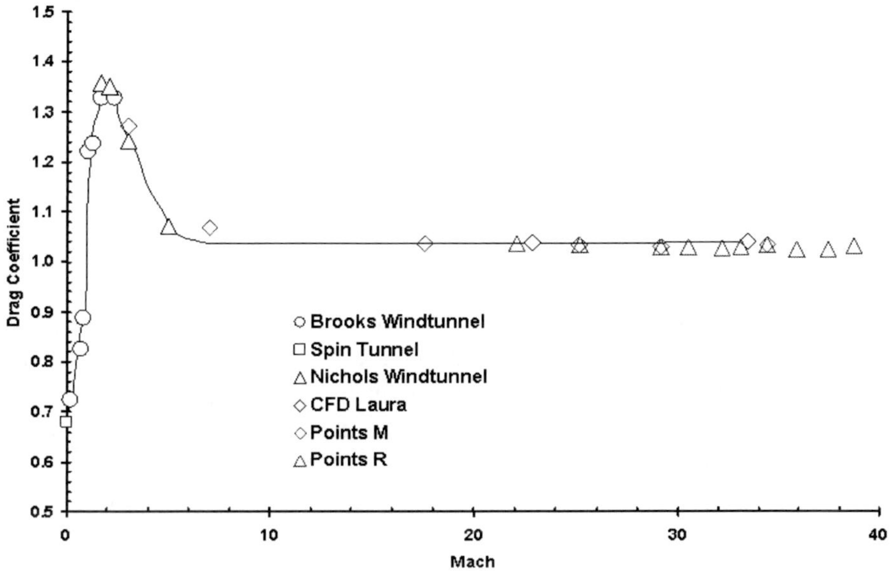

Figure 110. Drag coefficient of SRV comparison among present and literature results.

Comparison among present CFD results and literature data is shown in Figure 110 [1]. It presents an approximation of the drag coefficient across the Mach range for the 45-degree half-angle shape. The hypersonic value, 1.07, was computed at Mach 31.8 and 21.5 using the Langley Aerothermodynamic Upwind Relaxation Algorithm (LAURA) [1]. The subsonic value, 0.65, comes from tests conducted in the Langely 20-foot Vertical Spin Tunnel. The supersonic and transonic values are from Brooks and Nichols wind tunnel data on a similar geometry. For the purposes of this feasibility study, the important values are the hypersonic value, which affects the heat pulse, and the subsonic value that determines the impact velocity [1].

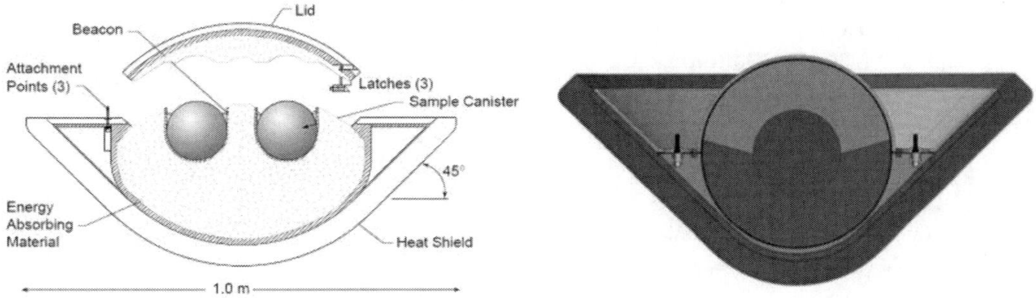

Figure 111. SRV configuration comparison.

It is worth to note that the important values are the hypersonic ones, which affect the heat pulse, and the subsonic value that determines the impact velocity [1]. As one can see, drag coefficients of the present evaluation compare rather well with literature data, especially at very high Mach number flow conditions. Below Mach 7, forebody-only CFD will not accurately predict the aerodynamics. The calculations must include the afterbody and wake. Therefore, differences are recognized at rather low Mach number. In this cases, the effect of base drag could be arose as one can see looking at the difference in back shell configuration between SRV and the capsule considered in the literature data shown in Figure 111.

CONCLUSION

In this work, numerical simulations have been undertaken for a superorbital re-entry capsule. A design analysis was performed to determine the aerothermal (i.e., both aerodynamic and aerothermodynamic) performances of an entirely passive Earth entry capsule for Sample Return mission. In particular, this work is useful to assess the aerodynamic performances and stability along the entry trajectory path, and to elaborate and validate an engineering tool that couples: aeroshape aerodynamic; trajectory and stability; aerothermal environments.

Present design result refers to both numerical and engineering-based analysis of a 1.1 m diameter spherically-blunted 45-deg half-angle forebody with a low-density, ablative heatshield.

Engineering-based design has been applied to assess flowfield regime and capsule aerodynamics; while Navier-Stokes simulations have been performed to evaluate flow field past the capsule to perform radiation analysis by means of PARADE code. Results of the analysis are presented and discussed.

REFERENCES

[1] A passive Earth-Entry Capsule for Mars Sample Return. *51st International Astronautics Federation Congress, Rio de Janeiro. 1998*. IAF-00-Q.3.04.

[2] Scott, C. D., Ried, R. C., Maraia, R. J., Li, C. P., and Derry, S. M., "An AOTV Aeroheating and Thermal Protection Study," H. F. Nelson (ed.), Thermal Design of Aeroassisted Orbital Transfer Vehicles, Vol. 96 of *Progress in Astronautics and Aeronautics*, AIAA, New York, 1985, pp. 198-229.

[3] Tauber M. E. and Suttont, K. "Stagnation-Point Radiative Heating Relations for Earth and Mars Entries". *Journal of Spacecraft and Rockets*. VOL. 28, NO. 1 (pag.40-42).

[4] *PLASMA RADIATION DATABASE PARADE* V2.2. Report TR28/96

[5] *USV_X-Phase A Aerodynamic and Aerothermodynamic Analysis*. Report CIRA-CF-06-1393.

[6] CLAE Project. *H3NS: Code Development Verification and Validation. Report* CIRA-CF-06-1017

[7] *BLAST Project: Input to ENG033 – Aerothermodynamic Database. Report* CIRA-CF-09-1271

[8] *Anderson, Hypersonic and High Temperature Gas Dynamics*. McGraw-Hill Book Company, New York 1989
[9] Review of Chemical Kinetic Problems of Future NASA Missions: Earth Entries. *Journal of Thermophysics and Heat Transfer* Vol 7. No 3. 1993.
[10] Bertin, Hypersonic Aerothermodynamics. *AIAA Educational Series,* Washington. 1994
[11] *Verification and Validation in Computational Science and Engineering*. Hermosa Publishers, Socorro, New Mexico. 1998
[12] *Notes on Earth Atmospheric Entry for Mars Sample Return Missions.* NASA/TP–2006-213486. 2006

In: Mars
Editor: Alberto G. Fairén

ISBN: 978-1-62618-102-1

Chapter 11

NEW TRENDS FOR HUMAN MISSIONS TO MARS

Jean Marc Salotti*

Laboratoire de l'Intégration du Matériau au Système,
Université de Bordeaux, France, and
Association Planète Mars, French Chapter of the Mars Society, France

ABSTRACT

In this paper, I present the new trends for human missions to Mars. Numerous scenarios have already been proposed. The first part of this chapter is dedicated to a brief survey of the main studies on the subject: NASA reference missions, Zubrin Mars Direct and others. Then, there is a specific focus on the driving parameters. I show that more attention should be paid to important parameters such as entry descent and landing choices and the number of astronauts, which are usually set at the beginning of the study. For that reason, numerous options still need to be examined. The last NASA report is not very optimistic but it does seem possible to reduce the complexity and risks of the mission. In the last part of the chapter, I examine the problems that will have to be addressed if it is decided to settle the red planet. I show that most problems concern the fields of ecology and sustainable development with evident links to similar problems on Earth.

INTRODUCTION

Human missions to Mars have been studied for a long time [6], [7], [8], [9], [10], [13], [14], [15]. From Von Braun's proposal to the last design reference architecture from NASA, a lot of work has been carried out. Many options have been examined, experts have been consulted and tests have been performed. However, there is still no consensus on the best scenario for such missions. There are so many possible variations around the main architectures and the relationships between the driving parameters are so complex that it is difficult to assess the feasibility and risks of all options. All-up versus pre-deploy, nuclear

* Corresponding author: jean-marc.salotti@ensc.fr

thermal, electric or chemical propulsion, LEO assembly or not, pressurized or unpressurized rovers, the issues are complex. Nevertheless, some trends are currently observed. These trends are presented and discussed here but also the elements of the mission around which there are still important debates.

A review of the main architectures that have been proposed for human missions to Mars is presented in the first section with a special focus on the Von Braun scenario, NASA reference missions, Zubrin Mars Direct concept and scenarios based on electric propulsion.

The second section is dedicated to recent works and trends for the design of human missions to Mars. It is divided into several subsections, in which the main driving parameters of the mission are presented and discussed.

The last section is more general and focuses on the links between the problems posed by Martian settlements and the problems we have on Earth in the domains of ecology and sustainable development. The use of renewable energies, for instance, will be a critical issue.

1. A Brief Survey of Human Missions to Mars

1.1. First Studies

As soon as humans were able to launch the first rockets, scientists and engineers made suggestions about the future of human spaceflight, starting with the moon but also exploring other planets. Werner Von Braun, specialist of the V2 rockets, started working on human missions to Mars in 1952, several years before the Sputnik launch [13]. Of course, he was one of the geniuses who made the men on the moon a reality but he never forgot Mars. The red planet was seen as one of the next logical steps because it was the closest body and its environment was not very harsh in comparison to other planets such as Venus. His proposals were strongly influenced by the enthusiasm and ever increasing performance in the rocket industry. It was the beginning of the space age and the technological frontiers were pushed back farther and farther each day. The Mercury program was first. It sent 7 Americans into low Earth orbit. Then the Gemini and the Apollo program followed and sent 12 Americans to the surface of the Moon. But the path was not easy and many technological difficulties had to be overcome. Noticeably, a giant launcher had to be built, in-space rendezvous had to be mastered and life support had to be operational several days or even weeks. Progress was so fast and important that the technological limits still seemed very far. The cost was not considered a problem. For those reasons, there was no doubt in Von Braun's mind that giant spaceships could be assembled in low Earth orbit to undertake a journey to Mars. Many other engineers probably had the same idea. The basic scenario was founded on the use of numerous giant launchers to send the pieces of giant spaceships into low Earth orbit. Once assembled, the spaceships would be directly sent to Mars. In his first proposal, the size of the crew and the size of the spaceships were exaggerated. In 1969, while the journey to the Moon was on the agenda, Von Braun was preoccupied by the next step of the space program. However, for the first time, politicians had some reticence about the cost. In August of the same year, just after the first steps on the Moon, Von Braun proposed a more realistic scenario. The main features were:

- After a long sequence of launches, 2 giant spaceships are assembled in low Earth orbit. Each one weighs 725 tons and the size of the crew is 6.
- Both spaceships are sent to Mars thanks to nuclear engines. During transit, there is no way to come back to Earth quickly. Each vehicle is therefore a possible rescue vehicle for the other. The transit time is 9 months.
- Both spaceships are placed in Mars orbit by means of propulsive maneuvers.
- 2 small landers are detached from the mother spaceships and land on the surface of Mars.
- 2 teams of astronauts explore the planet for 80 days.
- Each team uses the Mars ascent vehicle placed on top of the lander to come back to the mother spaceship.
- Both spaceships come back to Earth with a return trajectory that crosses Venus's orbit (1981 planetary configuration).

Despite his efforts to reduce the cost and some technological gaps, human missions to Mars were not considered a priority and all Martian projects were abandoned. It was the beginning of the space shuttle era. Space stations were built in low Earth orbit and space exploration was left to robots. In 1990, however, President Bush wanted to revive the endeavor for human space exploration. He commissioned a report on the best scenario for a human mission to Mars. A team of experts examined different options and provided a list of recommendations 90 days later. Their work was influenced by Von Braun's publications. The idea was to build a giant space station where giant spaceships could be assembled. Then the spaceships could be sent to the Moon and finally to Mars. The scenario was highly complex and the cost was clearly inacceptable. The proposal was rejected. For many space enthusiasts, it was the starting point of the main wind down of human space exploration. The ISS (International Space Station) was announced as the next logical step but there had already been several space stations in the past, such as Mir and Skylab and it was unclear how the work in the ISS could benefit the exploration of other planets.

1.2. Mars Direct

After the Moon adventure, many scientists carried on working on alternative scenarios for the human exploration of Mars. Since Von Braun's scenario was too complex and too expensive, the objective was to find new concepts to simplify the mission. There were several conferences on "The Case for Mars" during which innovative ideas emerged [20]. It was acknowledged that giant launchers like the Saturn V rocket had to be built. The key problem was to reduce the initial mass that had to be sent into low Earth orbit. In astronautics, it is well known that the heaviest part of a spaceship is the propellant required for the maneuvers. For a journey to Mars, some propellant is needed to escape the Earth's gravitational well and join Mars orbit. Another significant amount of propellant is needed for landing on the surface. And a huge amount of propellant is also needed for the launch from Mars and the return to the Earth. Two complementary strategies have been proposed to reduce that amount:

- Changing for a more efficient propulsion system that consumes less propellant.

– Producing the propellant for the return on Mars using local resources.

Robert Zubrin was one of the enthusiasts who worked hard to improve the scenarios for Mars exploration [20]. He advocated a combination of these two strategies, at least at the beginning of his study. He was indeed in favor of a new propulsion system for interplanetary transportation using nuclear energy (see part 2). However, David Baker convinced him that the use of nuclear energy could be a political disadvantage. Zubrin changed his mind and focused on the use of local resources to produce propellant for the return. Methane/oxygen is a good propellant for rockets [2]. Since carbon dioxide is present in the Martian atmosphere, it is possible to carry only hydrogen from Earth and to produce methane/oxygen by means of a small chemical unit and well-known chemical reactions. In 1990, one year after Bush obtained the 90-day report, he published his scenario with Baker [18]. His idea was to produce all the propellant needed for the return on Mars. There was a lot of mass savings and according to their calculations the trip could be undertaken with two spaceships. More importantly, a direct outbound trip and a direct return from the surface of Mars seemed possible for the crewed spaceship. It was the reason why they called that scenario "Mars Direct". The scenario followed a very specific sequence:

– A giant launcher is used to send a cargo vehicle to Mars.
– Eight months later, the vehicle lands on Mars and a chemical unit is automatically deployed together with a small nuclear reactor, which is required to supply energy.
– The chemical unit exploits carbon dioxide from the Martian atmosphere and hydrogen brought from Earth to produce methane and oxygen, which are stored in dedicated tanks.
– Two years after the first launch, the Earth return vehicle is ready.
– Another giant launcher is used to send another spaceship with a crew of four astronauts to the red planet.
– Eight months later, the vehicle lands on the surface of Mars close to the Earth return vehicle.
– The astronauts explore the surface over 500 days.
– The crew enters the Earth return vehicle and the engines are fired to send everyone back to Earth.

The idea seemed great. The scenario was much simplified and many space exploration enthusiasts were convinced. However, though many NASA experts considered the scenario very interesting, there were some doubts about several estimations about the masses involved, which could impact the feasibility. Surprisingly, there was neither report nor scientific paper in the literature explaining in detail why Mars Direct was not seriously considered by NASA. Some drawbacks could be surmised: The Earth return vehicle was very small and could only accommodate a crew of four with difficulty. The mass of the heat shield was probably underestimated. The mass of the propulsion system, engines and tanks, for the return was perhaps also underestimated. Fundamentally, because of the direct strategy, the Earth return vehicle was big and heavy. Launching a big vehicle from the surface of Mars might well be very difficult and risky. The technical reasons explaining NASA's rejection remained unclear however. Zubrin and Baker talked to several experts. They took into consideration their remarks and in 1993 they proposed a modified version of the scenario [19]. If a direct return from Mars is too difficult, it is possible to make a stop and split the inbound trip into two parts. In the Apollo program, the lunar module had to join the command module in lunar orbit. Similarly, a light Mars ascent vehicle can join an Earth return vehicle waiting in Mars orbit. This is the main feature of a new class of scenarios called Mars Semi-direct.

1.3. NASA Design Reference Mission

In 1994, a team of NASA personnel representing several field centers was formed to work on a reference mission for the human exploration of Mars [9]. They started almost from scratch to determine all the parameters of the mission. All domains were concerned: orbital mechanics, propulsion systems, life support, in situ resource utilization, energy, landing mechanisms, number and skills of astronauts, rovers, communication systems, development costs, organizational issues, etc. The problems were hard and they had to make some choices in the early stages of their work. The main principles were:

- Ambitious scientific objectives.
- Long stay on Mars as in Mars Direct or Semi-direct.
- Nuclear thermal propulsion for the transit between low Earth orbit and Mars.
- Based on Mars Semi-direct concept (in situ resource utilization for the production of propellant before the arrival of the crew and Earth return vehicle waiting in Mars orbit.)
- Backup options at critical steps of the mission.

The scenario was ambitious, perhaps too ambitious. The team was so confident on the feasibility of the scenario that the scientific objectives had been set very high. One of the key questions was the number of astronauts, their skills and their activities during transit and on the surface of Mars. It was found that every activity fell into one of the following categories: training, science and exploration, systems operations and maintenance, and planning. Considering operational constraints and the diversity of the skills required, it was decided that the best crew size was six. In addition, in order to maximize scientific productivity, three types of rovers were suggested for local, average distance and regional exploration and several tons of scientific payloads were proposed for three consecutive human missions. Safety issues were examined and many constraints and backup solutions were added, especially on the surface. In comparison with the original Semi-direct scenario proposed by Zubrin and Baker, the NASA reference mission became very complex. All in all, despite the use of nuclear thermal engines to minimize propellant consumption for the interplanetary transit, the initial mass in low Earth orbit was around 850 tons and four big spaceships had to be sent to Mars. Several launching configurations were possible. Although the payload was well above the Saturn V launch capability, a 200 ton-class launch vehicle was suggested. The other option was to split each interplanetary spaceship into two parts that would be assembled in orbit and to use a 100 ton-class launcher, but the strain on the launching sequence would have been very high. The main steps were finally:

- The first cargo with the Earth return vehicle is sent into Mars orbit.
- The second cargo with the Mars ascent vehicle, the chemical unit for propellant production and a small nuclear reactor are sent to the surface of Mars.
- Two years later, once the Mars ascent vehicle is ready, a third cargo is sent to Mars with a backup habitat and a scientific laboratory.
- At the same time, the first crewed vehicle is sent to the surface of Mars.
- Other cargos are sent the same year for the preparation of the next crew flight. They also provide backup solutions for the first crew.

NASA published the final report of the reference mission in 1997. The results were impressive but the initial mass in low Earth orbit was very high and so was the estimated cost. The scenario was too complex and politically unacceptable. Somewhere there was a paradox. In the original Mars Semi-direct scenario, there were only three spaceships sent to Mars. In the NASA reference mission, nuclear engines had been added to increase efficiency but four giant spaceships had to be sent to Mars. The scientific community was skeptical. Zubrin was convinced that his previous propositions were more appropriate and published a book with co-author David Weaver [20]. NASA carried on working on the scenario and tried to reduce the payload that had to be sent to Mars. In 1998, an addendum was published in which several significant modifications were made[6]:

- Scientific equipment is reduced.
- The TransHab concept is introduced: an inflatable habitat is suggested to reduce the mass of the structure.
- The mass of consumables is decreased.
- Elimination of the third cargo (backup habitat and dedicated laboratory).
- The 200 ton-class launcher is abandoned and replaced by an 80-ton class launcher called "Magnum". Two consecutive Magnum launches are necessary to assemble one nuclear thermal rocket in low Earth orbit.
- Repackaging of the Magnums' payload. The diameter is reduced to 7.5 meters instead of 10.
- The functions of in situ resource utilization systems are reexamined. It is suggested that they also supply oxygen for extravehicular activities.
- There are only small unpressurized rovers for the first crew. Pressurized rovers are sent to Mars in the next crew flights.
- The engines of the cargo lander have a dual use: they also are the engines of the Mars ascent vehicle.

Considerable efforts were made to reduce the overall mass budget for low Earth orbit. The new estimation was around 420 tons, which was half the estimation of the first study. In addition, several ideas were proposed for future investigations. First, nuclear energy can have a dual use: thermal for propulsion and electricity for the habitat. Power systems for interplanetary flight can therefore be reduced. Second, significant improvements are expected if high thrust solar electric propulsion can be mastered. Provided that in the near future some progress is made in several other domains, two futuristic "three magnums" scenarios were proposed (instead of six!). Without considering these possible extensions, NASA's proposal already seemed more realistic. However, it was not the right time for a new big project. A terrible accident occurred in 1986 with space shuttle Challenger, which had a bad impact on the perception of NASA's capabilities and it was the beginning of the construction of the international space station with important cost overruns.

1.4. NASA Design Reference Architecture 2009

In 2003, President Bush (the son of the former) asked NASA to make plans for a return to the Moon and to prepare for the next steps such as Mars exploration. The program was called "Constellation" and the great adventure of space exploration was back. Mars was not the priority but it was clear that a 100-ton class launcher had to be built and would help a lot in the planning of such a mission. In 2008, a new NASA team was formed to work on the details of the reference mission for the human exploration of Mars [7]. They were convinced that Constellation would be pursued until a mission to the Moon is undertaken. Ares I was supposed to replace the shuttle for sending astronauts to low Earth orbit and Ares V was the theoretical heavy launcher. As a consequence, they took into consideration Ares I and different configurations of Ares V to determine the details of the launching sequence. Concerning the scenario, the engineers did not want to reproduce the errors of the space shuttle. Robustness was mandatory and approximations had to be avoided. The payloads of the interplanetary spaceships were reexamined and margins were increased everywhere. The main steps of the new scenario were very similar to the ones of the 1997 reference mission but there were more launches, heavier vehicles and only oxygen was produced using Martian resources. More importantly, Braun and Manning were not much confident about the entry, descent and landing sequence. The mass of the heat shield was considerably increased and there were still several open questions. At the end of the study, they expressed their doubts about the feasibility of landing heavy vehicles on Mars. Moreover, loss of mission and loss of crew risks were scrutinized at every step of the mission. There were many uncertainties everywhere and the concluding remark was terrible. The risk level was in the order of 50%. There were some recommendations to reduce uncertainties and risks. However, in the world of information and communication very few people read the details of reports. Everything had to be summarized in one sentence. In the NASA report, one of the most striking sentences was:

"The general story that was constantly reiterated from the individual mission risk analyses was that current design philosophies and technologies would not provide an acceptable level of reliability for a Mars mission."

End of the story. As a consequence, for many space exploration enthusiasts, the dream was killed or reported to decades. The report was published in 2009 [7]. President Obama asked a commission of experts (Augustine Commission) to determine the best strategy for human space exploration. Many experts were interviewed and a human mission to Mars was not cited as a possible mission in a near future. Then Obama cancelled the Constellation program. A human mission to Mars was set as the priority goal but only for the long term.

Do we really have to wait decades for the first mission to Mars? Do we have to rely on the conclusion of the NASA report? Are there other scenarios that enable significant risk reductions? We try to answer these important questions in the next sections.

1.5. Electric Propulsion

Several authors proposed advanced propulsion systems to reduce the amount of propellant or to speed up the vehicle. If the velocity is sufficiently increased, transit times are minimized and short missions can be undertaken. That concept is supported by Chang Diaz who suggested using an efficient propulsion system that allows some variations of the specific impulse (VASIMR) [10].

As for all electric propulsion systems, the idea is to bring a dedicated power plant that supplies electricity to a specific device that accelerates particles (propellant). Particles can be accelerated to very high speeds but, for a given power, a trade-off has to be found between the speed of the particles and the number of particles that can be accelerated each second. High speeds correspond to high specific impulses (ISP) and the higher the ISP, the less propellant is consumed for the same ΔV requirement. But if few particles are accelerated each second, the thrust is low and it would take more time to reach that ΔV. In other words, efficiency plays against thrust.

If high ISP and high thrust are desired, higher electric power outputs are required. The problem is that power requirements quickly reach Megawatts, depending on the desired thrust and the mass of the vehicle. Let us consider the scenario proposed by Schmidt, Seboldt and Auweter-Kurtz in 2006 [16]:

- Type of mission: Short, but similar to NASA reference mission. Crew of 6. The Earth return vehicle and the Mars ascent vehicle are sent to Mars in two cargos prior to the crewed vehicle.
- Outbound trip for the crewed vehicle: 145 days.
- Surface stay: 30 days.
- Inbound trip: 270 days.
- Interplanetary propulsion system: ionic propulsion. Several options are proposed. Main features: Thrust: 100N, ISP: 3000s, conversion efficiency: 37%.
- Nuclear power plant: 4 MW, specific energy: 10kg/kW, mass: 42 tons.
- Initial mass in low Earth orbit around 600 tons.

The scenario is very promising. With electric propulsion systems, short missions seem possible with comparable initial mass in low Earth orbit. However, for such thrust values (100N), 4 MW are required.

For such power levels, a huge power plant is needed. The scenario may be realistic but the use of a huge nuclear power plant increases its complexity. Moreover, there are other constraints.

The thrust is so low that it would take 137 days to send each vehicle from low to high Earth orbit and the crew would have to be sent there separately. As in the NASA reference mission, the loss of mission risk is probably very high and it would take a long time to assess the technologies and implement the scenario. The same remarks can be made for all proposals suggesting the use of electric propulsion systems.

2. Driving Parameters

2.1. Usual Driving Parameters

In several studies dedicated to human missions to Mars, the classification of the scenarios is determined according to the same list of 5 driving parameters [6], [7]:

– Long surface stay / short surface stay:

According to the laws of orbital mechanics, different strategies can be found to send a vehicle from the Earth to Mars and back. However, a major constraint of the trip is the amount of propellant that is required to implement that strategy. That amount is directly linked to the ΔV budget according to Tsiolkovski equation and ΔV budgets are determined by orbital mechanics constraints. If a short surface stay is desired (< 30 days) to reduce the duration of the mission, the ΔV budget is high and the amount of propellant is tremendous because of the planetary configuration. Planet Earth indeed rotates around the sun much faster than Mars and if no effort is made to speed up the space vehicle, it will be totally out of reach when the vehicle lands on Mars. Electric propulsion systems may help in accelerating the vehicle without increasing much the required amount of propellant but at the expense of mission complexity and risks. For that reason, although longer surface stays require more consumables and more robust systems, they allow important mass savings and they are generally preferred to shorter ones. Following that strategy, once they are on Mars, astronauts have to wait 500 days before the next planetary configuration is appropriate for the return trip.

– Pre-deploy versus all-up:

An all-up strategy seems intuitively simpler. The idea is to send everything to Mars at the same time. It was Von Braun's strategy [13]. However, when ISRU is proposed to produce propellant on Mars, the threat is that the process will fail and the astronauts would not be able to come back. The pre-deploy strategy consists in the automatic deployment of ISRU systems in order to produce propellant and have the ascent vehicle ready before the astronauts are sent to Mars. In the last NASA report for the definition of the reference mission, the two strategies are compared and the risks are assessed [7]. Surprisingly, the risks are similar for both strategies. In the pre-deploy option, there are some concerns about the robustness of the systems that have to be maintained in a functional state during a long period of time. The pre-deploy option is nevertheless preferred but the main reason is that the first cargo of the next mission is sent together with the first crewed vehicle and that cargo brings another ascent vehicle, which is therefore a backup solution that does not exist in the all-up option. Since that reason is tightly linked to the specificity of the scenario and the risks are comparable, the all-up versus pre-deploy debate is still open. Our recent contribution suggests that other interesting scenarios based on the all-up strategy deserve investigation[15]. See also Section 2.3. Finally, an important conclusion is that the choice between pre-deploy and all-up is probably not a good driving parameter.

– Aerocapture versus propulsive:

Vehicles arriving from Earth have a hyperbolic trajectory. It is usually desired to send them to Mars orbit before any landing maneuver. There are two options to achieve that goal. The first one is to use the engines of the vehicle to reduce its velocity and place it in Mars orbit. That maneuver requires an important amount of propellant. The second option is to let the vehicle enter the upper layers of the Martian atmosphere and to use the atmospheric drag to reduce its velocity. A propulsive phase is required to circularize the orbit and avoid the next atmospheric entry but the mass of propellant needed for that maneuver is very small. That option is called aerocapture. From an energetic point of view, aerocapture is very interesting and enables important mass savings. However, the risks are higher. The trajectory has to be accurately controlled and the entire vehicle has to be protected against the heat caused by atmospheric frictions. According to NASA, aerocapture is not too risky for the cargos of the reference mission but it is not suggested for the crewed vehicle because of the complexity of its shape and the dangerousness of the maneuver [7]. Such a choice suggests that the question should be further examined. Since aerocapture enables important mass savings, what conditions have to be fulfilled to allow it also for a crewed vehicle? As far as we know, that problem has never been addressed. Smaller vehicles might help. A different scenario in which the capsule (Orion in the NASA reference mission) is integrated in the body of the mother spaceship instead of hanging on its side might also help a lot. Aerocapture could make the difference between two scenarios. In conclusion, aerocapture / no aerocapture is a driving parameter of the mission and there are still open questions to determine the best strategy, which might have an impact on the architecture of the mission.

– ISRU / no ISRU

This parameter has been identified as a key driving parameter for human missions to Mars. See Section 2.3 for a discussion on the subject.

– Nuclear thermal propulsion / Electric propulsion / Chemical propulsion

The choice of the propulsion system for interplanetary transfer has an important impact on the initial mass in low Earth orbit and especially on the mass of propellant. From an energetic point of view, the logical choice should be the propulsion system with highest specific impulse (ISP). High ISP values guarantee that less propellant is needed for the same ΔV requirement. However, the problem is more complex than just classifying the propulsion systems according to their ISP. When chemical propulsion is used, the energy of the system comes from the chemical links of the molecules. If that energy is not used, another type of energy has to be exploited to accelerate particles. Nuclear power plants or photovoltaic arrays are the two main options. In the case of nuclear power, the number of kilowatts it can supply for each kilogram of the system (specific power) is interesting but there are many constraints. First, the crew must be protected from the radiations of the plant. In the NASA reference mission, the solution is to add a shield that weighs 10 tons [7]. Second, there should be a complex assembly of the nuclear spaceship in low Earth orbit. Third, for security reasons, that orbit should be higher than for chemical propulsion options. Fourth, several years are needed for research, development and testing of the systems. And even after that period, the uncertainties will be high for the first missions to Mars. Moreover, for the electric propulsion option, the required power is in the order of several megawatts, which is tremendous. All in all, the use of nuclear power plants for nuclear thermal engines or electric propulsion is an interesting option, but mass savings have to be traded against the complexity of the option, the new risks and the costs. In the case of photovoltaic arrays, the technology is mature and recent technological developments suggest that their mass could be quite low [17]. However, if the vehicles are heavy, power requirements would be very high and gigantic arrays would be needed. The deployment of such arrays is an important challenge and the question of their robustness remains unsolved.

We presented a list of usual driving parameters but other parameters are important. It is interesting to look at the risks of mission loss that have been highlighted by NASA in its assessment of the reference mission [7]. High risks have been associated with the assembly and the entry, descent and landing phase. Surprisingly, neither the assembly nor the entry, descent and landing phase has been considered a driving parameter of the mission. Both phases have been studied but the choices were driven by the other choices of the mission and never the other way around. For the assembly problem, a simple solution is to send several small vehicles instead of a big one. The impact on other parameters of the mission would be important but interesting new architectures might emerge [14]. For the entry, descent and landing phase, the options are examined in the next section.

2.2. Entry, Descent and Landing on Mars

According to many engineers, entry, descent and landing (EDL) is the most dangerous phase of the mission [1]. Surface payloads are usually determined first and EDL specialists have to find the best way to land everything safely. However, EDL risks are tightly linked to the mass and volume of the landing vehicles, which in turn depend on many other parameters of the mission. Different landing configurations may facilitate EDL and reduce the risks. For instance, it might be less risky landing two small vehicles instead of one that is twice as heavy. Another strategy could be to reduce the payload by reducing the size of the crew. Although not mentioned in the 2009 NASA report, the EDL strategy is clearly one of the main driving parameters. Let us give some more precise details.

In order to land a vehicle on Mars, different mechanisms have to be used at different moments. The vehicle enters the Martian atmosphere with a velocity in the order of 5 km/s. The heat shield protects the vehicle and dissipates a large amount of the kinetic energy. Once the velocity is sufficiently reduced, parachutes may be used to continue decelerating the vehicle. The end of the descent, until landing, is propulsive. Depending on the strategy, different types of parachutes may be used, typically at hypersonic and subsonic regimes or no parachutes at all. There are two main physical parameters, the ballistic coefficient, see equation (1), and the lift to drag ratio (L/D).

$$\beta = \frac{M}{CS} \tag{1}$$

M is the mass of the vehicle, C the drag coefficient and S the reference surface for atmospheric braking. The Martian atmosphere is very thin, only a few millibars. For that reason, the deceleration is late and not very efficient. If the ballistic coefficient is high, which is inevitable for entry masses in the 50-100 tons range, the vehicle might even touch the ground before the velocity is sufficiently reduced for parachutes opening. The problem is illustrated in Figure 112.

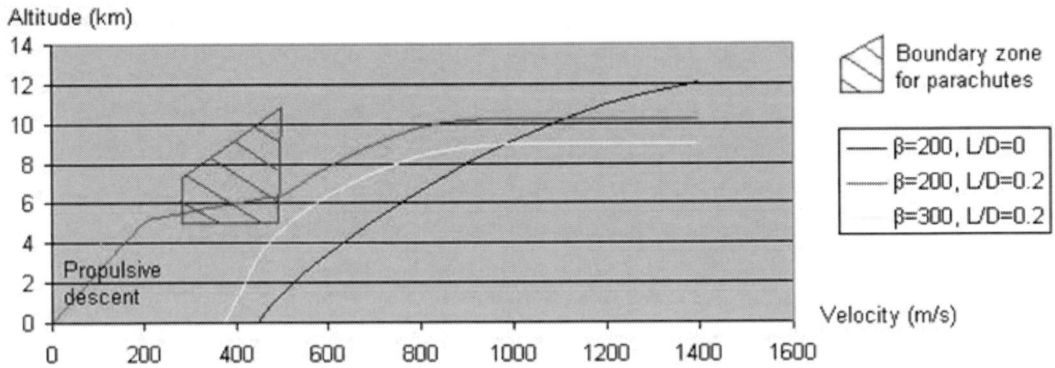

Figure 112.Illustration of the EDL constraints. The curves are approximated according to graphs found in Braun and Manning's paper [1].

The propulsive descent may start earlier without parachutes. However, many additional tons of propellant would be required and while the vehicle is still falling with hypersonic velocity a few seconds before a theoretical impact, the heat shield would have to be ejected quickly and the vehicle positioned in the appropriate position before firing the engines. When parachutes are used, the vehicle continues decelerating to subsonic velocity, it is automatically stabilized and the heat shield is easily pushed away thanks to gravitational and drag forces. To date, all robotic landers used parachutes but much bigger parachutes would be required for larger vehicles [1]. Is it realistic to deploy 100 meter wide parachutes? This is doubtful. Another option is to add some lift. However, the more lift is given, the more complicated the control of the descent. The orientation of the vehicle has indeed to be stabilized with high accuracy and a last delicate maneuver is needed to position the engines in opposition to the drag direction. From a risk standpoint, that option is probably acceptable if and only if the technologies are assessed and validated by means of numerous tests operated in the Martian atmosphere. Such tests would take a long time and would be very costly. There are not many alternatives. The current trend is to find a way to reduce the ballistic coefficient. There are two main options:

a) Using large inflatable heat shields. This option is very interesting but there might be a loss in terms of guidance accuracy. Moreover, adding a critical technology to the EDL sequence increases the risk.
b) Landing light vehicles. Such an option has never been regarded as a potential important classification parameter for human missions to Mars. The light or heavy property of a vehicle seems to be imposed by the payload that has to be landed on the surface. Nevertheless, there are at least two solutions to cope with that constraint. The first one is to reduce the number of astronauts, which has an important impact on

the mass of consumables. And the second consists in the replacement of a heavy landing vehicle by several smaller ones. Such an option might have a large impact on the architecture of the mission, allowing or disallowing certain strategies. The 2-4-2 concept, that was recently proposed, is one of them [15].

In conclusion, there are different options for landing vehicles on Mars and some of them may have an impact on other important choices. Considering the EDL strategy as one of the main driving parameters of human missions to Mars clearly offers new perspectives for the next studies on the subject.

2.3. In Situ Resource Utilization

In situ resource utilization (ISRU), especially for the production of propellant, is a key idea to reduce the total payload mass that has to be landed on Mars. Different types of propellant can be manufactured: hydrogen/oxygen, methane/oxygen and ethanol/oxygen are some examples. According to several authors, the best trade-off between efficiency and complexity is obtained with the methane/oxygen couple [2], [20]. The specific impulse is around 370s for a ratio close to 1/3.6. There are several ways to produce methane and oxygen. The main component of the Martian atmosphere is carbon dioxide and some water can be found in the ground. Water electrolysis can provide hydrogen and oxygen and the Sabatier reaction can be used to transform hydrogen and carbon dioxide into water and methane. In order to obtain the right amount of oxygen, it is possible to use more water or to reduce carbon dioxide to carbon monoxide. See Figure 113.

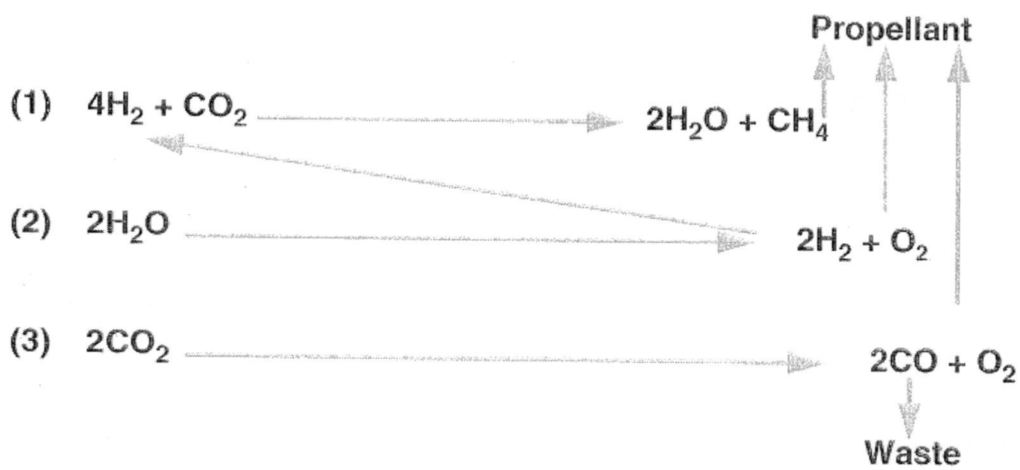

Figure 113. Production of methane and oxygen from local resources.

Such chemical reactions are well known and mastered but much energy is required, especially for water electrolysis, and catalysts have to be used. The amount of propellant that is needed is in the order of 50 tons and the total mass of ISRU and power systems is around 10 tons. Theoretically, the concept is therefore very efficient and allows important mass savings. However, specific tools are required to extract carbon dioxide from the atmosphere and water from the ground. The complexity of the systems is quite high and they have to be operational over a long period of time, typically several months. In the last NASA reference architecture, it is proposed to produce only oxygen and to bring methane from the Earth [7]. Oxygen is indeed the heaviest part of the propellant. It is recognized that the best option would be also to manufacture methane by means of ground excavators and water extraction units but there are numerous uncertainties about the robustness of such technologies. Can we rely on robotic capabilities? In NASA architecture, ISRU systems are pre-deployed and all the propellant for the ascent has to be produced before the departure of the first crew. However, if the threat is a robotic failure, what about sending the astronauts at the same time to help robots for the deployment and the maintenance of the systems? The entire architecture of the mission could be impacted [14]. The trade-off is therefore very complex and the debate is still open.

In addition, another debate exists for the choice of ISRU powering systems. Recent studies suggest the use of ultra-light solar panels instead of a nuclear power plant for power supply on the surface of Mars [3]. Solar panels alone would not be sufficient but even with complementary power systems such as fuel cells, the option might perform better than nuclear power in terms of specific energy (the estimated mass for the production of the same amount of energy). Once again, however, the deployment and maintenance of giant fields of solar panels might be a difficult task for robots and the robustness of that option is questionable. If humans were present on the surface, the robustness might be improved but the pre-deploy strategy would have to be abandoned.

In conclusion, the use of local resources for the production of propellant is a key idea. The strategy can be to minimize the payload that has to be landed on the surface or to maximize the robustness. As was the case for the EDL strategy, the trade-off is not obvious because the entire architecture of the mission is impacted.

2.4. Number of Astronauts

In the first NASA reference mission, the choice for a crew of 6 is justified by specific criteria that take human factors into consideration [9]. The objective was to determine a number of astronauts that is relatively optimal for the work in the field and the activities in the habitat. Complementary skills have been identified and 6 astronauts were considered a good trade-off to maximize the expertise of the group, the efficiency of the work and at the same time minimizing the size of the crew. However, the number of astronauts is an evident driving parameter for human missions to Mars. In all NASA studies, it is acknowledged that a modification of that number would have an important impact on many parameters of the mission but the size of the crew has never been called into question. Why? It might be that the size of the crew has not really been considered a driving parameter. And the other reason is that there are so many parameters that some choices have to be made early.

Since the complexity and risks of the current NASA reference mission are too high, an interesting question is the minimum crew size that would be acceptable. We already addressed that problem in a recent work [14]. Considering human factors and risks, the strict minimum is probably 3 [5], [14]. If the NASA reference mission were revised with a crew of 3 as a new assumption, there would be important mass savings. The vehicles would be much lighter, EDL risks would be reduced and the assembly phase could perhaps be suppressed.

Another idea is to send 2 vehicles with 2 astronauts in each. Like in the Von Braun scenario, each vehicle would be a possible backup vehicle for the other. In addition, during transit, as it is illustrated Figure 114, both vehicles could be attached with a cable and a rotation of the system could provide a kind of artificial gravity. These ideas have already been proposed [15]. The concept is called 2-4-2 because there are 2 astronauts in each small vehicle, there are 4 on the surface and there are 2 again in each vehicle for the return.

Figure 114.Illustration of artificial gravity in the context of the 2-4-2 concept.

3. Colonization and Sustainable Development

3.1. How to Settle the Red Planet?

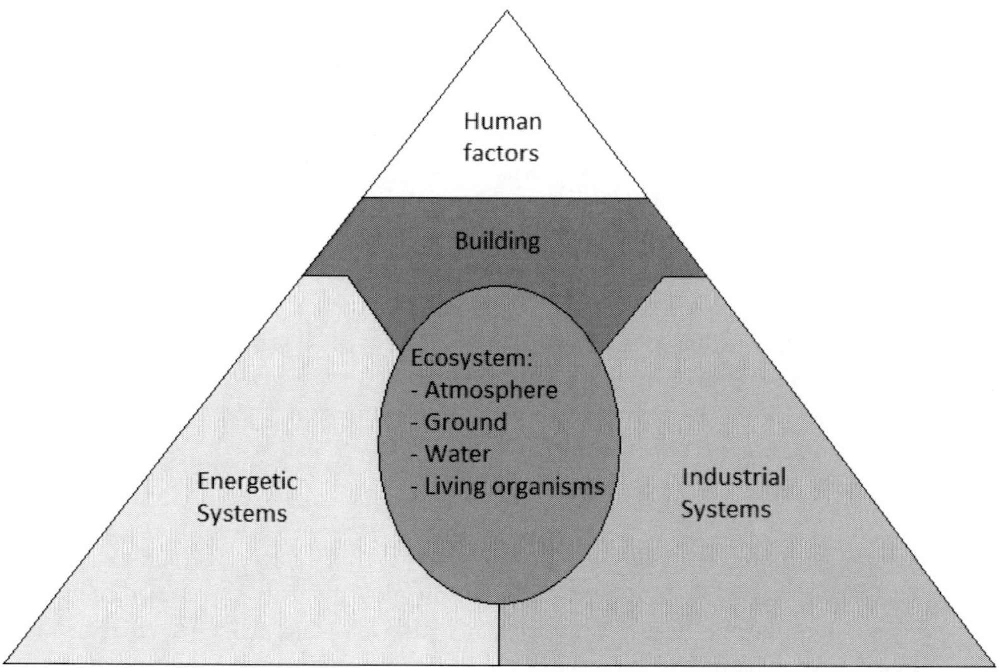

Figure 115. There will be 5 classes of problems for a Martian settlement.

Many studies have already addressed the problem of a Martian settlement [20]. The basic concept is to send a small colony to the red planet and to build a base, in which people could live, have children and develop new infrastructures. At first glance, the problem appears similar to the settlement of the colonies that occurred some centuries ago on the American continent. There were few people, few resources, no industry and the one-way trip from the European continent to America was long and risky. On Mars, however, the problems are more numerous. There is an atmosphere, but it is very cold and not breathable. There is some water in the ground and some ice at the poles but no rivers. There is ground with interesting minerals, but it is not appropriate for growing plants. On the other hand, the settlers will benefit from 21st century know-how and modern technologies. When addressing the key questions of a Martian settlement, what is striking is the link that can be found with sustainable development on Earth. And it is perhaps in that domain that the most relevant spin-offs, technology transfers and new organizational principles are expected.

Some examples are further developed in the following sections. As is illustrated in Figure 115, there are 5 important classes of problems, which can be presented according to a pyramidal hierarchy. At the root of the pyramid, the key problems concern energy production and industrial systems. In the center, impacted by every human action, there is the ecosystem and the needs to preserve the habitability of the base. Then we have the problems linked to the architecture, the building elements and the environment. At the top of the pyramid, there are human factors problems, organizational issues, social issues, wellbeing etc.

3.2. Energy

Supplying energy to the Martian base will be one of the most critical problems of the colony. Energy is required for heating, pumping, transporting and powering almost all machines of industrial processes. Food is also a kind of energy. On Earth, energy is abundant. However, neither fossil fuels nor uranium ores are renewable energy sources. They are limited in supply and will soon be depleted. One day or another, all society will have to depend solely on renewable energy sources. The problem is that if the world demand is maintained at its current level, the required energy will be huge and energy transformation will be very difficult. On Mars, since there is neither fossil fuel nor uranium ores, the Martian colony will have to face the problem from the start. An interesting question arises: if there is no revolution in energy technologies (nuclear fusion for instance), what solutions might emerge for Mars and will they be similar to the ones on Earth in a few decades or centuries?

On Mars, like on Earth, the primary energy source is the sun. There are different ways to transform photons into other types of energy:

- Photovoltaic cells convert solar power into electricity
- Solar concentrators convert solar power in heat, which can be directly exploited for heating purposes or converted into mechanical energy and electricity. Giant concentrators can be used for heating any kind of materials and replace industrial furnaces. An illustration is given in Figure 116. Such furnaces are not used in the industry nowadays because concentrating sunlight does not work during cloudy days and electricity is very cheap. Their main advantage is the autonomy they provide for the industrial process. If energy supply becomes a major problem of the future world, that reason might become essential for their use. Meanwhile, on Mars, since energy supply will be a major problem from the start, solar concentrators might well be used

in the early stages of the development of the colony, for instance for the production of glasses, ceramics or metallic objects.

- There are other methods to exploit solar power. Plants use photosynthesis to grow. They can be used in different ways, especially as food sources or as different types of biomass.

Figure 116. The first solar furnace with double reflexion was built in Mont-Louis, France, in1949.

There are other energy sources on Earth, like wind power and hydropower. They are both derived from solar power. On Mars, wind power will probably be neglected. Hydropower is non-existent but it could be exploited for energy storage. Energy storage is indeed a complementary problem that will have to be seriously tackled. Photovoltaic arrays supply electric power only during daylight. Algae fuel or bioethanol are interesting solutions for energy storage. However, in the case of electricity production, burning organic compounds is not an efficient way to convert solar power into energy. Although rivers are non-existent on Mars, it is possible to create underground artificial reservoirs at different levels and to push water up by means of pumping stations powered by photovoltaic arrays. Hydroelectricity would then be exploited at night. Such solutions already exist on Earth but they are not implemented on a large scale. The problem is complex and the best strategy for electricity production and energy storage remains unknown.

Whatever the strategy of the Martian colonizers for power supply, similar options will have to be found on Earth in the near future. The solutions for the sustainable development of a colony on Mars might even become examples to follow for societies on Earth.

3.3. Industry

In the early stages of the settlement, industry will be very small. Only simple objects will be manufactured on Mars, the others will have to be imported in limited quantities from

Earth. With such constraints and so many objects to be manufactured for the development of the colony, important principles will have to be followed:

- No pollution: In a confined environment such as a Martian base, every chemical substance present in the atmosphere may rapidly reach dangerous levels. Windows cannot be opened as on Earth. As a consequence, there will be strong constraints on the choice of materials to avoid any kind of toxicity.
- Robustness: When it is hard to replace old objects buy new ones, robustness is fundamental.
- Reparability: Objects may eventually break down but their reparability using simple tools is essential.
- Recycling: When raw materials are not abundant or hard to exploit, it is important to recycle all materials that can be extracted from the wastes.

More than just trying to recycle everything, objects will have to be designed from the start with materials and subparts that facilitate their maintenance, their reparability, their recycling and that avoid any toxicity. Ecological conception will be at the heart of all industrial processes on the Martian base. Such ideas have already been suggested on Earth but they compete with other ideas that prioritized financial returns [12].

3.4. Ecosystem

In a confined environment, atmosphere, water, ground, animals and plants can quickly evolve in a direction that is prejudicial for the survival of the colony. In the Biosphere II experiment that took place in Arizona a few years ago, there was a problem with the oxygen rate that decreased to dangerous levels [4]. Important lessons have been learned from that experience. First, even a small ecosystem is complex. There are many parameters impacting its stability and their interdependence is not always well understood. And second, artificial devices might be indispensable for the control of these parameters. On Earth, even on a large scale, it is recognized that human activities are the cause of catastrophic events with terrible consequences for the ecosystem. However, if there are important damages in a specific location, humans can easily move to another place. On Mars, moving will not be an option.

Water is very important for living organisms. It is also useful for washing and it can be used in many chemical reactions such as electrolysis for oxygen production. On Mars, water ice is present at the pole and underground mostly in the northern hemisphere. Extracting the rare water molecules from the ground (3%) is possible but the process is complex. Water will always remain a rare Martian resource for the settlers. As a consequence, water will have to be used in a parsimonious way and recycled as much as possible. Water recycling is a well-known problem in the space community. A water recycling unit is present in the International Space Station. Grey water generated from domestic activities can be recycled very efficiently. At ESA, the MELISSA project aims at recovering food, water and oxygen from grey water and organic wastes by means of plants and microorganisms [11]. Interestingly, several patents have been published and industrial applications have been implemented on Earth. The links between space activities and ecology have therefore already been established.

3.5. Building and Architecture

It is interesting to look at possible materials and architectures for the construction of Martian habitats, greenhouses and other buildings. Since the outside pressure is very low, their design will be very different from their counterparts on Earth. However, similar problems exist in the energy domain and similar solutions might be found. Outside temperatures are indeed very low and energy supply will be very constrained. As on Earth, a good thermal insulation of buildings will be a key issue. Another interesting idea, if radiation levels are acceptable, is to use the greenhouse effect through panels made of glass or other transparent materials. On Mars, there is no rain, hail or snow. Such panels can therefore be placed on the ceiling to maximize the effect. If insulation efforts and the greenhouse effect are sufficient, no complementary energy system would be necessary at all. Such options are very interesting because they minimize energy needs and they increase the robustness of the solution. In the case of a blackout and total loss of power supply, the impact on the survivability in the Martian habitats would be weak. At night, temperatures drop to very low levels. The low Martian atmospheric pressure acts as a very good thermal insulator but it would not be sufficient. The solution could be to store the heat in thermal accumulators such as water reservoirs. The architecture of the base could be impacted. Dust storms might be another problem, but even during such events, the sun is still visible and the greenhouse effect would not be totally absent. The best strategy remains unknown but what is interesting is that such problems are very similar to the problems we have on Earth to minimize the energy needs of buildings.

3.6. Human Factors

In a confined environment, every human action can have a terrible impact on the survival of the colony. The ecosystem is fragile and responsibilities are very high. In order to develop the colony in a sustainable way, humans will have to adopt eco-behaviors. Let us take some examples.

- Water will have to be recycled. However, the complexity and the efficiency of the recycling process greatly depend on the chemical products that are present in the water. Adopting an eco-behavior in that domain, so that water consumption and pollution are rather limited, is essential.
- Similarly, recycling in general is efficient only if household waste sorting is correctly performed. On Earth, there are not numerous classes of wastes. On Mars, however, the classification will probably be very strict. Organic wastes, for instance, are very important because they are almost the only resources to enrich the soil in organic compounds. A specific attention will be paid to nitrogen, which is not abundant on Mars but is vital for all living organisms. The objective will certainly be to recycle nitrogen with a rate close to 100%.
- Another important problem will be to maximize the lifetimes of all objects, buildings, machines, tools, clothes etc. Behavioral efforts will have to be made to use the objects cautiously and to carry out regular maintenance.

CONCLUSION

Two main conclusions can be drawn:

The last NASA study on human missions to Mars tends to suggest that such an enterprise is currently too risky to be undertaken any time soon. However, NASA did not investigate numerous options that could bring important simplifications to the scenario and significantly reduce the risks. The crew size can be decreased, small landers can be used to minimize EDL risks and the entire architecture can be optimized to simplify the scenario and to go even further in the reduction of the risks. The trend in that domain will probably be to work on the new driving parameters of the mission, especially crew size and EDL choices and to re-examine all other options with the new constraints. Finally, if a new consensus were to emerge, a human mission to Mars might be planned sooner than expected.

As soon as human missions to Mars are mastered, the logical next step could be the colonization of Mars. It would be an extraordinary challenge. What is interesting is that such a challenge addresses the key problems of ecology and sustainable development. Martians will have to live in perfect harmony with the ecosystem and all the constraints of a small colony. Let us conclude with an allusion to the Greek Mythology: Ares (Mars) and Aphrodite gave birth to a goddess. They had an incredible inspiration. They named that goddess … “Harmonia”!

REFERENCES

[1] Braun R.D. and Manning R.M., Mars Entry, Descent and Landing Challenges, *Journal of Spacecraft and Rockets,* Vol. 44 (2), 310–323, Mar-Apr, 2007.

[2] Burkhardt, H., Sippel, M., Herbertz, A. and J. Klevanski, Comparative Study of Kerosene and Methane Propellant Engines for Reusable Liquid Booster Stages, proceedings of the 4th International Conference on Launcher Technology *"Space Launcher Liquid Propulsion",* Liège (Belgium), 3-6 December 2002.

[3] Cooper C., Hofstetter W. Hoffman J.A. and Crawley E., Assessment of architectural options for surface power generation and energy storage on human Mars missions, *ActaAstronautica*, vol. 7-8, pages 1106-112, 2010.

[4] Dempster W.F., Tightly closed ecological systems reveal atmospheric subtleties – experience from Biosphere 2. *Advanced Space Research* 42 (12), 1951-1956, 2008.

[5] Dudley-Rowley M., Whitney S., Bishop S., Caldwell B. and Nolan P.D., Crew Size, Composition, and Time: Implications for Habitat and Workplace Design in Extreme Environments, SAE 2001-01-2139, *31st International Conference on Environmental Systems, Orlando, FL*, July 2001.

[6] Drake B.G., ed., Reference Mission Version 3.0 Addendum to the Human *Exploration of Mars: The Reference Mission of the NASA Mars Exploration Study Team* - EX13-98-036 - Exploration Office, NASA Johnson Space Center, June, 1998. (DRM 3.0)

[7] Drake B.G., ed., *Mars Architecture Steering Group, Human Exploration of Mars, Design Reference Architecture 5.0* (and addendum), NASA Johnson Space Center, 2009. (DRA 5.0)

[8] SA HUMEX team, *Study on the Survivability and Adaptation of Humans to Long-Duration Interplanetary and Planetary Environments,* Technical Note 1, Definition of Reference Scenarios for a European Participation in Human Exploration and Estimation of the Life Sciences and Life Support Requirements, HUMEX-TN-001, 2000.

[9] Hoffman S.J. and Kaplan D.I., eds., *Human Exploration of Mars: The Reference Mission of the NASA Mars Exploration Study Team* - NASA SP-6107 - NASA Johnson Space Center, July 1997. (DRM 1.0)

[10] Ilin A.V., Cassady L.D., Glover T.W., Carter M.D. and Chang Diaz F.R., *A Survey of Missions using VASIMR for Flexible Space Exploration*, JSC report 65825, 2010.

[11] Lasseur C., Tan G. and Hovland S., Overview of Life Support activities in ESA, *International symposium on Closed Habitation Experiment and Material Circulation Technology,* Rokkasho-mura, Aomori, Japan, September 28-30, 2004.

[12] McDonough W. and Braungart M, *Cradle to Cradle: Remaking the Way We Make Things*, North Point Press, 2002.

[13] Portree D., Humans to Mars: Fifty Years of Mission Planning, 1950 - 2000, *NASA Monographs in Aerospace History* Series, no 21, February 2001.

[14] Salotti J.M., Simplified scenario for manned Mars missions, *ActaAstronautica*, vol. 69, 266–279, 2011.

[15] Salotti J. M., Revised Scenario for Human Missions to Mars, ActaAstronautica, vol. 81, 273-287, 2012.

[16] Schmidt T.D., Seboldt W. and Auweter-Kurtz M., Flexible piloted Mars mission using continuous electric propulsion, *Journal of Spacecraft and Rockets,* vol. 43 (6), 2006.

[17] White S., Spence B., Trautt T. and Cronin P., ULTRAFLEX-175 on Space Technology 8 (ST8) – *Validating the Next Generation in Lightweight Solar Arrays, proceedings of NASA Science and Technology* Conference, University of Maryland, June 19-21, 2007.

[18] Zubrin R. and Baker D.A., Mars Direct: Humans to the Red Planet by 1999, *proceedings of the 41st Congress of the International Astronautical Federation,* 1990.

[19] Zubrin R. and Weaver D., *Practical Methods for Near-Term Piloted Mars Missions, AIAA 93-2089 AIAA/SAE 29th Joint Propulsion Conference,* Monterey CA, 1993.

[20] Zubrin R. and Wagner R., *The Case for Mars, The Plan to Settle the Red Planet and Why We Must,* Free Press, Touchstone Ed 1996.

INDEX

A

B

C

D

G

H

M

N

O

P

T

U

V

W

Y

Z